Biomass and Waste Conversion: Latest Advances and Prospects

Biomass and Waste Conversion: Latest Advances and Prospects

Editor

Agata Mlonka-Mędrala

MDPI

Basel • Beijing • Wuhan • Barcelona • Belgrade • Novi Sad • Cluj • Manchester

Editor
Agata Mlonka-Mędrala
Faculty of Energy and Fuels
AGH University of Science
and Technology
Krakow
Poland

Editorial Office
MDPI AG
Grosspeteranlage 5
4052 Basel, Switzerland

This is a reprint of articles from the Special Issue published online in the open access journal *Energies* (ISSN 1996-1073) (available at: https://www.mdpi.com/journal/energies/special_issues/Biomass_and_Waste_Conversion).

For citation purposes, cite each article independently as indicated on the article page online and as indicated below:

Lastname, A.A.; Lastname, B.B. Article Title. *Journal Name* **Year**, *Volume Number*, Page Range.

ISBN 978-3-7258-1997-3 (Hbk)
ISBN 978-3-7258-1998-0 (PDF)
doi.org/10.3390/books978-3-7258-1998-0

Contents

About the Editor

Agata Mlonka-Mędrala

Agata Mlonka-Mędrala (Ph.D) is currently an Associate Professor at the Faculty of Energy and Fuels of AGH University of Kraków. She is the author of more than 80 papers in peer-reviewed scientific journals and conference proceedings. She has participated in many international research projects and scholarships (Queen's University Ionic Liquid Laboratory in the United Kingdom; the Technical University of Denmark in Denmark; the Royal Institute of Technology (KTH) in Sweden; Lund University in Sweden; DBFZ German Biomass Research Center gGmbH in Germany; and Xi'an Jiaotong University in China). Agata Mlonka-Mędrala has received numerous awards for her research excellence, including recognition from the Rector of AGH University of Kraków, the Foundation for Polish Science, the Ministry of Science and Higher Education of Poland, and others.

Her research interests include high- and low-temperature corrosion in industrial processes, thermochemical conversion of biomass and alternative fuels, new advanced carbon materials' synthesis, analysis, and description using advanced instrumental techniques, and laboratory-scale corrosion studies in simulated environments.

energies

MDPI

Article

Properties of Animal-Origin Ash—A Valuable Material for Circular Economy

Izabella Maj *, Sylwester Kalisz and Szymon Ciukaj

Department of Power Engineering and Turbomachinery, Faculty of Energy and Environmental Engineering, Silesian University of Technology, 44-100 Gliwice, Poland; sylwester.kalisz@polsl.pl (S.K.); szymon.ciukaj@polsl.pl (S.C.)
* Correspondence: Izabella.maj@polsl.pl

Abstract: In the presented paper, two types of animal-origin biomass, cow dung and chicken litter, are characterized in terms of combustion-related problems and ash properties. It was found that these parameters strongly depend on the farming style. Whether it is cow dung or chicken litter, free-range raw materials are characterized by higher ash contents than industrial farming ones. Free-range samples contain chlorine at lower levels, while industrial farming samples are chlorine rich. Free-range samples are characterized by the predominant content of silica in the ash: 75.60% in cow dung and 57.11% in chicken litter, while industrial farming samples contain more calcium. Samples were classified by 11 "slagging indices" based on the ash and fuel composition to evaluate their tendencies for slagging, fouling, ash deposition and bed agglomeration. Furthermore, an assessment was made against the current EU law regulations, whether the ashes can be component materials for fertilizers. The phosphorus concentration in the investigated ashes corresponds to 4.09–23.73 wt% P_2O_5 and is significantly higher in industrial chicken litter samples. The concentrations of Hg, Cu, As, Ni, Cd and Pb in all samples are below the limits of the UE regulations. However, concentrations of Cr in all samples and Zn in industrial chicken litter exceed these standards.

Keywords: biomass; animal-origin biomass; chicken litter; cow manure; biomass ash; slagging; chlorine corrosion; SEM; ash characterization

Citation: Maj, I.; Kalisz, S.; Ciukaj, S. Properties of Animal-Origin Ash—A Valuable Material for Circular Economy. *Energies* **2022**, *15*, 1274. https://doi.org/10.3390/en15041274

Academic Editors: Agata Mlonka-Mędrala and David Chiaramonti

Received: 16 January 2022
Accepted: 7 February 2022
Published: 10 February 2022

1. Introduction

Animal-origin biomass is mainly a solid animal waste (manure) that is produced during animal breeding. Various types of animal biomass were under investigation when it comes to its energy potential and usability: cattle manure [1,2], poultry litter [3,4], turkey litter [5], goat dung [6], piggery waste [7,8], horse dung [9], deer manure [10] or even elephant dung [11].

Despite this diversity, the greatest potential of animal-origin biomass comes from poultry and cattle breeding. The daily production of droppings per bird depends on the chicken type and ranges from 150–160 g of droppings for an adult chicken to 65–110 g for pullets [12]. Poultry breeding results in massive litter production, which consists not only of manure, but waste bedding, food and feathers. The total amount of poultry litter generated per bird over the entire production cycle depends on the bedding change interval and is reported to be from 1.5 to 5.7 kg [13]. Similar to the poultry litter, the annual production of cow manure depends mainly on the bedding system. An adult cow produces 12.4–26.0 t of waste per year for deep bedding and 11.6–22.0 t/year for a bedding-less system [14].

The safe removal and utilization of animal litter is a key issue since it is considered to be a problematic type of waste. Both poultry litter and cow manure are rich in plant nutrients, such as nitrogen (N), phosphorus (P) and potassium (K) hence they have been utilized for soil conditioning as a fertilizer. However, in recent years, their effectiveness and safety as a fertilizer have become questionable. Animal manure is characterized by low carbon/phosphorous and nitrogen/phosphorous ratios. Nitrogen and phosphorous have the potential for leaching [15] and may result in the contamination of groundwater and eutrophication of water bodies [16] since soil and water are directly linked [17]. When fresh manure is spread on land, nitrous oxide and ammonia, potent greenhouse gases, can be emitted into the atmosphere. Another issue is the pathogens present in animal waste that are a possible threat to human and animal health [18]. Alternative strategies of litter utilization assume anaerobic digestion or thermal treatment, such as pyrolysis, gasification, direct combustion and co-combustion.

Tańczuk et al. determined the technical energy potential of chicken manure for four energy conversion variants taking into account the energy degradation during the production of useful energy, e.g., heat, and pre-processing of the litter (e.g., drying) [19]. The biggest energy loss was found for anaerobic digestion, while combustion demonstrated the most efficient scenario. Furthermore, direct combustion is a convenient and cost-effective method since it can be applied locally on a small scale in the farm neighborhood. Animal-origin biomass, similarly to plant-origin biomass and lignite, may be used in a processed form, for example, torrefied or as pellets and briquets [20–22]. Szymajda et al. [1] determined the quality of cow manure pellets (kinetic durability, bulk density and particle density), as well as flue gas composition during their combustion.

Animal litter combustion is favorable not only due to its energy potential but the possibility of nutrient recycling. If the chemical composition of ash is within the European Union (EU) standards, it can be applied in agriculture as a component of fertilizes. The EU allowed the marketing of fertilizer products of various origins, including ash from biomass combustion, within the scope of solid inorganic macronutrient fertilizers. The recirculation of ash into the soil is claimed to be the most sustainable disposal method according to the circular economy idea [23]. Ash from the combustion of animal-origin biomass may be a valuable by-product since it contains residual phosphorus (P) and potassium (K), which are excellent plant nutrients and could be processed into fertilizer [24]. The use of ashes for soil conditioning can be a part of the soil remediation process since it is expected to enhance the growth of phytoremediation plants. Such action may be beneficial to avoid further land degradation and promote land restoration hence these issues are important in terms of Sustainable Development Goals (SDGs) [25,26]. The studies of biomass ashes in terms of fertilizer usefulness and environmental safety were conducted mainly for plant-origin biomass [27,28], and only limited studies for animal-origin biomass can be found in the literature. The majority of these studies relate to poultry litter ashes from combustion and gasification processes, while cow manure ash has been less studied.

Ash from poultry litter combustion is germ-free and easy to transport [29]. It contains phosphorus and potassium that shows good bioavailability during field and pot tests and can be directly applied into a field as a P source [30], applied in the form of hydrated ash or P can be recovered by extraction/elution [31,32]. However, poultry litter ash may contain heavy metals and metalloids, such as Fe, Mn, As, Zn, Cu, I, Se and Co, which are used in the breeding process to prevent deficits and defects, improve mass gain and elevate egg production [33]. Faridullah et al. [34] presented increased concentrations of metals (Cu, Mn, Zn, Pb and Ni) with increasing combustion temperature and higher amounts of chicken litter ash than duck litter ash. Fiameni et al. [35] proposed a strategy for phosphorous and silica recovery from rice husk poultry litter ash. The proposed method aims to maximize the P extraction using hydrochloric acid and minimize the possible contamination by leachable heavy metals, such as zinc.

The elemental composition of ash, including metals, is crucial when considering its potential fertilizing application since it is highly dependent on local law regulations. In the EU, the limits of Zn, Hg, Cu, Cr, Cd, Ni, Pb and As for organic and mineral fertilizers and soil improvement materials have been regulated by the recent EU Fertilising Products Regulation (EU) 2019/1009, which was approved on 5 June 2019. When the requirements of the Regulation are fulfilled, the ash is no longer considered as waste within the principles of Directive 2008/98/EC, and it can be used as a component for fertilizing products. The products containing or consisting of such recovered ash are allowed to access the European market.

Apart from the use as a fertilizer, the chemical composition of biomass ash should be well recognized before the combustion process since it may cause severe problems in the furnace. Especially animal-origin biomass, whose composition differs significantly from coal and plant-origin biomass, is likely to cause critical combustion-related issues, and this fact determines its usability in the power sector [29,36]. The determination of the vital ash properties includes the characteristic ash fusion temperatures (AFT), chlorine content and the presence of alkalis. The AFTs for animal-origin biomass ashes are usually lower than for coal ashes. Vankát et al. determined the ash-softening temperature of animal manure to 1110–1170 °C and the flow point to 1140–1230 °C [37]. Animal feces ash is usually characterized by a relatively high chlorine content, which can exceed 10% [38]. Fahimi et al. investigated poultry litter ash that was calcium, phosphorous, potassium and sulfur-rich (>29 g/kg) [39]. The ash with high chlorine and alkali metal (K, Na) contents can lead to numerous undesirable issues, such as high-temperature corrosion, slagging, fouling, the formation of deposits on heating surfaces of the boiler and bed agglomeration in CFB boilers [40,41].

The presence of Cl in the ash deposits is vital, especially for high-temperature corrosion, since it leads to the development of low-melting mixtures containing metal chlorides. It favors the mobility of alkali compounds, which may form inorganic mixtures with silica and lower the fluid temperatures from around 1700 °C to around 750 °C. Therefore, chlorine-induced corrosion occurs in a combustion chamber according to the multi-step active oxidation model as it is the most widespread principle of high-temperature corrosion [42].

Many various indicators for classifying a fuel in terms of ash behavior are in use [43–47]. Such methods allow prediction of ash agglomeration, slagging, fouling or corrosion potential. They are based mainly on ash composition, ash fusion temperatures and fuel analysis. They were originally established for coal ashes. Nevertheless, their applicability for biomass ashes is under investigation. Garcia-Maraver et al. [43] classified 104 various biomass types by 10 commonly used slagging indices. For most cases, the coefficients show mixed results when applied to plant-origin biomass: the same fuel was categorized to have low, medium, high or extremely high slagging risk, depending on the index used for classification. However, among 104 various biomass fuels included in this research, only two animal-origin biomasses were investigated: chicken litter and meat-bone meal. Differently from other biomass types, for chicken litter, the categorization seems to be accurate: it was categorized as highly or extremely highly problematic by 7 out of 10 indices. This may suggest that these indices may be applicable to the animal-origin biomass due to its specific composition, different from plant-origin biomass. Lachman et al. [46] presented a compendium of slagging and fouling indices and their applicability to biomass fuels. Nevertheless, not enough attention was paid to animal-origin feedstock. These facts indicated the big knowledge gap when it comes to animal-origin biomass combustion problems. Hence, the presented paper covers a novel and unique research field. The problem of animal-origin biomass ash behavior prediction is still very poorly recognized and existing studies need to be supplemented.

The presented research aims to determine the ash properties of chicken litter and cow manure in terms of ash-related issues and potential use of ashes as an inorganic macronutrient fertilizer. Four types of cow manure and five types of chicken litter were investigated. The material was collected from different breeding styles: industrial farming and free-range to understand how the breeding system affects the feedstock and ash properties. According to the authors' best knowledge, there are no studies where the influence of the breeding system on fuel properties and ash composition is investigated. The ashes were examined by eleven so-called "slagging indices" to evaluate their potential for slagging, fouling and bed agglomeration. The possible application of such indices was determined for the investigated ashes since there is very limited data referring to animal-origin biomass ash behavior prediction. Further, the possible use of ashes for fertilizing purposes was assessed together with the potential risk associated with their introduction into the environment. The ash composition was evaluated against the criteria in the current

EU Fertilising Products Regulation (EU) 2019/1009, taking into account the limits of heavy metals: Zn, Hg, Cu, Cr, As, Cd, Ni and Pb in inorganic macronutrient fertilizers. The limits of other elements, such as Sb, Se, Sn, V, Mo and Co, were determined as well and compared with their average concentrations in European and North American soils. The presented research can improve the technologies of animal-origin biomass energy conversion.

2. Materials and Methods

The following types of animal-origin biomass were considered in this study:

- Cow manure from free-range farming CD1_FR, CD2_FR
- Cow manure from industrial farming CD3_IF, CD4_IF
- Chicken litter from free-range farming CL1_FR
- Chicken litter from industrial farming CL2_IF, CL3_IF, CL4_IF, CL5_IF

CD1_FR, CD2_FR and CL1_FR samples were collected from middle-size free-range animal farms located in Southern Poland.

CD3_IF and CL2_IF were collected from large industrial farms located in Southern Poland.

CD4_IF was investigated by the authors and research partners in previous research [36] and originated from a large industrial farm located in Eastern Poland.

CL3_IF, CL4_IF and CL5_IF samples were collected from a large poultry farm located in Ukraine by the Ukrainian research partner. They were collected within a time of one year to ensure seasonal diversity.

2.1. Feedstock and Ash Analysis

The biomass samples were chopped in a laboratory knife mill and stored at an ambient temperature. A small batch of feedstock was placed into a ceramic crucible, heated up to 550 °C and incinerated in a constant temperature zone in an electric muffle furnace. As a result, chemically stable ash with a minimal amount of unburned carbon (UBC) was obtained for further analysis.

Proximate and ultimate (elemental) analysis of biomass samples was conducted. For the ash, the oxide ash composition and ash fusion temperatures (AFT) were determined since they are the most common procedures for ash characterization in terms of combustion-related issues. To assess the potential risk associated with its introduction into the environment, the Zn, Hg, Cu, Cr, As, Co, Ni, Pb, Sb, Se, Sn, V, Mo and Cd concentrations were determined.

Feedstock analysis was conducted according to European standards for solid fuels: ash content PN-EN ISO 18122:2016-01, moisture content PN-EN ISO 18134-2:2017-03, Lower Calorific Value (LHV) and Higher Calorific Value (HHV) PN-EN ISO 18125:2017-07. A portion of fuel was incinerated at a constant volume in a calorimeter calibrated by the combustion of benzoic acid. C, H and N contents were determined by Infrared (IR) analyzer according to PN-EN ISO 16948:2015-07, Cl and S contents by the Ion Chromatography (IC) method according to PN-EN ISO 16994:2016-10.

The ash composition was determined by Inductively Coupled Plasma-Optical Emission Spectroscopy (ICP-OES). Metal and metalloid concentrations were determined according to PN-EN ISO 16968:2015-07 and PN-EN ISO 11885:2009. The ash fusion temperatures were determined by the microscope-photographic method according to standard CEN/TS 15370-1:2007. The procedure covers the identification and recording of initial deformation temperature (IDT), softening temperature (ST), hemisphere temperature (HT) and flow temperature (FT). The specific, characteristic shapes of the ash cylinders were recorded by a digital system. The procedure assumes a maximum temperature of 1500 °C and both oxidizing and reducing conditions. The only exceptions are samples CL3_IF–CL5_IF, which were collected and tested by the Ukrainian research partner. For these samples, the complete AFT investigation is not available, and only IDT and ST temperatures in oxidizing conditions are reported. For the same reason, the chlorine content and SEM analysis of these samples are not available.

The scanning electron microscope (SEM) analysis of ash samples was performed to investigate their morphology. The Zeiss Supra 35 microscope with the Trident XM4 series (EDX) X-ray spectrometer was used. The electron high tension (EHT) voltage was set to 10 kV. The pictures were taken with two types of magnification for each sample: 1000× and 250/500×, depending on the sample morphology.

2.2. Ash Deposition, Slagging and Fouling Prediction

The ash behavior indices used in this study are presented below, together with the evaluation of their values. An index can display low, moderate, high or extremely high slagging hazards [43,48–50].

The base-to-acid ratio:

$$B/A = \frac{Fe_2O_3+CaO+MgO+Na_2O+K_2O+P_2O_5}{SiO_2+Al_2O_3+TiO_2} \tag{1}$$

<0.5 low
0.5–1.0 moderate
1.0–1.75 high

B/A may be used in a simplified form:

$$B/A_{Simpl.} = \frac{Fe_2O_3 + CaO + MgO}{SiO_2} \tag{2}$$

The Bed Agglomeration Index (BAI):

$$BAI = \frac{Fe_2O_3}{Na_2O+K_2O} \tag{3}$$

<0.15 high

Babcock Index Rs (B/A ratio enhanced with sulfur content in fuel):

$$Rs = B/A \cdot S^d \tag{4}$$

<0.6 low
0.6–2.0 moderate
2.0–2.6 high
>2.6 extremely high

Fouling index Fu (B/A ratio enhanced with the alkali metals):

$$Fu = B/A \cdot (Na_2O + K_2O) \tag{5}$$

<0.6 low
0.6–40 moderate
>40 high

Slag Viscosity Index Sr:

$$Sr = \frac{SiO_2 \cdot 100\%}{SiO_2+Fe_2O_3+CaO+MgO} \tag{6}$$

>72 low
65–72 moderate
<65 high

Fuel alkalinity index μ_{alk}:

$$\mu_{alk} = \frac{1}{LHV \cdot A^d \cdot (Na_2O+K_2O)} \tag{7}$$

0.17–0.34 moderate
>0.34 high

Ash Fusibility Index (AFI):

$$\begin{gathered} AFI = \frac{4IT+HT}{5} \\ >1342 \text{ low} \\ 1232\text{–}1342 \text{ moderate} \\ 1052\text{–}1232 \text{ high} \\ <1050\text{extremely high} \end{gathered} \quad (8)$$

The indices are supplemented by the silica content in the ash SiO_2 (<20 low, 20–25 moderate, >25 high), the chlorine content in the fuel Cl (<0.2 low, 0.2–0.3 moderate, 0.3–0.5 high) and AFTs in oxidation conditions: initial deformation temperature IDT (>1100 °C low, 900–1100 °C moderate, <900 °C high) and softening temperature ST (>1390 °C low, 1250–1390 °C moderate, <1250 °C high).

3. Results and Discussion

3.1. Feedstock and Ash Characteristics

Proximate analysis, elemental (ultimate) analysis and heating values of the feedstock samples are presented in Table 1. The ash content of cow manure from free-range (CD1_FR, CD2_FR) was found to be higher than that of cow manure from industrial farming (CD3_IF, CD4_IF): 21.4% and 22.06% vs. 16.99% and 13.86%. Similarly, the ash content of free-range chicken litter (CL1_FR) is significantly higher than that of industrial farming chicken litter (CL2_IF–CL5_IF): 30.10% vs. 7.10–17.09%. This is reflected in low calorific values of free-range chicken litter. Nevertheless, the HHV of all samples, regardless of the breeding system, is within the range provided in the literature for poultry waste [51]. The moisture content of industrial farming chicken litter is visibly higher than for other samples. The chlorine content in free-range cow manure (0.086–0.33%) is lower than in industrial farming (0.54–1.02%). The chlorine content in free-range chicken litter (0.11%) is greatly below the content of the industrial-farming samples (0.66–0.99%). The chlorine content of free-range animal waste investigated in this study is comparable to the chlorine content of plant-origin biomass, while for industrial breeding, it can be considered as undesirably high. The presented analysis shows a big impact of farming style on the feedstock properties.

The chemical composition of ashes and their AFTs are summarized in Table 2. The elemental composition of ash is a feature of the specific biomass type, but the contents of ash-forming elements can be various even within a certain fuel type [52]. Therefore, the analyzed ashes show different quantitative chemical compositions and the farming style influenced the composition of ashes as strong as animal species. All the samples contain silica, calcium and phosphorus, but for free-range samples, the content of silica is significantly higher than for industrial farming. The massive difference in SiO_2 content can be observed for both cow manure (59.63–75.60% for free-range and 18.3–33.60% for industrial) and chicken litter (57.11% for free-range and 3.66–13.26% for industrial). The high concentration of silica in free-range animal-manure ash can be caused by contamination with sand and soil since animals can freely access the outdoor environment and consume grass directly from the ground.

Table 1. Proximate and elemental (ultimate) analysis of feedstock samples (a.r.—as received, d.b.—dry basis).

Parameter	Basis	Unit	Cow Manure				Chicken Litter				
			CD1_FR	CD2_FR	CD3_IF	CD4_IF [36]	CL1_FR	CL2_IF	CL3_IF	CL4_IF	CL5_IF
Moisture	a.r.	wt%	11.4	8.4	11.1	15.5	12.10	11.1	26.7	21.9	38.8
Ash	d.b.	wt%	21.4	22.06	16.99	13.86	30.10	17.09	9.31	10.70	7.10
HHV	d.b.	MJ/kg	17.26	16.93	17.91	19.04	12.22	17.22	15.60	16.90	16.60
	a.r	MJ/kg	15.49	15.5	15.92	16.09	10.97	15.31	11.40	13.20	10.20
LHV	d.b.	MJ/kg	15.78	15.86	16.72	17.84	11.32	16.02	15.20	15.80	16.10
	a.r.	MJ/kg	13.98	14.32	14.59	14.69	10.08	13.97	10.60	11.86	8.38
Cl	d.b.	wt%	0.086	0.33	1.02	0.54	0.11	0.99	0.96	0.66	0.82
C	d.b.	wt%	41.94	38.93	44.07	45.26	31.19	41.85	39.10	40.3	37.7
H	d.b.	wt%	5.38	4.89	5.45	5.53	3.91	5.5	5.10	5.40	5.20
N	d.b.	wt%	2.59	1.61	2.5	2.79	2.90	4.89	4.70	4.80	4.70
S	d.b.	wt%	0.34	0.32	0.47	0.32	0.50	0.97	0.73	0.75	0.31

Table 2. Ash analysis (dry basis) and ash fusion temperatures.

Parameter	Unit	Cow Manure				Chicken Litter				
		CD1_FR	CD2_FR	CD3_IF	CD4_IF [36]	CL1_FR	CL2_IF	CL3_IF	CL4_IF	CL5_IF
SO_3	wt%	0.88	1.18	4.45	2.63	1.02	9.68	0.94	1.01	0.82
K_2O	wt%	3.19	8.64	18.6	5.56	2.61	25.20	13.04	20.01	16.54
SiO_2	wt%	59.63	75.60	33.60	18.30	57.11	3.66	10.30	7.13	13.26
Fe_2O_3	wt%	1.52	1.11	1.24	1.06	1.94	0.92	4.10	4.11	1.84
Al_2O_3	wt%	4.28	2.65	1.95	1.31	4.15	0.48	1.66	1.11	2.81
Mn_3O_4	wt%	0.18	0.12	0.16	0.51	0.13	0.63	1.75	1.92	1.86
TiO_2	wt%	0.21	0.24	0.15	0.09	0.23	0.04	0.63	0.51	0.43
CaO	wt%	11.85	2.11	13.6	30.60	13.55	18.3	34.07	28.18	26.61
MgO	wt%	2.72	1.56	5.55	8.14	2.15	7.45	6.73	6.48	5.72
P_2O_5	wt%	8.21	4.09	10.8	17.50	7.81	21.00	19.23	22.49	23.74
Na_2O	wt%	3.57	0.73	2.66	3.20	3.21	3.87	6.80	6.26	5.74
BaO	wt%	0.02	0.04	0.05	0.03	0.02	0.03	0.15	0.17	0.14
SrO	wt%	0.02	0.02	0.03	0.04	0.02	0.05	0.34	0.39	0.31
Cl	wt%	0.65	7.56	6.55	2.57	0.90	5.67	-	-	-
						Ash Fusion Temperatures in reducing/oxidizing atmosphere				
initial deformation temperature (IDT)	°C	910/1020	1270/1260	1140/1130	1160/1230	1060/1140	1330/1400	-/1357	-/1254	-/1303
softening temperature (ST)	°C	1150/1240	1310/1290	1170/1170	1210/1270	1170/1210	1380/1470	-/1500	-/1500	-/1439
hemisphere temperature (HT)	°C	1370/1340	1460/1420	1200/1200	1320/1300	1300/1330	1430/>1500	-	-	-
flow temperature (FT)	°C	1410/1370	>1500/>1500	1310/1310	1430/1440	1320/1360	1490/>1500	-	-	-

Silica together with alkalis favors the formation of low-melting eutectics [45,53]. This is reflected in high AFTs for CL2_IF–CL5_IF samples that are low in silica. Lower AFTs of other samples are likely to be a result of a high silica-to-alumina (S/A) ratio. Liu et al. showed that the AFTs decrease with the increasing S/A ratio [54], and the S/A ratios of CD1_FR–CL1_FR samples are significantly higher than for other ones.

On the other hand, CL2_IF–CL5_IF ashes are characterized by higher concentrations of alkaline compounds, such as Ca (up to 34.07%) and K (up to 25.20%), which are mainly responsible for slagging and fouling. Contrarily, CD2_FR is characterized by the lowest concentration of calcium, 2.11%, and CL1_FR is characterized by the lowest concentration of potassium, 2.61%. The alumina content differs significantly as well and is higher for free-range samples than for industrial ones.

The analysis found that some samples contain chlorine at extremely high levels, up to 7.56% (CD2_FR), which may result in an active oxidation process and lead to severe corrosion damage in the furnace. On the other hand, the chlorine content in CD1_FR was only 0.65%.

The ash characteristics of both cow manure and chicken litter cannot be considered advantageous and demonstrate their high potential for ash deposition, slagging, fouling and high-temperature corrosion. If these problems appear, they can be minimized by using aluminosilicate fuel additives, such as halloysite, kaolin or bentonite. Their positive influence on plant-origin biomass and coal combustion has already been studied and successfully proven [55–57].

3.2. Ash Morphology

The SEM pictures of the analyzed ashes are shown in Figure 1. They are presented at the magnitude of 1000× (left column) and 250/500× (right column) for optimal morphology characterization. Numerous fibrous structures can be observed in the majority of the investigated ashes (Figure 1a,c,g,i). Most probably, they come from remaining straw bedding or plants consumed by the animals. The only exception is the CD3_IF sample, whose structure is much finer, without fibrous particles (Figure 1e,f). In all free-range ashes, particles of SiO_2 are present (Figure 1b,d,h). They are likely to be sand particles, as sand can be unintentionally consumed by the animals together with grass, plants, etc., in the outdoor environment. This fact can explain the high silica content in all free-range ashes.

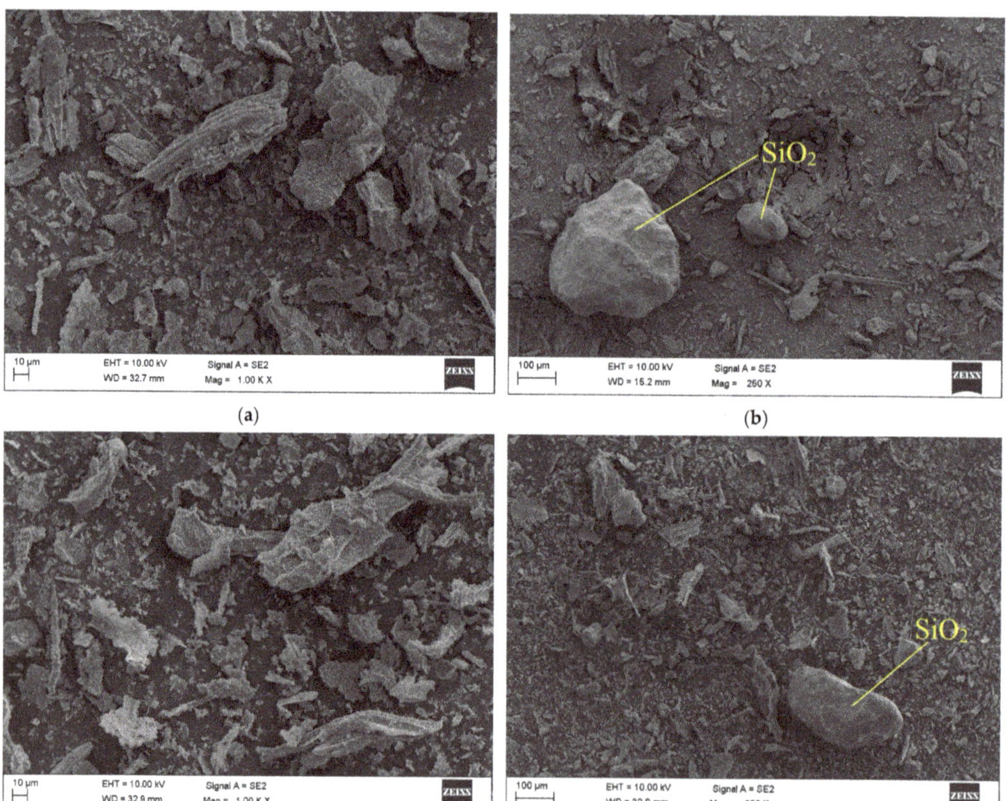

(a) (b)

(c) (d)

Figure 1. *Cont.*

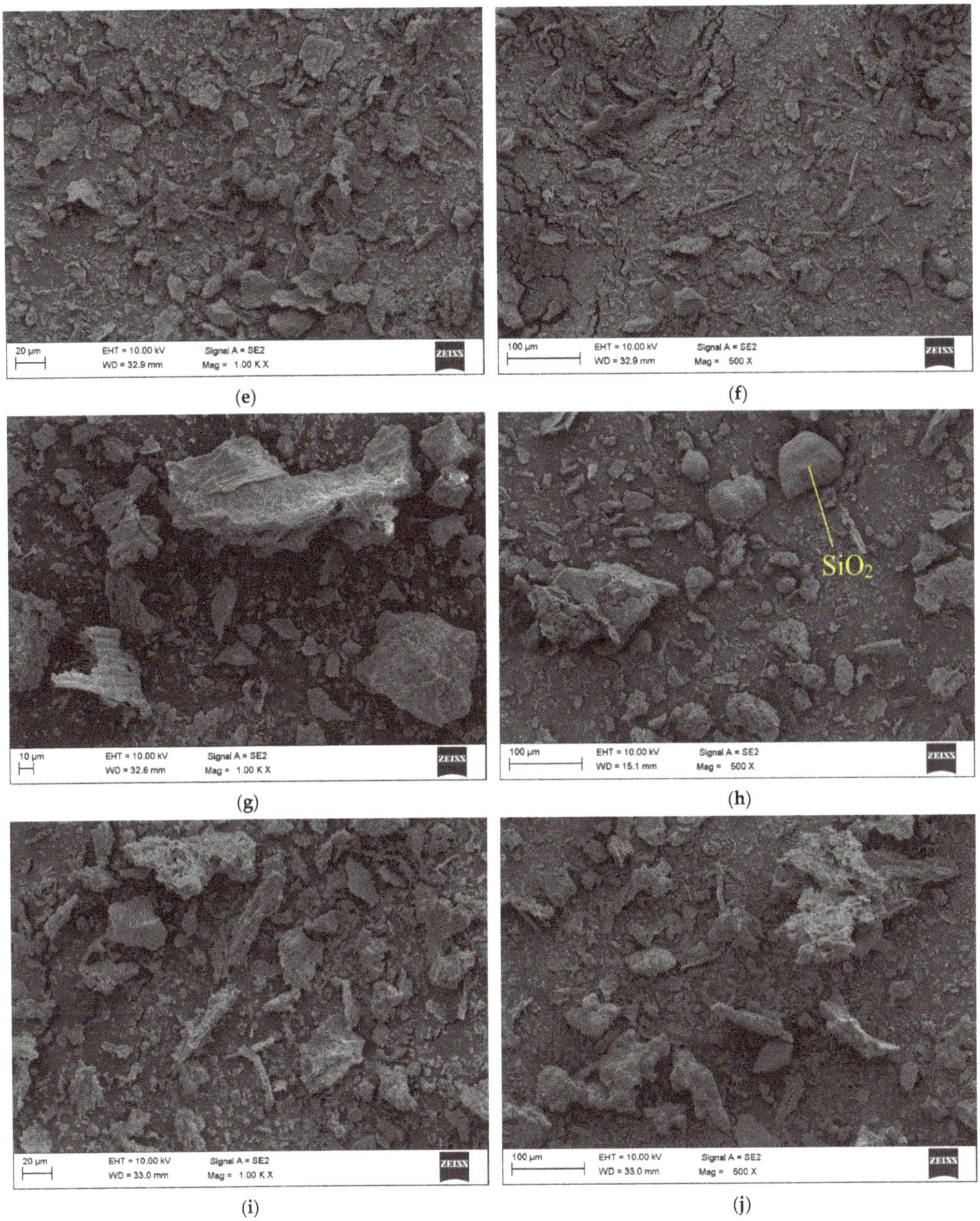

Figure 1. SEM pictures of ash samples. (**a**) CD1_FR; (**b**) CD1_FR; (**c**) CD2_FR; (**d**) CD2_FR; (**e**) CD3_IF; (**f**) CD3_IF; (**g**) CL1_FR; (**h**) CL1_FR; (**i**) CL2_IF; (**j**) CL2_IF.

3.3. Ash Deposition Tendencies

Ash deposition is one of the most unfavorable issues connected to solid biomass fuels. According to the deposition indices, the particular fuel can be evaluated as low, moderately, highly or extremely hazardous [48–50]. The results of classification for all the samples are presented in Table 3.

Table 3. Deposition hazard evaluation according to 11 prediction indices (d.b.—dry basis).

	Index	Unit	CD1_FR	CD2_FR	CD3_IF	CD4_IF [36]	CL1_FR	CL2_IF	CL3_IF	CL4_IF	CL5_IF
1	SiO_2	%	59.63	75.6	33.6	18.3	57.11	3.66	10.3	7.13	13.26
2	Cl	% d.b.	0.086	0.33	1.02	0.54	0.11	0.99	0.96	0.66	0.82
3	B/A	-	0.48	0.23	1.47	3.35	0.50	18.36	6.67	10.00	4.86
4	BAI	-	0.22	0.12	0.06	0.12	0.33	0.03	0.21	0.16	0.08
5	Rs	-	0.16	0.07	0.69	1.07	0.25	17.81	4.87	7.50	1.51
6	Fu	-	3.27	2.18	31.23	29.37	2.94	533.69	132.32	262.79	108.28
7	Sr	-	78.75	94.05	62.23	31.50	76.40	12.07	18.66	15.53	27.96
8	μ_{alk}	kg/GJ	0.44	0.31	0.17	0.46	0.50	0.13	0.36	0.23	0.39
9	Initial deformation temperature IT	°C	1020	1260	1130	1230	1140	1400	1357	1254	1303
10	Softening temperature ST	°C	1240	1290	1170	1270	1210	1470	1500	1500	1439
11	AFI	°C	1084	1292	1144	1244	1178	1420	1086	1003	1042

Though the melting point of silica is 1700 °C, its combination with alkali, K and Na favors the formation of low-melting eutectics [45]. Thus, the index based on the SiO_2 content displays a high tendency of ash deposition for most free-range ashes that are rich in silica and low for most industrial farming ashes that are low in silica.

Chlorine is a key compound in ash fusibility since its presence leads to the forming of inorganic mixtures with Si and reducing their fluid temperatures to about 750 °C. The molten mixtures of alkali chlorides can merge, form larger droplets and thus initiate the formation of slag. Moreover, alkali vapors that contain chlorine are prone to condensation more than non-chlorinated volatile alkalis [58]. The free-range cow manure and chicken litter examined in this study contain chlorine at a relatively low level, comparable to plant-origin biomass. These two samples are classified as having low deposition risk. Contrarily, very high Cl content in other samples resulted in their classification as extremely dangerous.

For the B/A ratio, the results obtained in this study and in studies of Garcia-Maraver et al. concerning biomass fuels [43] disagree with the results presented in other studies, where a decrease in the B/A ratio is followed by an increase of HT and FT and results in the reduction of the slagging tendencies [59]. All industrial farming chicken litter samples, whose AFTs are high, are categorized as having high slagging risks. The CD1_FR sample, whose AFTs were determined to be lower than other samples, was classified to have a low slagging tendency.

When the B/A index is supplemented with the sulfur content, as stands in the Rs index, the results transform. Sulfur is expected to form sulfates during the combustion of fuels where alkalis are present; however, not when they are bound as silicates [60]. In this case, samples with lower SiO_2 contents are more prone to deposit formation.

Alkalis play a key role in the formation of deposits, especially with a low presence of chlorine [60]. For this reason, the B/A index can be complemented with the sum of Na_2O and K_2O, resulting in the Fu index. In this case, industrial-farming chicken litter samples are classified to have high deposition risk, while free-range ones and cow manure to be moderately risky.

According to the BAI index, samples show mixed results. The clear influence of the farming style on the Fe_2O_3 content in the tested samples is not observed.

The Sr index assumes that silica is one of the least responsible components of deposit formation. Hence, when applying this index to silica-rich free-range ashes, they are classified as low risk, while all industrial farming ashes are classified as highly dangerous.

The indices that are based on ATFs (IT, ST and AFI) classified industrial-farming chicken litter ashes as less risky than cow manure ashes. Especially the CL2_IF sample, whose AFTs can be considered as very high, is classified to have low deposition risk.

3.4. Metals and Metalloids Concentration

EU legislation on heavy metal concentrations regulates the agricultural applications of inorganic macronutrient fertilizers. The concentrations of selected elements in the analyzed samples, together with their regulation limits, are presented in Table 4. For phosphorus concentration, an assumption was made that all phosphorous in the ashes is in the form of P_2O_5 [9]. The recalculation was made with the following conversion factor according to EU Regulation 2019/1009:

$$\text{phosphorus (P)} = \text{phosphorus pentoxide } (P_2O_5) \times 0.436$$

Table 4. Concentration of selected elements in ash samples (dry basis) confronted with current UE limits.

	Unit	CD1_FR	CD2_FR	CD3_IF	CL1_FR	CL2_IF	CL3_IF	CL4_IF	CL5_IF	EU Regulation 2019/1009 [1]
P_2O_5	wt%	8.21	4.09	10.8	7.81	21.00	19.23	22.49	23.74	
P [2]	wt%	3.58	1.78	4.71	3.41	9.16	8.38	9.81	10.35	
Zn	mg/kg	980	493	938	846	2787	3400	2700	2700	1500 [3]
Hg	mg/kg	<0.05	<0.05	<0.05	<0.05	<0.05	<0.05	<0.05	<0.05	1
Cu	mg/kg	138	35	129	109	612	232	321	280	600 [3]
Cr	mg/kg	22	178	52	24	36	73	65	44	2 [4]
As	mg/kg	8.86	1.78	1.38	9.99	<1.0	<1.0	<1.0	<1.0	40
Cd	mg/kg	4.35	2.23	2.43	<0.05	2.93	0.19	0.13	0.17	3 [5]
Cd^{recalc}	mg/kg P_2O_5	52.98	54.52	22.50	<0.64	13.95	0.99	0.58	0.72	60 [6]
Ni	mg/kg	18.1	14.1	22.6	16.0	41.8	99.0	75.0	58.0	100
Pb	mg/kg	40.00	16.40	18.80	49.30	5.24	4.26	2.37	3.39	120

[1] The EU Fertilising Products Regulation 2019/1009 of heavy metals concentration in inorganic macronutrient fertilizers. [2] An assumption was made that all P in the ashes is in the form of P_2O_5. The conversion factor was used according to EU Regulation 2019/1009: phosphorus (P) = phosphorus pentoxide (P_2O_5) × 0.436. [3] These limit values shall not apply where copper (Cu) or zinc (Zn) has been intentionally added to an inorganic macronutrient fertilizer for the purpose of correcting a soil micronutrient deficiency. [4] The limit value is 2 mg/kg dry matter is for Cr (VI) while in this table the total Cr concentrations are given. [5] Where an inorganic macronutrient fertilizer has a total phosphorus content of less than 5% P_2O_5: 3 mg/kg dry matter. [6] Where an inorganic macronutrient fertilizer has a total phosphorus content of 5% P_2O_5: 60 mg/kg P_2O_5.

Phosphorus is the desired element, and therefore, it is important to evaluate its concentration in ashes from various sources and breeding systems. From Table 4 it is evident that industrial farming chicken litter ashes have greatly more phosphorus than other samples.

Mercury and copper contents in all samples are greatly below the limits of the EU regulations. They are the most volatile of the heavy metals hence even if present in the biomass, they volatilize during the combustion process, which is the reason for their very low concentrations in the ashes.

For cadmium concentration in fertilizers, the EU regulation has two different limits depending on the P_2O_5 concentration in the investigated material. When the P_2O_5 concentration is below 5 wt%, the limit is 3 mg Cd per kg dry matter. When the P_2O_5 concentration is above 5 wt%, the limit is 60 mg per kg P_2O_5. Evaluating the Cd levels in the ashes against the EU limit needs an assumption that all phosphorus in the ashes is bound in the form of P_2O_5. With the following assumption, the concentration of Cd is recalculated in Table 4 (Cd^{recalc} expressed in mg per kg P_2O_5), which indicates that the ashes have Cd concentrations below the EU limit.

Arsenic and lead concentrations are below the limits, and none of the samples is over the limit of nickel concentration as well; however, the CL3_IF sample is close to the limit.

The concentrations of zinc in industrial farming chicken litter ashes are beyond the limits of UE regulations, reaching 3400 mg/kg. For cow manure and free-range chicken litter ashes, the limit is not exceeded. A possible explanation of the high Zn content is its addition as a food supplement. Zinc is used in industrial poultry breeding to improve reproduction, increase mass gain and boost egg production. High Zn content is, in general, an issue of animal-origin biomass. Nordin et al. [9] determined the Zn concentration in ash from the combustion of a mixture of horse manure and sewage sludge to 1400 mg/kg. Zhang et al. reported the Zn concentration in unprocessed chicken manure reaching up to 1063.32 mg/kg [61].

The elevated Cr concentration of chromium in all samples is likely to be a result of unregulated Cr limits in animal feed. Chromium concentration in chicken manure was reported to range up to 2402.95 mg/kg for a flock size of 2000–20,000 birds [61]. However, due to significant phosphorus concentrations, ashes with exceeded chromium levels can be used in a smaller amount as an additive for fertilizing products.

The limits of other metals and metalloids, such as Sb, Se, Sn, V, Mo and Co, in fertilizers are not controlled under EU regulations. In Table 5, the concentration of these elements in ash samples (dry basis) are presented together with their average concentrations in European or North American soils to assess the potential risk of their introduction into the environment.

Table 5. Concentration of selected elements in ash samples (dry basis) together with their concentrations in European/North American soils.

	Unit	CD1_FR	CD2_FR	CD3_IF	CL1_FR	CL2_IF	CL3_IF	CL4_IF	CL5_IF	Concentration in European/USA Soils
V	mg/kg	<0.05	29.7	15.1	34	13.4	88	49	36	median 60 [62]
Sb	mg/kg	<5.00	<5.00	<5.00	<5.00	<5.00	<5.00	<5.00	<5.00	0.02–31.1 median 0.60 [63]
Se	mg/kg	<5.00	<5.00	<5.00	<5.00	<5.00	<5.00	<5.00	<5.00	up to 600 [64]
Sn	mg/kg	<5.00	<5.00	<5.00	<5.00	<5.00	<5.00	<5.00	<5.00	<2–106 median 3.0 [63]
Mo	mg/kg	5.59	10.90	11.80	6.25	46.50	43.00	31.00	40.00	<0.1–17.2 median 0.62 [63]
Co	mg/kg	6.07	2.97	4.57	7.61	3.66	31.00	10.70	7.03	0.1–7.0 [65]

The knowledge of vanadium behavior in soils is poor compared with other heavy metals, such as Cu, Pb and Zn. However, the median value of total vanadium concentration in European surface soils is 60 mg/kg, with maximum values up to 500 mg/kg. Most toxicity-based limits for unacceptable risks range from 90 to 500 mg/kg for those EU members that have established limitations for vanadium concentration in soils [62]. Thus, the content of vanadium in the tested ashes can be considered harmless.

The content of antimony, tin and molybdenum in the European surface soils was found to be in the ranges <0.02–31.1, <2–106 and <0.1–17.2 mg/kg, respectively [63]. Selenium was found in soils with concentrations up to 600 mg/kg [64]. The normal range of cobalt concentration in agricultural soils are 0.1–7.0 mg/kg [65]. Among these elements, only molybdenum is present in the investigated ashes in amounts exceeding the typical values of soil. However, molybdenum is a micronutrient necessary for the proper development of plants. Its deficit leads to the appearance of light flaws on the leaves, the death of buds and difficulty in the formation of leaf blades, and hence it was under investigation as a component of fertilizers [66]. On this basis, it can be concluded that the tested ashes do not pose a threat to the environment in terms of examined metal and metalloid contents.

4. Conclusions

It was found that the properties of animal-origin biomass strongly depend on the farming style, not only on the animal species. Whether it is cow manure or chicken litter, the free-range raw materials showed different fuel and ash characteristics than industrial farming ones.

Free-range materials are characterized by higher ash contents. The great difference can be observed in chlorine concentration: Free-range samples of both cow manure and chicken litter contain chlorine at lower levels, comparable to plant-origin biomass, while industrial farming samples are chlorine-rich (up to 1.02% of Cl). Free-range samples are characterized by the predominant content of silica in the ash, up to 75.6% in cow manure and 57.11% in chicken litter. The high concentration of silica in free-range ashes is caused by the presence of sand particles that were found in all free-range samples by SEM analysis.

The samples were classified by 11 "slagging indices". Similar to plant-origin biomass, the coefficients display mixed results when applied to animal-origin biomass. The ash fusion temperatures of industrial farming chicken litter can be considered very high since these samples were classified to have low deposition risk by indices that are based on AFTs (IT, ST and AFI). On the other hand, based on chlorine concentration, all industrial ashes were classified as highly risky.

The content of phosphorous, metals and metalloids was determined in ash samples to assess their potential for agricultural application together with the risk of their introduction into the environment. Ashes from industrial chicken litter contain more phosphorus than other ones. The concentrations of Hg, Cu, As, Ni, Cd and Pb in all samples are below the limits established in the EU Fertilising Products Regulation of heavy metals concentration in inorganic macronutrient fertilizers. The concentrations of Cr in all samples and Zn in industrial farming chicken litter are beyond the limits of UE Regulation. A possible explanation is an intentional addition of Zn into the poultry feed as a food supplement. Cr can be present in the feed as well; however, its concentration in animals' food is not regulated in the EU. Such ashes can be processed as an additive to mineral fertilizers in small amounts, taking into account their elevated Zn and Cr concentration.

Author Contributions: Conceptualization, I.M. methodology, I.M.; validation, I.M. and S.K.; formal analysis, I.M. and S.K.; investigation, I.M.; resources, I.M. and S.C.; writing—original draft preparation, I.M. and S.K.; writing—review and editing, I.M. and S.K.; visualization, I.M.; supervision, S.K.; project administration, S.K. and S.C.; funding acquisition, S.K. All authors have read and agreed to the published version of the manuscript.

Funding: The research was carried out as part of the project "Process optimisation and valorisation of combustion by-products in the transition to a circular economy (UPS-Plus)" (www.ccf.polsl.pl, accessed on 1 January 2022), grant number POIR.04.04.00-00-31B4/17-00 financed by the TEAM-TECH Core Facility by Foundation for Polish Science.

Institutional Review Board Statement: Not applicable.

Informed Consent Statement: Not applicable.

Data Availability Statement: The data is available on request from the corresponding author.

Conflicts of Interest: The authors declare no conflict of interest. The funders had no role in the design of the study; in the collection, analyses, or interpretation of data; in the writing of the manuscript, or in the decision to publish the results.

References

1. Szymajda, A.; Łaska, G.; Joka, M. Assessment of Cow Dung Pellets as a Renewable Solid Fuel in Direct Combustion Technologies. *Energies* **2021**, *14*, 1192. [CrossRef]
2. Marin-Batista, J.D.; Villamil, J.A.; Qaramaleki, S.V.; Coronella, C.J.; Mohedano, A.F.; Rubia, M.A. de la Energy Valorization of Cow Manure by Hydrothermal Carbonization and Anaerobic Digestion. *Renew. Energy* **2020**, *160*, 623–632. [CrossRef]
3. Atimtay, A.; Yurdakul, S. Combustion and Co-Combustion Characteristics of Torrefied Poultry Litter with Lignite. *Renew. Energy* **2020**, *148*, 1292–1301. [CrossRef]
4. Bhatnagar, N.; Ryan, D.; Murphy, R.; Enright, A.-M. Trace Element Supplementation and Enzyme Addition to Enhance Biogas Production by Anaerobic Digestion of Chicken Litter. *Energies* **2020**, *13*, 3477. [CrossRef]
5. Williams, A.G.; Leinonen, I.; Kyriazakis, I. Environmental Benefits of Using Turkey Litter as a Fuel Instead of a Fertiliser. *J. Clean. Prod.* **2016**, *113*, 167–175. [CrossRef]
6. Pan-in, S.; Sukasem, N. Methane Production Potential from Anaerobic Co-Digestions of Different Animal Dungs and Sweet Corn Residuals. *Energy Procedia* **2017**, *138*, 943–948. [CrossRef]
7. Theofanous, E.; Kythreotou, N.; Panayiotou, G.; Florides, G.; Vyrides, I. Energy Production from Piggery Waste Using Anaerobic Digestion: Current Status and Potential in Cyprus. *Renew. Energy* **2014**, *71*, 263–270. [CrossRef]

8. Orlando, M.-Q.; Borja, V.-M. Pretreatment of Animal Manure Biomass to Improve Biogas Production: A Review. *Energies* **2020**, *13*, 3573. [CrossRef]
9. Nordin, A.; Strandberg, A.; Elbashir, S.; Åmand, L.-E.; Skoglund, N.; Pettersson, A. Co-Combustion of Municipal Sewage Sludge and Biomass in a Grate Fired Boiler for Phosphorus Recovery in Bottom Ash. *Energies* **2020**, *13*, 1708. [CrossRef]
10. Wang, H.; Xu, J.; Sheng, L.; Liu, X.; Zong, M.; Yao, D. Anaerobic Digestion Technology for Methane Production Using Deer Manure Under Different Experimental Conditions. *Energies* **2019**, *12*, 1819. [CrossRef]
11. Stępień, P.; Świechowski, K.; Hnat, M.; Kugler, S.; Stegenta-Dąbrowska, S.; Koziel, J.A.; Manczarski, P.; Białowiec, A. Waste to Carbon: Biocoal from Elephant Dung as New Cooking Fuel. *Energies* **2019**, *12*, 4344. [CrossRef]
12. Dobrzański, Z. The Relationship between Modern Poultry Production Systems and the Protection of Natural and Productive Environment. First Agricultural Portal. 2002. Available online: http://www.ppr.pl/artykul-ppr-2924.php?_resourcePK=2924 (accessed on 1 May 2021). (In Polish).
13. Dalólio, F.S.; da Silva, J.N.; de Oliveira, A.C.C.; Tinôco, I.D.F.F.; Barbosa, R.C.; de Oliveira Resende, M.; Albino, L.F.T.; Coelho, S.T. Poultry Litter as Biomass Energy: A Review and Future Perspectives. *Renew. Sustain. Energy Rev.* **2017**, *76*, 941–949. [CrossRef]
14. The Agency for Restructuring and Modernization of Agriculture (Poland). Available online: https://www.arimr.gov.pl/fileadmin/pliki/kontrole/2018/Zal_06.pdf (accessed on 1 May 2021).
15. Vandecasteele, B.; Reubens, B.; Willekens, K.; de Neve, S. Composting for Increasing the Fertilizer Value of Chicken Manure: Effects of Feedstock on P Availability. *Waste Biomass Valorization* **2014**, *5*, 491–503. [CrossRef]
16. Szogi, A.A.; Vanotti, M.B. Prospects for Phosphorus Recovery from Poultry Litter. *Bioresour. Technol.* **2009**, *100*, 5461–5465. [CrossRef] [PubMed]
17. Keesstra, S.; Sannigrahi, S.; López-Vicente, M.; Pulido, M.; Novara, A.; Visser, S.; Kalantari, Z. The Role of Soils in Regulation and Provision of Blue and Green Water. *Philos. Trans. R. Soc. B Biol. Sci.* **2021**, *376*, 1–8. [CrossRef]
18. Billen, P.; Costa, J.; van der Aa, L.; van Caneghem, J.; Vandecasteele, C. Electricity from Poultry Manure: A Cleaner Alternative to Direct Land Application. *J. Clean. Prod.* **2015**, *96*, 467–475. [CrossRef]
19. Tańczuk, M.; Junga, R.; Kolasa-Więcek, A.; Niemiec, P. Assessment of the Energy Potential of Chicken Manure in Poland. *Energies* **2019**, *12*, 1244. [CrossRef]
20. Olugbade, T.O.; Ojo, O.T. Binderless Briquetting Technology for Lignite Briquettes: A Review. *Energy Ecol. Environ.* **2021**, *6*, 69–79. [CrossRef]
21. Olugbade, T.; Ojo, O.; Mohammed, T. Influence of Binders on Combustion Properties of Biomass Briquettes: A Recent Review. *BioEnergy Res.* **2019**, *12*, 241–259. [CrossRef]
22. Olugbade, T.O.; Ojo, O.T. Biomass Torrefaction for the Production of High-Grade Solid Biofuels: A Review. *BioEnergy Res.* **2020**, *13*, 999–1015. [CrossRef]
23. Dróżdż, D.; Wystalska, K.; Malińska, K.; Grosser, A.; Grobelak, A.; Kacprzak, M. Management of Poultry Manure in Poland —Current State and Future Perspectives. *J. Environ. Manag.* **2020**, *264*, 110327. [CrossRef]
24. Lynch, D.; Henihan, A.M.; Bowen, B.; Lynch, D.; McDonnell, K.; Kwapinski, W.; Leahy, J.J. Utilisation of Poultry Litter as an Energy Feedstock. *Biomass Bioenergy* **2013**, *49*, 197–204. [CrossRef]
25. Keesstra, S.; Mol, G.; de Leeuw, J.; Okx, J.; Molenaar, C.; de Cleen, M.; Visser, S. Soil-Related Sustainable Development Goals: Four Concepts to Make Land Degradation Neutrality and Restoration Work. *Land* **2018**, *7*, 133. [CrossRef]
26. Novara, A.; Pulido, M.; Rodrigo-Comino, J.; di Prima, S.; Smith, P.; Gristina, L.; Gimenez-Morera, A.; Terol, E.; Salesa, D.; Keesstra, S. Long-Term Organic Farming on a Citrus Plantation Results in Soil Organic Carbon Recovery. *Cuad. Investig. Geográfica* **2019**, *45*, 271. [CrossRef]
27. Zając, G.; Szyszlak-Bargłowicz, J.; Gołębiowski, W.; Szczepanik, M. Chemical Characteristics of Biomass Ashes. *Energies* **2018**, *11*, 2885. [CrossRef]
28. Jarosz-Krzemińska, E.; Poluszyńska, J. Repurposing Fly Ash Derived from Biomass Combustion in Fluidized Bed Boilers in Large Energy Power Plants as a Mineral Soil Amendment. *Energies* **2020**, *13*, 4805. [CrossRef]
29. Kelleher, B.P.; Leahy, J.J.; Henihan, A.M.; O'Dwyer, T.F.; Sutton, D.; Leahy, M.J. Advances in Poultry Litter Disposal Technology —A Review. *Bioresour. Technol.* **2002**, *83*, 27–36. [CrossRef]
30. Codling, E.E.; Chaney, R.L.; Sherwell, J. Poultry Litter Ash as a Potential Phosphorus Source for Agricultural Crops. *J. Environ. Qual.* **2002**, *31*, 954–961. [CrossRef]
31. Sugiyama, S.; Wakisaka, K.; Imanishi, K.; Kurashina, M.; Shimoda, N.; Katoh, M.; Liu, J.-C. Recovery of Phosphate Rock Equivalents from Incineration Ash of Chicken Manure by Elution-Precipitation Treatment. *J. Chem. Eng. Jpn.* **2019**, *52*, 778–782. [CrossRef]
32. Kaikake, K.; Sekito, T.; Dote, Y. Phosphate Recovery from Phosphorus-Rich Solution Obtained from Chicken Manure Incineration Ash. *Waste Manag.* **2009**, *29*, 1084–1088. [CrossRef]
33. Bolan, N.S.; Szogi, A.A.; Chuasavathi, T.; Seshadri, B.; Rothrock, M.J.; Panneerselvam, P. Uses and Management of Poultry Litter. *World's Poult. Sci. J.* **2010**, *66*, 673–698. [CrossRef]
34. Faridullah, F.; Irshad, M.; Yamamoto, S.; Honna, T.; Eneji, A.E. Characterization of Trace Elements in Chicken and Duck Litter Ash. *Waste Manag.* **2009**, *29*, 265–271. [CrossRef]
35. Fiameni, L.; Fahimi, A.; Marchesi, C.; Sorrentino, G.P.; Zanoletti, A.; Moreira, K.; Valentim, B.; Predeanu, G.; Depero, L.E.; Bontempi, E. Phosphorous and Silica Recovery from Rice Husk Poultry Litter Ash: A Sustainability Analysis Using a Zero-Waste Approach. *Materials* **2021**, *14*, 6297. [CrossRef] [PubMed]

36. Maj, I.; Kalisz, S.; Szymajda, A.; Łaska, G.; Gołombek, K. The Influence of Cow Dung and Mixed Straw Ashes on Steel Corrosion. *Renew. Energy* **2021**, *177*, 1198–1211. [CrossRef]
37. Vankát, A.; Krepl, V.; Kára, J. Animal Dung as a Source of Energy in Remote Areas of Indian Himalayas. *Agricult. Trop. Subtrop.* **2009**, *43*, 140–142.
38. Tran, Q.T.; Maeda, M.; Oshita, K.; Takaoka, M. Phosphorus Release from Cattle Manure Ash as Soil Amendment in Laboratory-Scale Tests. *Soil Sci. Plant Nutr.* **2017**, *63*, 369–376. [CrossRef]
39. Fahimi, A.; Bilo, F.; Assi, A.; Dalipi, R.; Federici, S.; Guedes, A.; Valentim, B.; Olgun, H.; Ye, G.; Bialecka, B.; et al. Poultry Litter Ash Characterisation and Recovery. *Waste Manag.* **2020**, *111*, 10–21. [CrossRef]
40. Vassilev, S.v.; Baxter, D.; Andersen, L.K.; Vassileva, C.G. An Overview of the Composition and Application of Biomass Ash. *Fuel* **2013**, *105*, 19–39. [CrossRef]
41. Díaz-Ramírez, M.; Frandsen, F.J.; Glarborg, P.; Sebastián, F.; Royo, J. Partitioning of K, Cl, S and P during Combustion of Poplar and Brassica Energy Crops. *Fuel* **2014**, *134*, 209–219. [CrossRef]
42. Kofstad, P. *High Temperature Corrosion*; Elsevier: London, UK, 1988.
43. Garcia-Maraver, A.; Mata-Sanchez, J.; Carpio, M.; Perez-Jimenez, J.A. Critical Review of Predictive Coefficients for Biomass Ash Deposition Tendency. *J. Energy Inst.* **2017**, *90*, 214–228. [CrossRef]
44. Weber, R.; Poyraz, Y.; Beckmann, A.M.; Brinker, S. Combustion of Biomass in Jet Flames. *Proc. Combust. Inst.* **2015**, *35*, 2749–2758. [CrossRef]
45. Miles, T.R.; Miles, T.R.; Baxter, L.L.; Bryers, R.W.; Jenkins, B.M.; Oden, L.L. Boiler Deposits from Firing Biomass Fuels. *Biomass Bioenergy* **1996**, *10*, 125–138. [CrossRef]
46. Lachman, J.; Baláš, M.; Lisý, M.; Lisá, H.; Milčák, P.; Elbl, P. An Overview of Slagging and Fouling Indicators and Their Applicability to Biomass Fuels. *Fuel Processing Technol.* **2021**, *217*, 106804. [CrossRef]
47. Mlonka-Mędrala, A.; Magdziarz, A.; Kalemba-Rec, I.; Nowak, W. The Influence of Potassium-Rich Biomass Ashes on Steel Corrosion above 550 °C. *Energy Convers. Manag.* **2019**, *187*, 15–28. [CrossRef]
48. Jenkins, B.M.; Baxter, L.L.; Miles, T.R.; Miles, T.R. Combustion Properties of Biomass. *Fuel Processing Technol.* **1998**, *54*, 17–46. [CrossRef]
49. Vamvuka, D.; Zografos, D. Predicting the Behaviour of Ash from Agricultural Wastes during Combustion. *Fuel* **2004**, *83*, 2051–2057. [CrossRef]
50. McLennan, A.R.; Bryant, G.W.; Bailey, C.W.; Stanmore, B.R.; Wall, T.F. Index for Iron-Based Slagging for Pulverized Coal Firing in Oxidizing and Reducing Conditions. *Energy Fuels* **2000**, *14*, 349–354. [CrossRef]
51. Qian, X.; Lee, S.; Soto, A.; Chen, G. Regression Model to Predict the Higher Heating Value of Poultry Waste from Proximate Analysis. *Resources* **2018**, *7*, 39. [CrossRef]
52. Xing, P.; Mason, P.E.; Chilton, S.; Lloyd, S.; Jones, J.M.; Williams, A.; Nimmo, W.; Pourkashanian, M. A Comparative Assessment of Biomass Ash Preparation Methods Using X-ray Fluorescence and Wet Chemical Analysis. *Fuel* **2016**, *182*, 161–165. [CrossRef]
53. Nutalapati, D.; Gupta, R.; Moghtaderi, B.; Wall, T.F. Assessing Slagging and Fouling during Biomass Combustion: A Thermodynamic Approach Allowing for Alkali/Ash Reactions. *Fuel Processing Technol.* **2007**, *88*, 1044–1052. [CrossRef]
54. Liu, B.; He, Q.; Jiang, Z.; Xu, R.; Hu, B. Relationship between Coal Ash Composition and Ash Fusion Temperatures. *Fuel* **2013**, *105*, 293–300. [CrossRef]
55. Sobieraj, J.; Gądek, W.; Jagodzińska, K.; Kalisz, S. Investigations of Optimal Additive Dose for Cl-Rich Biomasses. *Renew. Energy* **2021**, *163*, 2008–2017. [CrossRef]
56. Kalisz, S.; Ciukaj, S.; Mroczek, K.; Tymoszuk, M.; Wejkowski, R.; Pronobis, M.; Kubiczek, H. Full-Scale Study on Halloysite Fireside Additive in 230 t/h Pulverized Coal Utility Boiler. *Energy* **2015**, *92*, 33–39. [CrossRef]
57. Wejkowski, R.; Kalisz, S.; Tymoszuk, M.; Ciukaj, S.; Maj, I. Full-Scale Investigation of Dry Sorbent Injection for NOx Emission Control and Mercury Retention. *Energies* **2021**, *14*, 7787. [CrossRef]
58. Yang, T.; Ma, J.; Li, R.; Kai, X.; Liu, F.; Sun, Y.; Pei, L. Ash Melting Behavior during Co-Gasification of Biomass and Polyethylene. *Energy Fuels* **2014**, *28*, 3096–3101. [CrossRef]
59. Munir, S. Potential Slagging and Fouling Problems Associated with Biomass-Coal Blends in Coal-Fired Boilers. *J. Pak. Inst. Chem. Eng.* **2010**, *38*, 1–11.
60. Tortosa Masiá, A.A.; Buhre, B.J.P.; Gupta, R.P.; Wall, T.F. Characterising Ash of Biomass and Waste. *Fuel Processing Technol.* **2007**, *88*, 1071–1081. [CrossRef]
61. Zhang, F.; Li, Y.; Yang, M.; Li, W. Content of Heavy Metals in Animal Feeds and Manures from Farms of Different Scales in Northeast China. *Int. J. Environ. Res. Public Health* **2012**, *9*, 2658–2668. [CrossRef]
62. Carlon, C. *Derivation Methods of Soil Screening Values in Europe. A Review and Evaluation of National Procedures towards Harmonisation*; European Commission, Joint Research Centre: Ispra, Italy, 2007.
63. De Vos, W.; Tarvainen, T. *Geochemical Atlas of Europe. Part 2*; Geological Survey of Finland: Espoo, Finland, 2006.
64. Fernández-Martínez, A.; Charlet, L. Selenium Environmental Cycling and Bioavailability: A Structural Chemist Point of View. *Rev. Environ. Sci. Bio/Technol.* **2009**, *8*, 81–110. [CrossRef]
65. Kabata-Pendias, A. *Trace Elements in Soils and Plants*, 3rd ed.; CRC Press: Warsaw, Poland, 2000; ISBN 9780429191121.
66. Anderson, A.J. *Molybdenum as a Fertilizer*; Commonwealth Scientific and Industrial Research Organization: Canberra, Australia, 1956.

energies MDPI

Article

Intermediate Pyrolysis of Brewer's Spent Grain: Impact of Gas Atmosphere

Artur Bieniek [1,*], Wojciech Jerzak [1], Małgorzata Sieradzka [1], Łukasz Mika [2], Karol Sztekler [2] and Aneta Magdziarz [1]

1 Faculty of Metals Engineering and Industrial Computer Science, AGH University of Science and Technology, Al. Mickiewicza 30, 30-059 Cracow, Poland; wjerzak@agh.edu.pl (W.J.); msieradz@agh.edu.pl (M.S.); amagdzia@agh.edu.pl (A.M.)
2 Faculty of Energy and Fuels, AGH University of Science and Technology, Al. Mickiewicza 30, 30-059 Cracow, Poland; lmika@agh.edu.pl (Ł.M.); sztekler@agh.edu.pl (K.S.)
* Correspondence: artbie@agh.edu.pl

Abstract: This work focuses on the impact of carrier gas on the quantity and quality of pyrolytic products received from intermediate pyrolysis of the brewer's spent grain. In this study, three types of carrier gases were tested: argon, nitrogen, and carbon dioxide at three temperatures of 500, 600, and 700 °C. On the basis of the process conditions, the yield of products was determined. The ultimate analysis of the char was performed, and for selected chars, the combustion properties were determined. Gas chromatography of the organic fraction of oil was performed, and the compounds were determined. Additionally, microscale investigation of the spent grain pyrolysis was performed by thermogravimetric analysis. The results showed that there were no significant differences in product yields in various atmospheres. Char yield changed only with temperature from 28% at 500 °C up to 19% at 700 °C. According to ultimate analysis, the char from CO_2 pyrolysis was approximately 2% richer in carbon and this fact did not influence on the combustion properties of the char. The oil fraction was characterized mainly by acids with a maximum content of 68% at 600 °C in an argon atmosphere and the acid concentration depended on the carrier gas as follows line: Ar > N_2 > CO_2.

Keywords: Brewer's spent grain; intermediate pyrolysis; carrier gas; fixed bed reactor

Citation: Bieniek, A.; Jerzak, W.; Sieradzka, M.; Mika, Ł.; Sztekler, K.; Magdziarz, A. Intermediate Pyrolysis of Brewer's Spent Grain: Impact of Gas Atmosphere. *Energies* **2022**, *15*, 2491. https://doi.org/10.3390/en15072491

Academic Editor: Changkook Ryu

Received: 28 February 2022
Accepted: 25 March 2022
Published: 28 March 2022

1. Introduction

Current energy and ecological policy promotes the reuse and management of all types of municipal and industrial solid wastes in technological processes [1,2]. The main goal of this promotion is to reduce the amount of waste in landfills and also search for alternative solutions for fossil fuels. The use of municipal solid waste in the power engineering sector allows a reduction the amount of fossil fuels involved in energetic processes. Especially organic wastes are worth considering. This is important because fossil fuels have limited resources and can be regenerated over long periods of time [3,4]. Another reason is the fact that global demand for primary energy is dynamically increasing, and it leads to intensified consumption of all kinds of fuel, especially fossil fuels such as natural gas, coal, or oil [5].

In 2000, the usage of industrial and municipal solid waste in the energy sector allowed to produce approximately 1.24 EJ of primary energy, and in 2019, the production of primary energy from waste was 2.59 EJ [6]. It was above two times the increase in waste share in the primary energy generation structure. Most wastes are characterized by an organic chemical structure, and these wastes can be classified as biomass sources, according to the definition of biomass included in the directive of the European Parliament and of the Council [7]. Biomass as a source of energy belongs to renewable energy sources. From this point of view, the reuse of waste in the energy process has a lot of energetic, economical, and ecological benefits [8].

Today, the power engineering sector uses organic wastes primarily in combustion, where the chemical energy is converted to heat that could be used in other processes [9–11]. Another method is the thermal conversion of solid waste into valuable fuels by a process such as pyrolysis. The primary feedstock is thermally degraded in the absence of an oxygen atmosphere, where, in the results, long polymer structures are broken into smaller chemical compounds. This process is called pyrolysis and usually operates in the temperature range of 350 °C to 750 °C [12–14]. Pyrolysis always delivers three types of products: solid residue, condensable fraction of heavy hydrocarbons, and light-weight gases. Solid residues are built with fixed carbon and ash included in feedstock, while volatile matter and noncondensable gases are gaseous products of the degradation of long structure of polymers.

An example of food industrial solid waste is brewer's spent grain (BSG). Brewer's spent grain is organic residue generated after the beer production process. In 2020, around 32 billion L of beer were produced in the European Union [15]. Production of 100 L of beer was estimated to require 20–22 kg of barley malt, which is the primary organic component [16]. According to the presented data, it can be observed that a large amount of BSG could be obtained and reused in the energy sector. Nowadays, BSG waste is converted to animal feed due to high content of nutrients, even after the brewing process, or is stored in landfills [16,17].

BSG as a waste material could be involved in the power engineering sector. Because of its organic nature, it is classified as a biomass, and its use could increase the impact of green energy on the structure of primary energy demand. The wet BSG waste directly from the lauter tun contains around 77–81% water, and for further processes, the BSG should be dried to increase its energetic potential [18,19]. BSG is considered as fuel in the combustion process where the included chemical energy is transformed into heat [20,21]. It is characterized by a higher calorific value of 17.8 up to 19.1 $MJ \cdot kg^{-1}$ and a relatively small ash content of 1.7% to 5.4% [22]. On the other hand, many scientific works present BSG as a good example of feedstock material in pyrolysis or other chemical processes of solid material conversion [23–27]. Because of these methods, valuable chemical compounds can be evolved from BSG. Čater et al. [23] investigate a method of to produce biogas from brewer's spent grain by bioaugmentation enhanced with anaerobic hydrolytic bacteria. The results of the researchers have shown that 17.8% of the methane from BSG can be produced with the bioaugmentation of pseudobutyrivibrio xylanivorans. Yinxin et al. [24] investigated a co-pyrolysis of BSG and sewage sludge in a limited oxygen atmosphere (<5%). The results have shown that the higher the content of sewage sludge in the primary mixture, the higher the pH and the lower the ash content in the biochar received. The conversion of BSG to calorific fuel has been reported in [25], where intermediate pyrolysis combined with catalytic steam reforming was investigated by Mahmood et al. That research has shown that higher temperature affects gaseous product production, and the heating value of gas was from 10.8 to 25.2 MJ/m^3, depending on the conditions. The microscale of BSG pyrolysis was described by Borel et al. [26]. Researchers used thermogravimetry analysis to investigate the thermal degradation of a BSG in the presence of nitrogen atmosphere. On the basis of the curves received from the TG analysis, scientists evaluated the independent parallel reaction model of the pyrolysis of basic components of BSG.

This paper provides a study of the intermediate pyrolysis of the brewer's spent grain in a laboratory scale fixed-bed reactor. This work focuses mainly on the impact of a gaseous atmosphere on the pyrolysis process. Based on the literature review, it was found that only some investigations presented the influence of sweeping gas on different types of pyrolysis, mainly in fast pyrolysis, with diversified biomass feedstocks [28–31]. Thus, the main goal of this paper was to compare the quality and quantity of products obtained under three kinds of atmosphere (Ar, N_2, and CO_2). In [28], Zhang et al. investigated a fast pyrolysis of corncob in a fluidized reactor under five different carrier gases: nitrogen, carbon dioxide, carbon monoxide, methane, and hydrogen. The results suggested that the smallest amount of bio-oil would be obtained under the CO atmosphere, and it was 49.6%, while the other gases gave similar results, and the liquid fraction was around

58%. Similar experiments were carried out by Liu et al. [29]. Researchers investigated the pyrolysis process of binder cold-briquetted lignite (BCBL) in a fixed bed reactor under carbon monoxide, carbon dioxide, methane, hydrogen, and nitrogen atmosphere. Scientists received the specific surface area of semichars in the range of 1.71–2.53 m^2/g where the lowest value was received for the CO atmosphere and the highest for N_2. Investigation of the impact of gaseous atmosphere on the pyrolysis process could also be found in [30], where Zhang and Zhang conducted experiments concerning a catalytic microwave-induced fast pyrolysis of medicinal herb residue under carbon dioxide, carbon monoxide, hydrogen, and nitrogen, and their studies were focused on determining the pyrolytic product yields and oil chemical compositions. The received results showed the dependency between the maximum content of the oil fraction and the gas atmosphere in the following line $N_2 > CO_2 > H_2 > CO$. These results confirm the results of the work [28].

In this work, studies concerning the impact of the gaseous atmosphere on the pyrolytic products from intermediate pyrolysis have been presented. This article focuses on the quality and quantity analysis of the pyrolytic products received under various conditions. The aim of this article was to provide an answer to the following questions: Are there crucial differences in the quality of products obtained under different gases in intermediate pyrolysis? Is it possible to reduce the amount of N-containing compounds in the oil if commonly used nitrogen is replaced by argon?

Moreover, intermediate pyrolysis compared to fast and slow pyrolysis characterizes the same yields of char, gas, and oil. This distribution of yields is desirable in the production process. As reported in the newest paper [32] that intermediate pyrolysis with polygeneration may play important an role in the reduction of CO_2 emissions up to 2050. The additional advantage of intermediate pyrolysis is the fact that oil has less oxygen content than that from fast pyrolysis. Moreover, the feedstocks used in intermediate pyrolysis can contain significant amounts of moisture, and hence, they do not require a pretreatment process. Thus, the motivation of this work was the investigation of the impact of carrier gas on intermediate pyrolysis. This fact makes this study novel, and the presented results may enhance the knowledge about another kind of pyrolysis.

2. Materials and Methods

The experiments were carried out in a horizontal fixed-bed reactor with a nonlinear heating rate of the sample. Three different carrier gases were selected: argon and nitrogen, which are considered non-reactive gases with solid feedstock, and carbon dioxide, which at higher temperatures could react with carbon [33]. The tests were carried out at three temperatures of 500, 600, and 700 °C to determine the effect of temperature on the received results.

Pyrolytic products were divided into three groups: solid residue (char), condensable fraction (oil), and light-weight gases. For every investigated case, the yield of products was determined. Chars were analyzed via ultimate analysis, and carbon, nitrogen, and hydrogen contents were determined. For selected chars obtained under a nitrogen and carbon dioxide atmosphere, combustion characteristics were plotted, and ignition and burnout temperatures were determined. The organic fraction of oil was analyzed by gas chromatography technique. Organic compounds which form the structure of oil have been determined and classified into chemical groups. Furthermore, the main experiments have been supplied by the microscale study involving a thermogravimetric analysis, where the pyrolysis behavior of BSG under considered gases was recorded.

2.1. Characterization of Feedstock

Brewer's spent grain (BSG) was selected as a feedstock material. This feedstock is the main organic residue after the beer production process. The BSG for the tests was supplied from the Polish brewery industry located in AGH Cracow. The material was collected directly after the brewing process, and the feedstock was immediately packed in hermetic bags and stored at 4 °C to prevent microbial fermentation during delivery. The raw feedstock contained a lot of moisture and for this reason, the BSG was dried at 40 °C for several days. Then, the air-dried material was milled, and for further experiments, nonuniform size particles (less than 1 mm) were selected. It was expected that the non-uniform size of the particles would be characterized by a lower bulk density. Consequently, the biomass sample should have more spaces to release volatile matter, and carrier gas should have easier sweeping into the sample.

2.2. Experimental Setup

Intermediate pyrolysis experiments of the brewer's spent grain were carried out in a horizontal fixed bed reactor. The reactor scheme is presented in Figure 1.

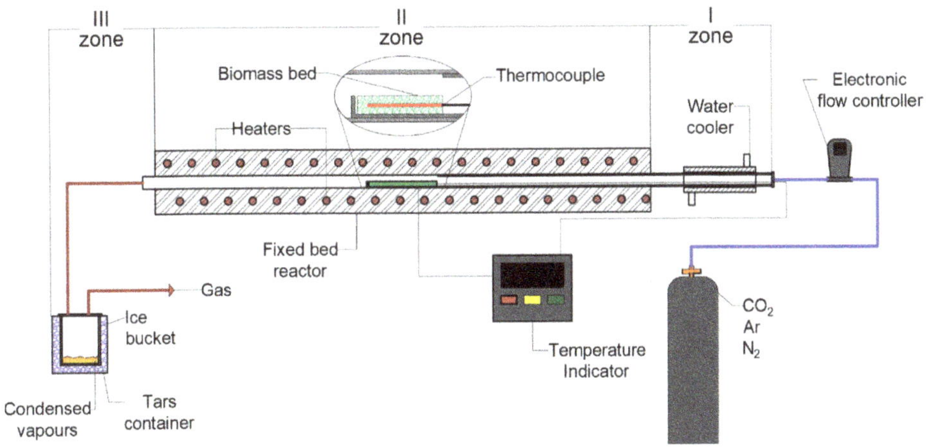

Figure 1. Experimental setup for the intermediate pyrolysis process.

The fixed-bed reactor is built with two concentric positioned stainless-steel pipes: external and internal. The external pipe is motionless and is surrounded by electric heaters. The external pipe with the heaters is enclosed by ceramic insulation. The internal pipe is placed inside the external pipe, and this pipe is able to move by push. In the internal pipe, a longitudinal hole was created, and in this hole, the feedstock was placed. The biomass sample was inserted into the reactor when the internal pipe was pushed.

The presented laboratory setup consists of three main sections. In the first zone, on the right side of Figure 1, a biomass sample was prepared for the experiment, and the solid residues were cooled after pyrolysis. BSG sample (1 g) was applied to the reactor. The sample was pyrolyzed for 3 min, and after this time, the sample was cooled. Inside the internal pipe, the sweeping gas was flowing. Nitrogen, argon, and carbon dioxide were chosen, and for each case, the volume flow rate was set at 300 mL/min. The purity of each gas was 99.9999%. The flow rate was monitored by an electronic regulator. Additionally, in a first zone, the water cooler was installed which allowed to cool a solid residue after pyrolysis.

In the second zone, the fixed-bed reactor was placed where intermediate pyrolysis was conducted. Heat was supplied by electrical heaters. The reactor temperature was measured on the wall of the external pipe using a K thermocouple. In this study, the temperature was set at 500 °C, 600 °C, and 700 °C.

The temperature was controlled by an autotransformer, and the temperature value was displayed in the indicator. During the experiment, in the initial time, a temperature drop of approximately 30–40 °C was observed due to the insertion of a cold internal pipe with the sample. Then, the temperature returned to the initial temperature. In this laboratory setup, cold feedstock was inserted into the preheated reactor up to the set temperature of the process, and the sample was heated with the non-linear heating curve. The detailed description of the non-linear heating rate can be found in previous research [34].

The last part of the setup provided the capture of liquid and gaseous products. Water and heavy hydrocarbons from volatile matter were liquified in a container that had been placed in an ice bucket. The remaining gases left the reactor and were not captured for analysis.

2.3. *Analytical Methods of Feedstock and Products Characterization*

2.3.1. Ultimate, Proximate, and Fiber Analysis

The elemental composition of brewer's spent was determined using a Truspec CHNS 628 Leco analyzer. All solid products from intermediate pyrolysis of BSG under considered gaseous atmospheres were also analyzed by ultimate analysis involving the same device. This apparatus determines carbon, hydrogen, and nitrogen content using combustion methods. A pre-weighted sample was inserted to the furnace where it was combusted with pure oxygen to carbon dioxide (CO_2) and steam (H_2O). High temperatures also allowed to create NO_x. The species received on detectors determine the content of basic elements.

Proximate analysis was performed for air-dried brewer's spent grain only. Moisture, volatile matter and ash content were determined according to the standards EN ISO 18134-2:2017, EN ISO 18122:2015, and EN ISO 18122:2015, respectively.

Fiber structure analysis of BSG was conducted according to van Soest method [35] to enhance knowledge about cellulose, hemicellulose, and lignin content. All basic analyzes of feedstock properties were duplicated.

2.3.2. Thermogravimetric Analysis

Thermogravimetric analysis was involved to investigate the impact of carrier gas on the BSG pyrolysis and also to determine the combustion properties of selected chars.

The mass loss of the BSG under the respected gases was determined using a STA 449 F3 Jupiter thermal analyzer. Approximately 12 mg of the sample was loaded onto platinum crucibles with an Al_2O_3 inlay. The sample was heated from ambient temperature to 800 °C at a heating rate of 5 K/min with a volume flow rate of 70 mL/min. After reaching the final temperature, the sample was kept for 20 min under isothermal conditions. Then, it was cooled to the initial temperature again.

According to the results of ultimate analysis of solid residues from intermediate pyrolysis, the chars received under carbon dioxide and nitrogen atmosphere at the lowest and the highest temperatures were selected for thermogravimetry analysis. The reason for choosing only four cases for TG analysis was as an attempt to investigate the char received under one of the inert atmospheres and compare it to the char from the reactive atmosphere at a given temperature.

These experiments were performed using Mettler Toledo TGA/SDTA 851. Around 6 mg of char was placed in a corundum crucible. The sample was then combusted under air atmosphere with a volume flow rate of 40 mL/min. The char was heated at a heating rate of 10 K/min from ambient temperature to 800 °C. During the experiments, mass loss, time, and temperature were recorded, and combustion profiles were determined. The results of the TG analysis, and their graphical interpretation allowed the ignition temperature and the burnout temperature to be found [36].

The ignition temperature refers to the minimum temperatures at which fuels spontaneously ignite without external source of ignition. The ignition temperature was determined according to work [36]. The burnout temperature is the temperature at which 99% of the initial fuel mass was converted.

All experiments were performed twice to test the repeatability, and a good consistency, with the standard errors of ± 1 °C, was observed.

2.3.3. Products Yield

The products from intermediate pyrolysis were classified into three main groups: char (solid residue), oil fraction, (liquid fraction consisted with organic fraction of condensable vapors and water), and non-condensable gases. The yield of each pyrolysis product was computed according to Equations (1)–(3).

$$Y_{char} = \frac{M_{char}}{M_{biomass}} \cdot 100\% \quad (1)$$

$$Y_{oil} = \frac{M_{oil}}{M_{biomass}} \cdot 100\% \quad (2)$$

$$Y_{gas} = 100\% - Y_{oil} - Y_{char} \quad (3)$$

where Y_x is the yield of x and M_x refers to the mass of x.

2.3.4. Gas Chromatography–Mass Spectroscopy

The liquid fraction collected from the intermediate pyrolysis contained a condensable fraction of volatiles (tar) and water. For further investigations, water had been separated, and only the organic fraction was analyzed. The purpose of the tar analysis was to identify the quality and quantity of the compound that built the oil structure received under various conditions. Analysis was performed using gas chromatography mass spectroscopy (GC–MS) and an Agilent GC 7890 B apparatus equipped with an MS 5977A mass spectrometer and a flame ionizer detector. Before analysis, the tars were dissolved in acetone. Then, tar analysis was carried out according to the evaporation of compounds from the samples. The sample was heated from 40 to 300 °C at a rate of 3 °C/min and kept for 10 min. The relative content of the determined compound was calculated as a ratio of the peak area of the compound to the sum of all the compound areas of the peak detected. All samples were investigated two times.

3. Results

3.1. *Brewer's Spent Grain Properties*

The results of the ultimate, proximate, and component analysis of BSG are presented in Table 1. An air-dry brewer's spent grain lost approximately 70% of the water that was included in the as-received state [18]. BSG contains a large amount of volatiles, and this fact classifies it as a good candidate for pyrolysis, especially for oil production, where large amount of volatiles is desirable.

Table 1. Results of ultimate, proximate, and component analysis of BSG in an air-dry state.

Ultimate Analysis (wt%)	
C	47.57
H	6.78
N	4.08
[a] O	41.59
Proximate Analysis (wt%)	
Moisture (M)	6.23
Volatile matter (VM)	81.57
Ash (A)	3.35
[b] Fixed carbon (FC)	8.85
Component Analysis (wt%)	
Cellulose	17.18
Hemicellulose	34.16
Lignin	3.12
Extractives	45.54

[a] O: by difference, ash free. [b] FC = 100 wt%–M–VM–A.

The BSG is built mainly from extractives such as proteins, fats, and sugars. Lignocellulosic hardwood feedstock is mainly composed of three main components, lignin, cellulose, and hemicellulose, and cellulose is the main component. In the case of agricultural biomass such as BSG, extractives constitute the main substances. It means that BSG is still rich in nutrients and, for this reason, is reused for animal food.. The results proved that lignin has the lowest share as a component, which is the main contributor in char formation during pyrolysis [37]. Additionally, fixed carbon is only 8.85%, and these facts lead to less char production than in the case of hardwood feedstocks [38]. The results of BSG characterization are compatible with data from the literature review [25,39].

3.2. Results of Thermogravimetry Analysis of BSG

The impact of carrier gas on the pyrolysis process was also tested on a microscale involving thermogravimetric analysis. The main goal was to determine the conversion behavior of the feedstock during heating under respectable gases. The thermogravimetric curves for BSG pyrolysis under N_2, CO_2, and Ar atmospheres are presented in Figure 2. Based on TG analysis, the derivative thermogravimetric (DTG) curves were created and are presented in Figure 2.

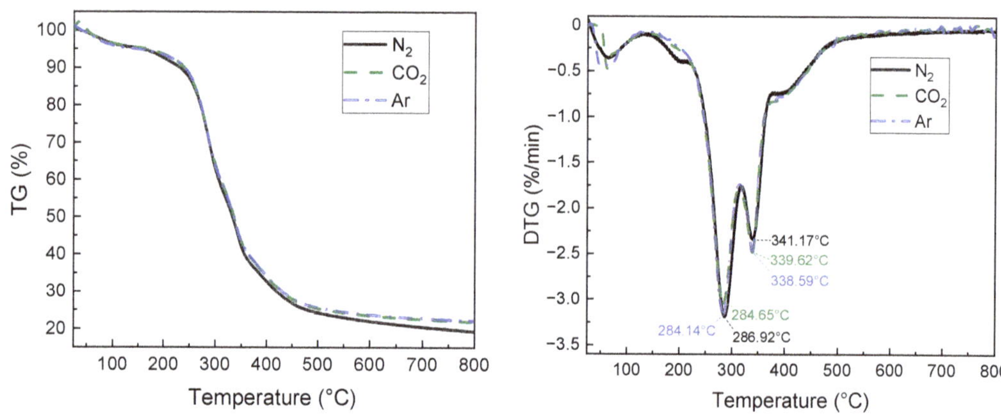

(**a**) (**b**)

Figure 2. TG (**a**) and DTG (**b**) curves of BSG pyrolysis under nitrogen, carbon dioxide, and argon atmospheres at a heating rate of 5 K/min.

From the presented figures, it can be noticed that feedstock material is pyrolyzed in the same way under each gas. There were no significant differences in the pyrolysis behavior. The first stage of the process between 25 °C and 105 °C referred to moisture evaporation. In the temperature range between 250 °C and 450 °C, BSG lost the highest amount of initial mass, and the mass dropped from around 91.5% to 30% of the initial mass. This zone refers to the main pyrolysis, where volatiles are released from biomass particles. It could be observed that this process was nearly the same for all the considered cases. For every test, there were two peaks where the mass conversion was the highest. The first peak could be observed at a temperature around 284 °C, where the mass loss was around 3.2%/min, and the second peak occurred at a temperature close to 340 °C with mass loss of around 2.4%/min.

After reaching 450 °C, the mass conversion was deaccelerated, and only residual volatiles were released from the particles. In this range of pyrolysis, it was also assumed that the reactive gas, carbon dioxide, had started to react with char. It was expected that the TG and DTG curves for inert gases would be nearly the same, while for carbon dioxide smaller amount of initial mass would be observed, especially for higher temperatures. From the presented results, the TG curves are closely the same for all gases. It could be concluded that, in this temperature range, carbon dioxide does not react with the char in a noticeable way. The received curves of BSG pyrolysis are compatible with the literature review [25].

3.3. Yield of Products from Intermediate Pyrolysis

The yield of pyrolytic products from the intermediate pyrolysis under an atmosphere of N_2, CO_2, and Ar at 500 °C, 600 °C, and 700 °C is presented in Figure 3. The oil represents the condensable fraction of volatiles and condensed water; the char represents the solid residue remained after the experiment, and the gas refers to light-weight gases which did not condensate in the ice bucket. The yield of products was calculated according to Equations (1)–(3).

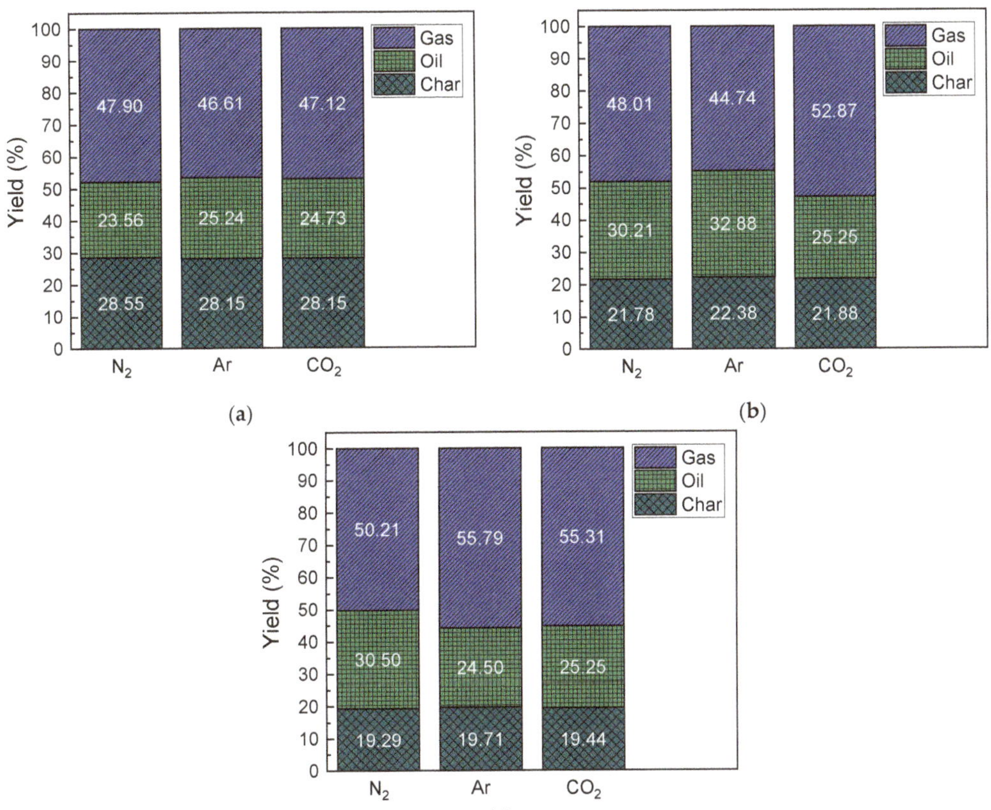

Figure 3. Products yields received from pyrolysis of BSG at (**a**) 500 °C, (**b**) 600 °C, and (**c**) 700 °C.

The presented results have confirmed that the carrier gases did not influence the products yields in a noticeable way. In particular, the char content remained at the same level and its yield changed only with temperature from around 28% at 500 °C to 19% at 700 °C. The char content was expected to be lower under the CO_2 atmosphere due to the heterogeneous reaction, and this reaction would be more intensive at higher temperature. In these cases, it could be concluded that the solid fraction did not react with the reactive atmosphere significantly, and the yield of char was nearly the same for CO_2 as for inert gases.

The yield of other products also did not change with the utilization of the carrier gas or even with the temperature. The oil fraction was in the range from 23.56% to 32.88%, and the gas yield was from 44.74% to 55.79% of the initial mass of BSG. A slight disproportion could be observed at 600 °C where the oil fraction yield was higher for inert gases than for carbon dioxide, and this difference was of around 7.5%.

Obtained results of products yield for intermediate pyrolysis are valid with literature review, and similar results could be found in work [40]. Scientists made an analysis of intermediate pyrolysis of 8 different feedstocks, and char content has been reported from 27–40%, oil yield was from 20% to 47.5%, and gas fraction consisted of 24.5–40.5% of the initial mass of sample. The results of the impact of carrier gas also agree with Reference [28], where difference between yield of char for various gases was around 1.5%.

3.4. Analytical Methods of Feedstock and Products Characterization

3.4.1. Ultimate Analysis of Char

The chars from the experiments were collected to analyze their properties, especially the elemental composition. Although the yields of char between various carrier gases were nearly identical, it was worth finding out about the composition of char from different cases. Table 2 presents the results of the ultimate analysis of the chars received under various conditions. Additionally, the atomic ratios of H/C and O/C were included in this table.

Table 2. Results of ultimate analysis of BSG char in an air-dry state.

Temperature (°C)	Gas Atmosphere	C (wt%)	H (wt%)	N (wt%)	[a] O (wt%)	H/C	O/C
500	N_2	65.00	4.25	6.36	24.39	0.785	0.281
	Ar	66.28	4.19	6.59	22.94	0.758	0.260
	CO_2	67.29	4.15	6.55	22.00	0.740	0.245
600	N_2	67.02	3.07	6.23	23.68	0.550	0.265
	Ar	67.18	3.14	6.30	23.38	0.561	0.261
	CO_2	68.60	3.00	6.65	21.75	0.524	0.238
700	N_2	68.44	2.48	5.94	23.14	0.435	0.254
	Ar	69.08	2.48	5.92	22.52	0.430	0.244
	CO_2	70.39	2.32	6.22	21.06	0.396	0.224

[a] O: by difference, ash free.

According to the received results, it can be observed that char from pyrolysis under carbon dioxide is slightly richer in carbon. This situation happens at all temperatures considered. The maximal difference in carbon content between chars is around 2%, and the following correlation was observed in every case: CO_2 > Ar > N_2. Taking into account other elements, it can be observed that the highest nitrogen content is received under CO_2 atmosphere except 500 °C, and this situation has been proven in [41]. Additionally, the smallest amount of oxygen in char is always obtained under a CO_2 atmosphere. From presented results, under CO_2 pyrolysis, chars are characterized by the lowest value of H/C and O/C ratios. The lower the H/C and especially the O/C ratios, the better the quality of the chars, and the char is close to being built with pure carbon, which is the desirable effect in solid fuels [42].

The small differences between the elemental composition can be confirmed according to Reference [43]. Gao et al. used pure nitrogen and a 50/50 mixture of nitrogen and carbon dioxide in lignite pyrolysis at 550, 700, and 800 °C. The differences in carbon content were around 1.2, 3.4, and 1.3%, respectively, at 500, 700, and 800 °C, but in all cases, the carbon chars from nitrogen pyrolysis were richer in carbon. In the work of Luo et al. [44], scientists investigated a pyrolysis of bituminous coal and platanus sawdust under a pure nitrogen and pure carbon dioxide atmosphere. The received results on the elemental composition of the char suggest the small difference in carbon content of 2% and, in this case, the chars obtained under carbon dioxide pyrolysis were also richer in carbon.

3.4.2. Combustion Characteristics of Selected Chars

From the results of the ultimate analysis, chars that were characterized by the lowest and highest carbon content at the lowest and highest investigated temperature were considered for further analysis. For this reason, the chars obtained from pyrolysis under nitrogen and carbon dioxide atmosphere at 500 and 700 °C were selected. The chars were analyzed by the thermogravimetric analysis and were combusted under air conditions. The results of char combustion behavior are presented in Figure 4. Based on these results, the ignition and burnout temperatures were determined, and results are presented in Table 3.

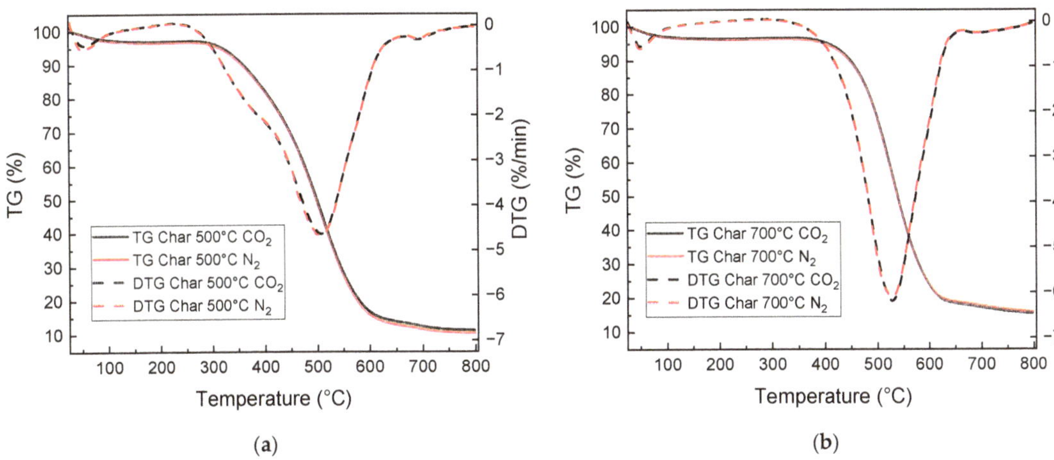

Figure 4. TG and DTG curves of char combustion which was obtained from pyrolysis under carbon dioxide and nitrogen atmosphere at (**a**) 500 °C and (**b**) 700 °C.

Table 3. Ignition and burnout temperatures for chars obtained at 500 and 700 °C under carbon dioxide and nitrogen atmosphere.

Sample Name	Ignition Temperature (°C)	Burnout Temperature (°C)
Char_500_CO_2	400.1	711.9
Char_500_N_2	397.0	710.8
Char_700_CO_2	466.3	739.5
Char_700_N_2	463.4	739.6

The small differences in the carbon content in the chars did not reflect the combustion behavior of the chars considered. The curves overlap by themselves, and for this reason, it could be concluded that the gaseous atmosphere of pyrolysis did not influence the combustion behavior of received chars. The conclusion can be confirmed by the similar ignition and burnout temperatures of chars. Only the temperature of pyrolysis had an impact on the values of the ignition and burnout temperatures. The ignition temperature of the chars was around 66 K higher for the char obtained at 700 °C. The same correlation was observed for the burnout temperature, but the difference was around 29 °C.

3.5. Characteristics of Organic Fraction of Oil

The oils collected from the experiments consisted of two separate phases: the aqueous phase (water + soluble compounds) and the organic phase. The phases had been separated, and for further analysis, only the organic fraction was involved. Gas chromatography was performed, where organic compounds were determined. The obtained results allowed the classification of the compounds on the basis of their functional group. The solutions of this classification are presented in Figure 5. Additionally, in Table 4 is the list of detected compounds and their relative content in the organic fraction of oil.

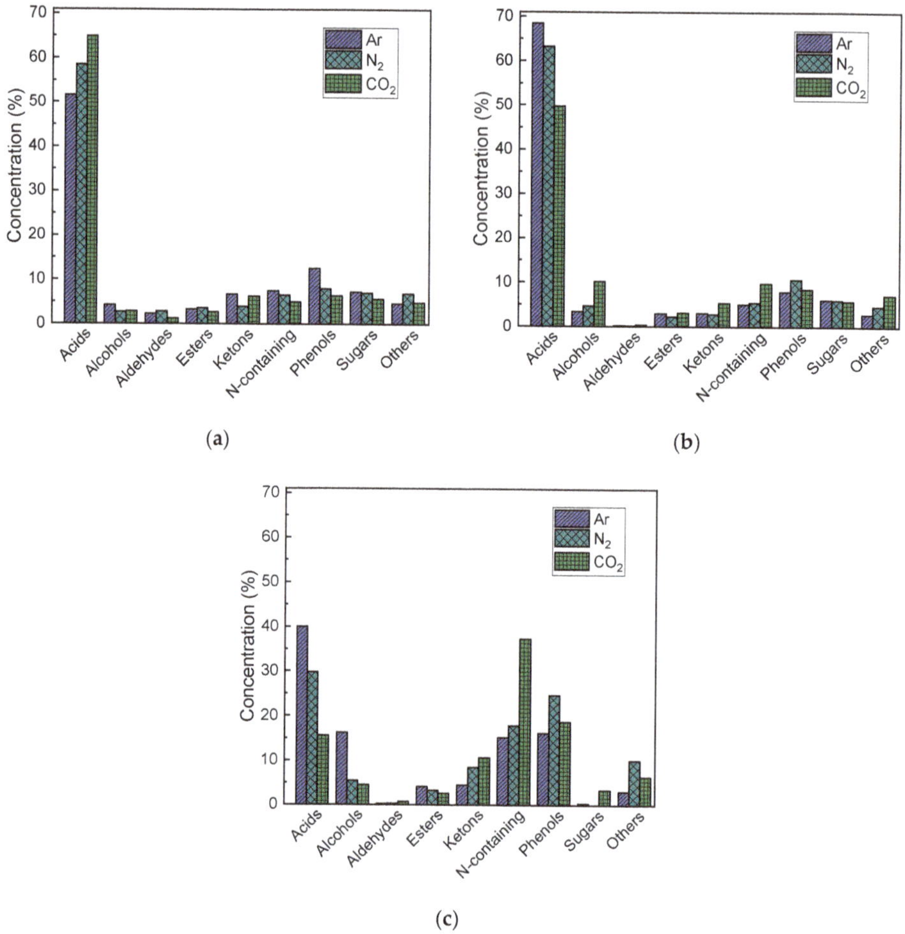

Figure 5. Relative content of different groups of chemicals in organic fraction of oil received under argon, nitrogen, and carbon dioxide at (**a**) 500 °C, (**b**) 600°C, and (**c**) 700 °C.

Despite the similar yields of the oil under respected gases, from Figure 5 and Table 4, it could be observed that a diversified structure of oil and a content of selected compound are different for every gas. The main group of compounds were acids, which was confirmed in work [25], and this group of compounds was dominant at every pyrolysis temperature and used carrier gas. At 600 °C and 700 °C, it was observed that under an argon atmosphere, the higher amount of acids evolved while the lowest amount was under carbon dioxide. The reverse situation occurred for pyrolysis at 500 °C, where at this temperature, carbon dioxide promoted the higher content of acids. A high amount of acids in the oil could be result of significant content of lipids up to even 13% in the BSG structure [45]. Among the acid compounds, the n-Hexadecanoic acid was dominant, and its concentration was dependent on the carrier used in the following line: Ar > N_2 > CO_2 at all temperatures. Another major compound among acidic group was 9,12-Octadecadienoic acid (Z,Z)-, and the highest content of 25.17% was determined at 500 °C under CO_2 atmosphere. For 600 °C and 700 °C, the same correlation was recorded as for the previous compound, but at 700 °C under CO_2, the compound was not detected. The last major compound identified was oleic acid, and its concentration was the highest for nitrogen atmosphere at 500 °C and 600 °C.

Another large group of compounds include nitrogen in their structure. The presence of these compounds in oil would be the result of the destruction of proteins that also have significant content in BSG up to 31% [46,47]. The N-containing compounds had the highest share in the oil structure at 700 °C and under CO_2, and their relative content was around 40%. For nitrogen and argon, N-containing compounds consisted around 18 and 15%, respectively, at the same temperature. The lower the temperature, the significant lower the number of N-containing compounds in the oil structure, and the relative content became the same for each gas. In this group, the main determined compound was 3-Octanamine received under carbon dioxide at 700 °C with a relative content of around 11.3%.

As the temperature of pyrolysis increased, the total phenol content grew, and the highest concentration was observed at 700 °C under a nitrogen atmosphere. In this atmosphere, the phenol content increased from around 8% at 500 °C to 30% at 700 °C. The dominant compound in this group was p-cresol, which was observed under all atmospheres at all temperatures. The highest concentration of p-cresol was determined under nitrogen at 700 °C. The presence of p-cresol could be the result of thermal degradation of lignin structures [48].

Analyzing Table 4 and Figure 5, it is worth also mentioning the content of derivatives of saccharides, and with it the oil, galacto-heptulose, that was determined. Its concentration is noticeable at 500 and 600 °C for all gases, and it was around 6–7%. At 700 °C, the compound was measured only in the case of carbon dioxide and was 3.45% of relative content. The presence of galcato-heptulose could be the result of saccharide degradation [49].

Presented analysis confirms that acids have the major share in oil structure with relative content of 52–65%, 49–68%, and 26–40%, respectively, at 500, 600, and 700 °C. The content of acids was dependent on used carrier gas. At higher temperatures, argon promoted higher concentration of acids where the lowest was observed for carbon dioxide. The diversified structure of oil received under various atmospheres at the same temperature was also confirmed in works [28,29,50].

Table 4. Chemical compositions of the oils obtained under argon, nitrogen, and carbon dioxide.

Compound	Formula	Relative Content at Different Conditions (%)								
		500 °C			600 °C			700 °C		
		Ar	N_2	CO_2	Ar	N_2	CO_2	Ar	N_2	CO_2
		Acids								
Propanoic acid	$C_3H_6O_2$	0.98	1.17	0.74	1.29	0.96	1.53	0.00	2.36	1.33
Propanoic acid, 2-methyl-	$C_4H_8O_2$	0.00	0.18	0.11	0.00	0.00	0.00	0.00	0.68	0.00
Butanoic acid, 3-methyl-	$C_5H_{10}O_2$	0.43	0.33	0.00	0.00	0.95	0.00	0.00	0.33	0.00
Undec-10-ynoic acid	$C_{11}H_{18}O_2$	0.00	0.00	0.00	0.47	0.33	0.52	0.55	0.32	0.67
n-Hexadecanoic acid	$C_{16}H_{32}O_2$	32.46	27.94	29.36	34.76	32.72	22.96	32.72	22.74	12.83
Oleic Acid	$C_{18}H_{34}O_2$	7.78	10.36	9.51	9.68	10.15	8.97	2.93	1.72	0.76
9,12-Octadecadienoic acid (Z,Z)-	$C_{18}H_{32}O_2$	10.02	18.49	25.17	22.21	18.10	15.71	3.76	1.59	0.00
		Alcohols								
2-Pentanone, 4-hydroxy-4-methyl-	$C_6H_{12}O_2$	0.70	0.59	0.83	1.20	0.47	0.88	14.33	2.71	3.02
2-Furanmethanol	$C_5H_6O_2$	0.98	0.75	0.60	0.97	0.72	1.84	0.18	2.13	0.00
5,7-Octadien-3-ol, 2,4,4,7-tetramethyl-, (E)-	$C_{12}H_{22}O$	0.19	0.23	0.19	0.00	0.00	0.00	0.00	0.31	0.00
1-Heptatriacotanol	$C_{37}H_{76}O$	2.34	1.12	1.22	1.17	3.50	7.42	0.66	0.00	0.93
3-Cyclohexen-1-ol, 1-methyl-	$C_{10}H_{18}O$	1.01	0.74	0.75	0.36	0.49	2.26	0.00	2.34	1.94
4,7-Methano-1H-inden-1-ol, 3a,4,7,7a-tetrahydro-	$C_{10}H_{12}O$	0.00	0.12	0.10	0.17	0.19	0.09	1.38	1.23	0.87
Bicyclo [2.2.1]heptan-2-ol, 7,7-dimethyl-, acetate	$C_{11}H_{18}O_2$	0.86	0.26	0.11	1.01	0.33	0.94	2.61	2.34	1.51
Tetracyclo [4.4.1.1(7,10).0(2,5)]dodec-3-en-11-ol	$C_{12}H_{16}O$	0.00	0.00	0.00	0.00	0.00	0.00	1.05	0.32	0.55
		Aldehydes								

Table 4. *Cont.*

Compound	Formula	Relative Content at Different Conditions (%)								
		500 °C			600 °C			700 °C		
		Ar	N_2	CO_2	Ar	N_2	CO_2	Ar	N_2	CO_2
2,3,4,5,6,7,8-Heptahydroxyoctanal	$C_8H_{16}O_8$	1.33	1.33	1.11	0.00	0.00	0.00	0.00	0.22	0.00
		Esters								
2-Propanone, 1-(acetyloxy)-	$C_5H_8O_3$	1.74	1.68	1.63	0.61	0.66	1.16	0.20	0.28	0.00
2,5-Octadecadiynoic acid, methyl ester	$C_{19}H_{34}O_2$	0.00	0.52	0.21	0.64	0.32	0.90	0.83	0.36	1.27
2-Octen-1-ol, 3,7-dimethyl-, isobutyrate, (Z)-	$C_{14}H_{26}O_2$	0.40	0.30	0.18	0.94	0.50	0.57	0.98	0.50	0.00
Propanoic acid, ethenyl ester	$C_5H_{10}O_2$	0.21	0.37	0.15	0.00	0.00	0.00	0.00	0.22	0.00
2-Acetoxy-1,2-dihydronaphthalene	$C_{12}H_{12}O_2$	0.95	0.71	0.54	0.81	0.78	0.52	2.17	1.90	1.41
		Eters								
Geranyl vinyl ether	$C_{12}H_{20}O$	0.88	0.78	1.25	0.54	0.24	2.46	0.64	1.57	0.94
		Ketons								
3-Hexen-2-one	$C_6H_{10}O$	0.17	0.00	0.27	0.00	0.00	0.25	0.27	1.01	3.97
2-Butanone, 4-cyclohexyl-	$C_{10}H_{18}O$	0.18	0.26	0.19	0.00	0.00	0.00	0.00	0.48	0.00
Cyclohexanone, 2-methyl-	$C_{10}H_{18}O$	1.46	1.26	1.51	1.04	1.38	0.96	0.00	0.96	0.94
2H-Pyran-2-one, tetrahydro-4-(2-methyl-3-methylene-1-buten-4-yl)-	$C_{11}H_{16}O_2$	0.33	0.28	0.64	0.46	0.47	0.91	0.25	0.16	1.40
3H-Cyclodeca[b]furan-2-one, 4,9-dihydroxy-6-methyl-3,10-dimethylene-3a,4,7,8,9,10,11,11a-octahydro-	$C_{15}H_{20}O_4$	0.91	1.48	0.16	0.28	0.19	0.40	0.22	0.12	0.72
Cyclohexanone, 2-(2-butynyl)-	$C_{10}H_{14}O$	0.93	0.00	1.72	0.00	0.00	0.00	0.00	0.00	0.00
		N-Containing								
2-Propen-1-amine, N,N-bis(1-methylethyl)-	C_4H_9NO	0.63	0.32	0.40	0.26	0.32	0.76	0.57	1.56	2.29
2-Pyrroline, 1,2-dimethyl-	$C_{11}H_{18}N_2O_2$	0.66	0.25	0.48	0.45	0.95	0.72	0.58	1.55	1.56
3-Octanamine	$C_{10}H_{19}N$	0.00	0.00	0.14	0.00	0.00	0.00	0.00	0.00	11.29
3-Pyrrolidinol	C_4H_9NO	0.47	0.17	0.14	0.00	0.00	1.54	0.00	0.00	0.00
4-(2,5-Dihydro-3-methoxyphenyl)butylamine	$C_{18}H_{35}NO$	0.00	0.13	0.00	0.37	0.38	0.42	0.72	0.26	0.70
4-(2,5-Dihydro-3-methoxyphenyl)butylamine	$C_8H_{19}N$	0.53	1.02	0.59	0.65	1.14	0.71	0.00	0.86	0.63
4-Piperidinone, 2,2,6,6-tetramethyl-	$C_9H_{19}N$	0.18	0.00	0.10	0.00	0.00	0.45	1.15	1.80	8.38
7-Azabicyclo [4.1.0]heptane, 1-methyl-4-(1-methylethyl)-	$C_{11}H_{19}NO$	0.50	0.00	0.00	0.16	0.00	0.24	0.00	0.86	0.54
9-Octadecenamide, (Z)-	$C_{11}H_{19}NO$	0.32	0.82	0.00	0.52	0.88	0.82	9.93	0.00	1.73
Acetamide, N-methyl-N-[4-(3-hydroxypyrrolidinyl)-2-butynyl]-	$C_{10}H_{13}N_3$	0.91	1.01	0.00	0.00	0.00	0.53	0.26	5.23	1.27
Debrisoquine	$C_9H_{17}NO$	0.32	0.00	0.11	0.00	0.00	0.00	0.00	1.14	0.00
Ethanimidic acid, ethyl ester	C_5H_5N	0.55	1.06	0.78	0.00	0.00	0.00	0.00	0.66	0.00
Hydrazinecarboxylic acid, phenylmethyl ester	$C_6H_{13}N$	0.15	0.16	0.19	0.36	0.22	0.35	0.59	1.08	0.61
N-[3-[N-Aziridyl]propylidene]-1-azacycloheptylamine	$C_6H_{11}N$	0.00	0.00	0.00	0.23	0.23	0.22	0.30	0.68	3.67
Piperidine, 1-methyl-	$C_{10}H_{21}N_3$	1.00	0.42	1.13	0.94	0.51	1.12	0.00	0.00	0.00
Pyridine	$C_8H_{10}N_2O_2$	0.91	0.91	0.70	0.59	0.79	1.01	0.37	0.00	2.37
Pyridine, 2-methyl-	C_6H_7N	0.32	0.28	0.30	0.56	0.00	0.88	0.74	2.21	2.29

Table 4. *Cont.*

Compound	Formula	Relative Content at Different Conditions (%)								
		500 °C			600 °C			700 °C		
		Ar	N_2	CO_2	Ar	N_2	CO_2	Ar	N_2	CO_2
Phenols										
Phenol	C_6H_6O	0.98	1.09	0.74	0.38	2.63	2.00	0.57	6.45	4.47
o-Cresol	C_7H_8O	0.16	0.19	0.14	0.54	0.64	0.39	1.00	1.87	1.32
p-Cresol	C_7H_8O	3.95	3.04	2.32	6.23	6.77	5.65	10.26	14.87	9.97
o-Ethylphenol	$C_8H_{10}O$	1.08	1.27	1.16	1.50	3.05	1.81	0.44	3.82	2.93
Mequinol	$C_7H_8O_2$	1.35	1.32	1.04	0.53	0.64	0.43	1.78	1.02	0.63
4-Ethyl-2-methoxyphenol	$C_9H_{12}O_2$	1.25	1.24	1.15	0.00	0.00	0.00	0.00	0.00	0.00
Phenol, 2,6-dimethoxy-	$C_8H_{10}O_3$	0.64	0.71	0.55	0.00	0.00	0.00	0.00	0.20	0.00
Phenol, 4-methoxy-3-(methoxymethyl)-	$C_9H_{12}O_3$	0.41	0.43	0.45	0.00	0.00	0.00	0.00	0.32	0.00
Androst-5,7-dien-3-ol-17-one	$C_{19}H_{26}O_2$	3.78	0.00	0.00	0.22	0.00	0.00	2.59	0.00	2.31
Sugars										
Galacto-heptulose	$C_7H_{14}O_7$	7.29	7.06	5.67	6.13	6.02	5.79	0.39	0.00	3.45
Other										
Cyclopropane, 2-(1,1-dimethyl-2-pentenyl)-1,1-dimethyl-	$C_{12}H_{22}$	0.36	0.36	0.30	0.00	0.00	0.00	0.00	0.60	0.00
R-Limonene	$C_{10}H_{16}$	1.44	0.75	0.47	0.00	0.00	0.00	0.00	0.62	0.00
Indene	C_9H_8	0.34	0.37	0.31	0.16	0.49	0.50	0.44	1.08	0.87
1,5,5-Trimethyl-6-methylene-cyclohexene	$C_{10}H_{16}$	0.00	0.36	0.38	0.36	0.61	0.34	1.40	0.00	0.65

4. Conclusions

In this work, intermediate pyrolysis of the brewer's spent grains was carried out in a horizontal fixed bed reactor at three temperatures of 500, 600, and 700 °C. The study focused on the analysis of the influence of carrier gas on the pyrolytic products received under the following conditions: argon, carbon dioxide, and nitrogen. The conducted analysis allows to draw the following conclusions:

1. According to thermogravimetric analysis, there were no significant differences in the pyrolysis behavior of brewer's spent grain under argon, nitrogen, and carbon dioxide with a heating rate of 5 K/min. The maximum conversion of samples was almost the same for each gas at 284 °C.
2. Intermediate pyrolysis of BSG delivered pyrolytic products independently of the carrier gas. Only temperature had an impact on product concentrations. Especially the char content dropped from around 28.5% to 19.5% as the temperature increased.
3. The ultimate analysis showed that the char received under carbon dioxide was richer in carbon by approximately 2% at all temperatures.
4. The selected chars received under various carrier gases had the same combustion properties as the ignition and burnout temperatures.
5. Analyzing oil composition, the most dominant group of organic compounds was acids with a maximum content of 68% at 600 °C under an argon atmosphere. The presented analysis showed the correlation between the content of acids and the carrier gas as follows: Ar > N_2 > CO_2. at 600 °C and 700 °C, less acids could be obtained in the case of carbon dioxide relative to inert gases.
6. The highest content in the oil was n-Hexadecanoic acid, and it could be a result of degradation of fats. Carbon dioxide promotes a lower concentration of this compound in the oil at 600 °C and 700 °C.
7. The inert atmosphere did not influence the yield of N-containing compounds in oil. at 600 °C and 700 °C, argon provided a higher amount of these compounds.
8. According to the results, it is worth considering the carbon dioxide as the carrier gas in the char production via intermediate pyrolysis. Char obtained under CO_2 is

characterized by the highest carbon content. Additionally, carbon dioxide is used, which indirectly may lead to decrease of CO_2 in the environment.

Author Contributions: Conceptualization, A.B., A.M. and W.J.; methodology, A.B, A.M. and W.J.; software, A.B.; validation, A.B., M.S. and A.M.; formal analysis, A.M and W.J.; investigation, A.B. and M.S; resources, A.B., A.M. and W.J.; data curation, A.B., K.S. and Ł.M.; writing—original draft preparation, A.B, K.S and Ł.M.; writing—review and editing, A.M., W.J. and M.S.; visualization, A.B, K.S. and Ł.M.; supervision, A.M and W.J.; project administration, A.M.; funding acquisition, A.M. All authors have read and agreed to the published version of the manuscript.

Funding: This work was supported by the Ministry of Science and Higher Education, Poland [grant AGH UST no. 16.16.110.663].

Institutional Review Board Statement: Not applicable.

Informed Consent Statement: Not applicable.

Data Availability Statement: Not applicable.

Conflicts of Interest: The authors declare no conflict of interest.

References

1. Directive 2009/30/EC Directive 2009/30/EC of the European Parliament and of the Council of 23 April 2009 Amending Directive 98/70/EC as Regards the Specification of Petrol, Diesel and Gas-Oil and Introducing a Mechanism to Monitor and Reduce Greenhouse Gas Emissions and Amend. 2009. Available online: https://eur-lex.europa.eu/legal-content/EN/TXT/?uri=celex%3A32009L0030 (accessed on 1 October 2021).
2. Nathan, A.J.; Scobell, A. Summary for Policymakers. *Foreign Aff.* **2014**, *91*, 1–30. [CrossRef]
3. Mohr, S.H.; Wang, J.; Ellem, G.; Ward, J.; Giurco, D. Projection of world fossil fuels by country. *Fuel* **2015**, *141*, 120–135. [CrossRef]
4. Poskart, A.; Skrzyniarz, M.; Sajdak, M.; Zajemska, M.; Skibiński, A. Management of Lignocellulosic Waste towards Energy Recovery by Pyrolysis in the Framework of Circular Economy Strategy. *Energies* **2021**, *14*, 5864. [CrossRef]
5. Anisimova, N. The capability to reduce primary energy demand in EU housing. *Energy Build.* **2011**, *43*, 2747–2751. [CrossRef]
6. Global Bioenergy Statistics. 2019. Available online: http://www.worldbioenergy.org/uploads/-191129%20WBA%20GBS%202019_HQ.pdf (accessed on 1 October 2021).
7. European Union. Directive 2009/28/EC Directive 2009/28/EC of the European Parliament and of the Council of 23 April 2009 on the promotion of the use of energy from renewable sources and amending and subsequently repealing Directives 2001/77/EC and 2003/30/EC. *Off. J. Eur. Union* **2009**, *5*, 2009.
8. Motghare, K.A.; Rathod, A.P.; Wasewar, K.L.; Labhsetwar, N.K. Comparative study of different waste biomass for energy application. *Waste Manag.* **2016**, *47*, 40–45. [CrossRef] [PubMed]
9. García, R.; Pizarro, C.; Álvarez, A.; Lavín, A.G.; Bueno, J.L. Study of biomass combustion wastes. *Fuel* **2015**, *148*, 152–159. [CrossRef]
10. Gunasee, S.D.; Carrier, M.; Gorgens, J.F.; Mohee, R. Pyrolysis and combustion of municipal solid wastes: Evaluation of synergistic effects using TGA-MS. *J. Anal. Appl. Pyrolysis* **2016**, *121*, 50–61. [CrossRef]
11. Wang, X.; Zhang, J.; Bai, S.; Zhang, L.; Li, Y.; Mikulčić, H.; Chen, J.; Wang, L.; Tan, H. Effect of pyrolysis upgrading temperature on particulate matter emissions from lignite semi-char combustion. *Energy Convers. Manag.* **2019**, *195*, 384–391. [CrossRef]
12. Kaczor, Z.; Buliński, Z.; Werle, S. Modelling approaches to waste biomass pyrolysis: A review. *Renew. Energy* **2020**, *159*, 427–443. [CrossRef]
13. Mlonka-Mędrala, A.; Evangelopoulos, P.; Sieradzka, M.; Zajemska, M.; Magdziarz, A. Pyrolysis of agricultural waste biomass towards production of gas fuel and high-quality char: Experimental and numerical investigations. *Fuel* **2021**, *296*, 120611. [CrossRef]
14. Glushkov, D.; Nyashina, G.; Shvets, A.; Pereira, A.; Ramanathan, A. Current Status of the Pyrolysis and Gasification Mechanism of Biomass. *Energies* **2021**, *14*, 7541. [CrossRef]
15. Eurostat. Happy International Beer Day! Products Eurostat News. Available online: https://ec.europa.eu/eurostat/web/products-eurostat-news/-/edn-20210805-1 (accessed on 1 October 2021).
16. Agro Industry Młóto Browarnicze–Stary Problem, Nowe Pomysły-Agroindustry.pl. Available online: https://www.agroindustry.pl/index.php/2021/01/12/mloto-browarnicze-stary-problem-nowe-pomysly/ (accessed on 28 May 2021).
17. Rachwał, K.; Waśko, A.; Gustaw, K.; Polak-Berecka, M. Utilization of brewery wastes in food industry. *PeerJ* **2020**, *8*, e9427. [CrossRef] [PubMed]
18. Mussatto, S.I.; Dragone, G.; Roberto, I.C. Brewers' spent grain: Generation, characteristics and potential applications. *J. Cereal Sci.* **2006**, *43*, 1–14. [CrossRef]
19. Mutlu, Ö.Ç.; Büchner, D.; Theurich, S.; Zeng, T. Combined Use of Solar and Biomass Energy for Sustainable and Cost-Effective Low-Temperature Drying of Food Processing Residues on Industrial-Scale. *Energies* **2021**, *14*, 561. [CrossRef]

20. Celaya, A.M.; Lade, A.T.; Goldfarb, J.L. Co-combustion of brewer's spent grains and Illinois No. 6 coal: Impact of blend ratio on pyrolysis and oxidation behavior. *Fuel Process. Technol.* **2015**, *129*, 39–51. [CrossRef]
21. Arranz, J.I.; Sepúlveda, F.J.; Montero, I.; Romero, P.; Miranda, M.T. Feasibility Analysis of Brewers' Spent Grain for Energy Use: Waste and Experimental Pellets. *Appl. Sci.* **2021**, *11*, 2740. [CrossRef]
22. Gil-Castell, O.; Mascia, N.; Primaz, C.; Vásqez-Garay, F.; Baschetti, M.G.; Ribes-Greus, A. Brewer's spent grains as biofuels in combustion-based energy recovery processes: Evaluation of thermo-oxidative decomposition. *Fuel* **2022**, *312*, 122955. [CrossRef]
23. Čater, M.; Fanedl, L.; Malovrh, Š.; Marinšek Logar, R. Biogas production from brewery spent grain enhanced by bioaugmentation with hydrolytic anaerobic bacteria. *Bioresour. Technol.* **2015**, *186*, 261–269. [CrossRef]
24. Yinxin, Z.; Jishi, Z.; Yi, M. Preparation and Application of Biochar from Brewery's Spent Grain and Sewage Sludge. *Open Chem. Eng. J.* **2015**, *9*, 14–19. [CrossRef]
25. Mahmood, A.S.N.; Brammer, J.G.; Hornung, A.; Steele, A.; Poulston, S. The intermediate pyrolysis and catalytic steam reforming of Brewers spent grain. *J. Anal. Appl. Pyrolysis* **2013**, *103*, 328–342. [CrossRef]
26. Borel, L.D.M.S.; Lira, T.S.; Ribeiro, J.A.; Ataíde, C.H.; Barrozo, M.A.S. Pyrolysis of brewer's spent grain: Kinetic study and products identification. *Ind. Crops Prod.* **2018**, *121*, 388–395. [CrossRef]
27. Olszewski, M.P.; Arauzo, P.J.; Maziarka, P.A.; Ronsse, F.; Kruse, A. Pyrolysis Kinetics of Hydrochars Produced from Brewer's Spent Grains. *Catalysts* **2019**, *9*, 625. [CrossRef]
28. Zhang, H.; Xiao, R.; Wang, D.; He, G.; Shao, S.; Zhang, J.; Zhong, Z. Biomass fast pyrolysis in a fluidized bed reactor under N2, CO_2, CO, CH_4 and H_2 atmospheres. *Bioresour. Technol.* **2011**, *102*, 4258–4264. [CrossRef]
29. Liu, J.; Li, G.Q.; Chen, L.; Wang, Y.; Xu, Y.; Qiao, X.X.; Zhang, Y.F. Effects of atmospheric gas on pyrolysis characteristics of briquetted lignite and surface properties of semi-char. *Fuel Process. Technol.* **2016**, *151*, 40–49. [CrossRef]
30. Zhang, B.; Zhang, J. Influence of Reaction Atmosphere (N_2, CO, CO_2, and H_2) on ZSM-5 Catalyzed Microwave-Induced Fast Pyrolysis of Medicinal Herb Residue for Biofuel Production. *Energy Fuels* **2017**, *31*, 9627–9632. [CrossRef]
31. Mellin, P.; Yu, X.; Yang, W.; Blasiak, W. Influence of Reaction Atmosphere (H_2O, N_2, H_2, CO_2, CO) on Fluidized-Bed Fast Pyrolysis of Biomass Using Detailed Tar Vapor Chemistry in Computational Fluid Dynamics. *Ind. Eng. Chem. Res.* **2015**, *54*, 8344–8355. [CrossRef]
32. Yang, Q.; Zhou, H.; Bartocci, P.; Fantozzi, F.; Mašek, O.; Agblevor, F.A.; Wei, Z.; Yang, H.; Chen, H.; Lu, X.; et al. Prospective contributions of biomass pyrolysis to China's 2050 carbon reduction and renewable energy goals. *Nat. Commun.* **2021**, *12*, 1698. [CrossRef] [PubMed]
33. Speight, J.G. Upgrading by Gasification. In *Heavy Oil Recovery and Upgrading*; Gulf Professional Publishing: Houston, TX, USA, 2019; pp. 559–614. [CrossRef]
34. Jerzak, W.; Bieniek, A.; Magdziarz, A. Multifaceted analysis of products from the intermediate co-pyrolysis of biomass with Tetra Pak waste. *Int. J. Hydrogen Energy* **2021**. [CrossRef]
35. Soest, P.J. Van Use of Detergents in the Analysis of Fibrous Feeds. II. A Rapid Method for the Determination of Fiber and Lignin. *J. Assoc. Off. Anal. Chem.* **1990**, *73*, 491–497. [CrossRef]
36. Li, Q.; Zhao, C.; Chen, X.; Wu, W.; Li, Y. Comparison of pulverized coal combustion in air and in O_2/CO_2 mixtures by thermo-gravimetric analysis. *J. Anal. Appl. Pyrolysis* **2009**, *85*, 521–528. [CrossRef]
37. Ding, T.; Li, S.; Xie, J.; Song, W.; Yao, J.; Lin, W. Rapid Pyrolysis of Wheat Straw in a Bench-Scale Circulating Fluidized-Bed Downer Reactor. *Chem. Eng. Technol.* **2012**, *35*, 2170–2176. [CrossRef]
38. Basu, P. *Biomass Gasification, Pyrolysis and Torrefaction: Practical Design and Theory*; Elsevier Inc.: Amsterdam, The Netherlands, 2013; ISBN 9780123964885.
39. Olszewski, M.P.; Arauzo, P.J.; Wądrzyk, M.; Kruse, A. Py-GC-MS of hydrochars produced from brewer's spent grains. *J. Anal. Appl. Pyrolysis* **2019**, *140*, 255–263. [CrossRef]
40. Tinwala, F.; Mohanty, P.; Parmar, S.; Patel, A.; Pant, K.K. Intermediate pyrolysis of agro-industrial biomasses in bench-scale pyrolyser: Product yields and its characterization. *Bioresour. Technol.* **2015**, *188*, 258–264. [CrossRef] [PubMed]
41. Aljaziri, J.; Gautam, R.; Alturkistani, S.; Fiene, G.M.; Tester, M.; Sarathy, S.M. On the effects of CO_2 atmosphere in the pyrolysis of *Salicornia bigelovii*. *Bioresour. Technol. Rep.* **2022**, *17*, 100950. [CrossRef]
42. Smoot, L.D. General Characteristics of Coal. In *Pulverized-Coal Combustion and Gasification*; Springer: Boston, MA, USA, 1979; pp. 123–132. [CrossRef]
43. Gao, S.P.; Zhao, J.T.; Wang, Z.Q.; Wang, J.F.; Fang, Y.T.; Huang, J.J. Effect of CO_2 on pyrolysis behaviors of lignite. *Ranliao Huaxue Xuebao/J. Fuel Chem. Technol.* **2013**, *41*, 257–264. [CrossRef]
44. Luo, Y.; Ben, H.; Wu, Z.; Nie, K.; Han, G.; Jiang, W. Impact of CO_2 on Pyrolysis Products of Bituminous Coal and Platanus Sawdust. *Polymers* **2019**, *11*, 1370. [CrossRef] [PubMed]
45. Lynch, K.M.; Steffen, E.J.; Arendt, E.K. Brewers' spent grain: A review with an emphasis on food and health. *J. Inst. Brew.* **2016**, *122*, 553–568. [CrossRef]
46. Ikram, S.; Huang, L.Y.; Zhang, H.; Wang, J.; Yin, M. Composition and Nutrient Value Proposition of Brewers Spent Grain. *J. Food Sci.* **2017**, *82*, 2232–2242. [CrossRef]

47. Santos, M.; Jiménez, J.J.; Bartolomé, B.; Gómez-Cordovés, C.; Del Nozal, M.J. Variability of brewer's spent grain within a brewery. *Food Chem.* **2003**, *80*, 17–21. [CrossRef]
48. Nonaka, H.; Yamamoto, R.; Funaoka, M. Selective conversion of hardwood lignin into syringyl methyl benzofuran using p-cresol. *Polym. J.* **2016**, *48*, 977–981. [CrossRef]
49. Moldoveanu, S.C. Chapter 16 Pyrolysis of Carbohydrates. *Tech. Instrum. Anal. Chem.* **2010**, *28*, 419–470. [CrossRef]
50. Tarves, P.C.; Mullen, C.A.; Boateng, A.A. Effects of Various Reactive Gas Atmospheres on the Properties of Bio-Oils Produced Using Microwave Pyrolysis. *ACS Sustain. Chem. Eng.* **2016**, *4*, 930–936. [CrossRef]

energies

MDPI

Article

Plastic Waste Management towards Energy Recovery during the COVID-19 Pandemic: The Example of Protective Face Mask Pyrolysis

Magdalena Skrzyniarz [1], Marcin Sajdak [2], Monika Zajemska [1], Józef Iwaszko [1], Anna Biniek-Poskart [3], Andrzej Skibiński [3], Sławomir Morel [1] and Paweł Niegodajew [4,*]

[1] Faculty of Production Engineering and Materials Technology, Czestochowa University of Technology, 19 Armii Krajowej Avenue, 42-200 Czestochowa, Poland; magdalena.skrzyniarz@pcz.pl (M.S.); monika.zajemska@pcz.pl (M.Z.); jozef.iwaszko@pcz.pl (J.I.); slawomir.morel@pcz.pl (S.M.)

[2] Department of Air Protection, Faculty of Energy and Environmental Engineering, Silesian University of Technology, 22 B Konarskiego Avenue, 44-100 Gliwice, Poland; marcin.sajdak@polsl.pl

[3] Faculty of Management, Czestochowa University of Technology, 19 B Armii Krajowej Avenue, 42-200 Czestochowa, Poland; anna.poskart@pcz.pl (A.B.-P.); andrzej.skibinski@pcz.pl (A.S.)

[4] Faculty of Mechanical Engineering and Computer Science, Czestochowa University of Technology, 21 Armii Krajowej Avenue, 42-200 Czestochowa, Poland

* Correspondence: pawel.niegodajew@pcz.pl; Tel.: +48-34-32-50-537

Abstract: This paper presents an assessment of the impact of the COVID-19 pandemic on the waste management sector, and then, based on laboratory tests and computer calculations, indicates how to effectively manage selected waste generated during the pandemic. Elemental compositions—namely, C, H, N, S, Cl, and O—were determined as part of the laboratory tests, and the pyrolysis processes of the above wastes were analysed using the TGA technique. The calculations were performed for a pilot pyrolysis reactor with a continuous flow of 240 kg/h in the temperature range of 400–900 °C. The implemented calculation model was experimentally verified for the conditions of the refuse-derived fuel (RDF) pyrolysis process. As a result of the laboratory tests and computer simulations, comprehensive knowledge was obtained about the pyrolysis of protective masks, with particular emphasis on the gaseous products of this process. The high calorific value of the pyrolysis gas, amounting to approx. 47.7 MJ/m^3, encourages the management of plastic waste towards energy recovery. The proposed approach may be helpful in the initial assessment of the possibility of using energy from waste, depending on its elemental composition, as well as in the assessment of the environmental effects.

Keywords: plastic waste; waste management; pyrolysis; thermal conversion; protective mask

Citation: Skrzyniarz, M.; Sajdak, M.; Zajemska, M.; Iwaszko, J.; Biniek-Poskart, A.; Skibiński, A.; Morel, S.; Niegodajew, P. Plastic Waste Management towards Energy Recovery during the COVID-19 Pandemic: The Example of Protective Face Mask Pyrolysis. *Energies* **2022**, *15*, 2629. https://doi.org/10.3390/en15072629

Academic Editor: Javier Fermoso

Received: 4 March 2022
Accepted: 31 March 2022
Published: 3 April 2022

1. Introduction

The appearance and rapid spread of the SARS-CoV-2 virus have caused an intensive increase in the demand for personal protective equipment, especially face masks, due to the obligation to wear them in public spaces [1]. The production of disposable food packaging has also significantly risen during the pandemic [2]. Quarantine, isolation, and the fear of a personal visit to the store have caused an increase in online sales of food, as well as other products necessary in everyday life [3,4]. For practical and hygienic reasons, many of these items are packed in disposable plastic packaging [5,6]. The necessity to ship the ordered products involves the use of cardboard boxes, packaging, cling film, bubble wrap, etc. to safely deliver the ordered products [7]. In addition, shoppers are encouraged to wear disposable gloves and to pack all fresh produce separately in plastic bags when shopping in stationery stores [8,9]. The consequence of this state of affairs is, unfortunately, an enormous increase in the amount of plastic waste, which, as of 22 November 2020, in the 25 countries with the highest incidence of COVID-19, reached approximately 54,000 tons [10]. This situation was also reflected in the production of the European plastics industry, especially in the second half of 2020.

The presented data show that the production of plastics before the COVID-19 pandemic increased until 2018 and with it the amount of plastic waste (Figure 1). In contrast, the production of the European plastics industry has decreased since 2018. However, after a sharp decline in production due to the restrictions resulting from the COVID-19 pandemic in the first half of 2020, in the second half of this year, the production of plastics in EU countries began to increase rapidly, according to the Association Plastics Manufacturers in Europe (Plastics Europe) and the European Association of Plastics Recycling and Recovery Organisations (EPRO).

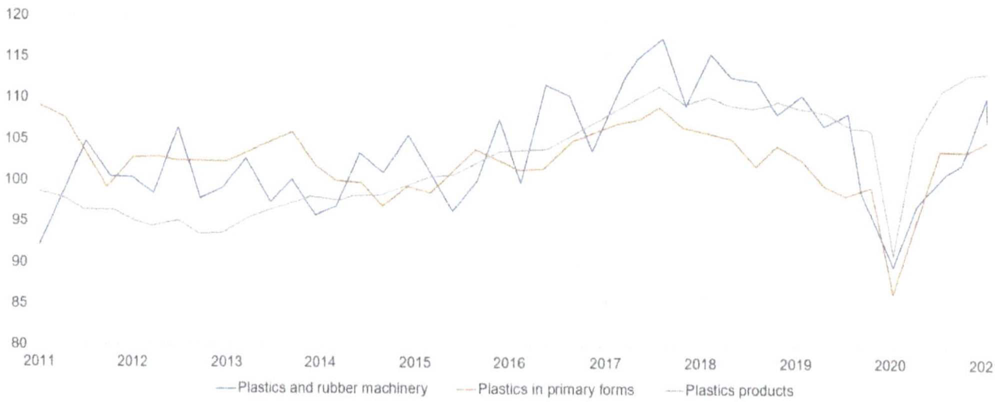

Figure 1. Plastic industry production in EU27 (index 2015 = 100, seasonally adjusted quarterly data) [11].

The pandemic has also had a negative impact on the economy and the proper functioning of local governments [12]. Municipalities and cities are struggling with an augmenting amount of municipal solid waste (MSW) [13] and medical waste, with a simultaneous lower budgetary inflow of funds as a result of the economic downturn [14]. Local governments cannot adequately manage the growing amount of hazardous waste placed in municipal landfills because of a lack of forces and resources at their disposal [15]. It should be noted that the challenges associated with plastics are largely owing to the fact that the ways in which such plastics are produced and utilised are not sustainable, because personal protective equipment (PPE) for hygiene reasons cannot be reusable, recycled, or are often made of single-use plastic. There is no doubt that the COVID-19 pandemic has changed the production and use of plastics [16]. These plastics significantly contribute to limiting the spread of the virus [12]. Nonetheless, according to the authors of this publication, the rapid increase in plastic waste due to the widespread use of masks and gloves, as well as changes in the manufacture and use of disposable products, could, in the short term, undermine EU efforts to reduce plastic pollution and move to more sustainable plastic handling [17,18].

The shock of the outbreak and expansion of the virus on a global scale poses enormous challenges to economies, public finances, and health systems with the threat of a possible recession [19]. The solid waste management sector is particularly susceptible, especially in EU countries, including Poland, which do not meet the requirement of recycling 50% of municipal waste (in 2020, recycling of MSW amounted to 26.7%, according to GUS data) [20,21]. Taking the example of Poland, it can be seen that during the pandemic there was a clear slowdown in the amount of processed waste, resulting in lower incomes in waste management treatment (WMT). Moreover, the situation in the secondary raw materials market had a negative impact on the entire waste industry. The prices of new commodities also fell, which resulted in their greater demand. On the other hand, the waste-to-energy industry (WTE) achieved good ratings and results. For example, the solid waste disposal sector in the Chinese market did not experience large deviations from the turnover recorded in 2019 [22].

The demand for personal protective equipment (PPE) has grown enormously worldwide since the start of the pandemic [23]. Such a high demand for PPE has led to the generation of huge amounts of plastic waste. For example, in Singapore, during 2 months of lockdown, an additional 1400 tons of plastic was generated [24]. Estimated calculations performed by Sazzadul Haque et al. [25] in Bangladesh have resulted in forecasts that the combined production of disposable face masks and other personal protective equipment is 3.4 billion units per month. This, in turn, translates into the production of 472.3 tons of plastic waste. According to estimates provided by Benson et al. [26], in Africa alone, assuming that people used at least one mask every day, it is estimated that the number of masks produced and disposed of during the day per person is around 412 million. Thus, more than 12 billion medical or textile masks are thrown away every month. Assuming that the average mass of the mask is 8.58 g, this gives the probability that about 105 Mt of masks per month only on the African continent may be discharged into the environment. According to Aragaw et al. [27], each month, around 129 billion masks and 65 billion disposable gloves are used and discarded worldwide.

The waste produced during the pandemic mainly consists of plastics such as polypropylene (PP), polyethylene (PE), polyester (PEs), polycarbonate (PC), polyethylene terephthalate (PET), polyvinyl chloride (PVC), and high-density polyethylene (HDPE), in addition to a small amount of paper, textiles, cotton, and natural rubber (Table 1) [28,29]. As can be seen from the literature review (Table 1), plastic waste can be a valuable fuel due to its high calorific value (38–46 MJ/kg).

Table 1. Proximate and ultimate analysis of plastic waste.

Plastic	Proximate Analysis (wt% Dry Basis)		Elemental Analysis (wt% Dry Basis)					LHV (MJ/kg)	Source
	Volatile Matter	Ash	C	H	N	S	O		
HDPE	98.2	1.8	84.5	13.8	0.1	0.1	1.5	43.0	[30]
LDPE	99.8	0.0	86.8	12.9	0.1	-	0.2	43.6	
PP	99.6	0.4	85.0	13.9	0.1	0.0	1.0	43.6	
PS	99.8	0.1	90.5	7.9	0.4	0.2	1.0	38.8	
PET	92.9	6.9	62.5	4.0	0.1	0.0	33.5	21.1	
PET-12	-	4.9	77.1	12.6	0.2	-	5.2	-	[31]
PET-28	-	12.1	67.2	9.7	0.1	-	10.9	-	
PET	86.75	6.83	63.94	4.52	0.01	0.04	31.49	-	[32]
PS	98.81	-	92.2	7.8	-	-	-	41.25	[33]
PE	99.96	-	85.5	14.5	-	-	-	-	
HDPE	91.88	3.9	83.4	12.71	1.08	0.002	2.8	46.48	[34]
PP	93.84	3.68	83.28	13.81	1.01	0.001	1.90	44.43	
PS	94.33	0.84	89.2	8.78	0.01	0.00	2.01	40.34	

"-" No data available.

Disposable face masks account for a significant share of the plastic waste stream generated during the pandemic [35]. As reported by Benson et al., the global production of disposable face masks is around 1.6 million tons/day, which means around 3.4 billion are thrown away daily due to the COVID-19 pandemic [26]. Protective masks are made of heterogeneous material, i.e., they consist mainly of several layers of PP and PE [36] (and additionally, they are equipped with an elastic band for the ears and a stiffening wire on the nose), which hinders their safe and quick recycling [37,38].

The most appropriate method of neutralising PPE should be material recycling, but due to the difficulty of separating the individual layers from each other, it would be a very time-consuming process [39]. Considering all the emerging difficulties that may arise in the recycling of PPE, one of the most effective methods of their disposal is the thermal transformation process [40–42]. Thus, effective waste management is a global concern that necessitates a reassessment of current technology and solutions. [43]. According to the introduced regulations, among others in Poland, waste generated during a pandemic and ending up in the mixed waste stream should be thermally neutralised by incineration, pyrolysis, or gasification [44,45]. In addition, mobile installations using the method of the thermal conversion of waste, i.e., pyrolysis, may occur to be an effective solution that responds to the needs of the waste industry during the pandemic. Mobile installations could be successfully utilised by entities collecting waste during the COVID-19 epidemic to neutralise hazardous (infectious) waste directly at the place of its collection, thus minimising the risk to public health and the environment. The above solution may also be used in the case of municipal waste contaminated with a biological agent from quarantine facilities and from places of isolation of patients at home. For the safe disposal of post-COVID-19 waste, a novel and sustainable approach is needed. As indicated in the literature, one of the most perspective methods of the thermal conversion of solid municipal waste, including plastic waste is pyrolysis, which can be a promising route to sustainable waste management. It is a thermochemical process carried out in the absence of oxygen in a pyrolysis reactor, in which waste decomposes at a temperature of 400–900 °C. This method enables the effective neutralisation of hazardous waste, but also leads to the formation of valuable solid (char) and gaseous (high-calorific pyrolysis gas) products that can be used, thus reducing the total cost of pyrolysis. The share and quality of individual products varies, depending, among others, on the type of pyrolysis reactor, type of waste, temperature, and residence time in the reactor [45].

The advantage of a pyrolysis installation compared with a waste incineration plant is its greater flexibility in terms of the amount of processed waste. These installations are economically justified also in the case of disposing of smaller amounts of waste than in the case of incineration plants, which gives opportunities, for example, to small municipal corporations operating, among others, in Polish conditions, for effective and profitable solid waste management. The proposed disposal method is also characterised by lower financial outlays and environmental fees, compared with incineration plants, as well as lower emissions of pollutants, i.e., SO_2 and NO_x. Hence, pyrolysis is considered to be a sustainable solution that may be economically profitable on a large scale and could minimise environmental concerns [46]. According to available data, with the limited number of possible solutions in this area, there is a great need for innovation enabling it to meet the key challenges in the field of plastic waste management in the pandemic era and to integrate new solutions in the field of thermal waste treatment technologies into the existing waste management system.

Taking into account above mentioned facts, pyrolysis seems to be the most promising thermal conversion method for such kind of waste into an environmentally inert, as well as valuable product from the energy point of view. Furthermore, considering the latest research results obtained by the authors of the article [47], in the case of RDF pyrolysis, it is justified to develop pyrolysis technology with simultaneous management of the products of this process.

Pyrolysis gas's high calorific value (30 MJ/m^3) encourages its use in both heating and the iron and steel industry. The gas produced by RDF pyrolysis can be co-incinerated with natural gas in industrial heating chambers, reducing the consumption of that fossil fuel. Pyrolysis gas can also be a substitute for coke oven gas for heating furnaces. This gas could be a viable alternative to conventional fuels, helping to reduce the role of waste storage while also improving environmental protection. The liquid fraction, in turn, is a potential source of valuable chemicals such as benzene and toluene. It can also be used as liquid fuel with properties similar to fuel oil or diesel fuel. Moreover, the preliminary technological tests carried out by the authors have shown the possibility of using the solid pyrolysis product in the production of insulating building materials, which increases the economic attractiveness of pyrolysis compared with other methods, e.g., combustion.

Bearing in mind the above, the authors of this study undertook this study to assess the impact of the COVID-19 pandemic on the waste management sector and to identify an effective method of managing selected waste generated during the pandemic.

2. Materials and Methods

The subject of the research was non-medical, three-layer, biologically uncontaminated disposable face masks presented in Figure 2, produced by Quanzhou Ruoxin Hygiene Products Co. Ltd. Caicuo, Luoyang, Huian Country, Quanzhou City, Fujian Province, China, manufactured according to the GB/T 32610-2016 standard, model NS-2020.

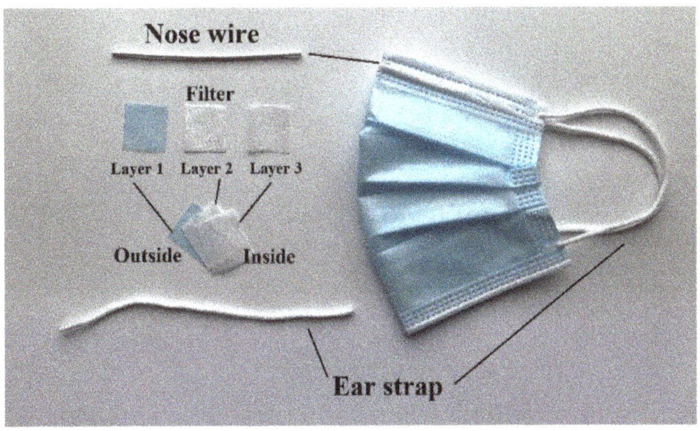

Figure 2. View of disposable protective mask used for studies.

As part of the research, both laboratory experiments and computer simulations were carried out. The research scheme is shown in Figure 3.

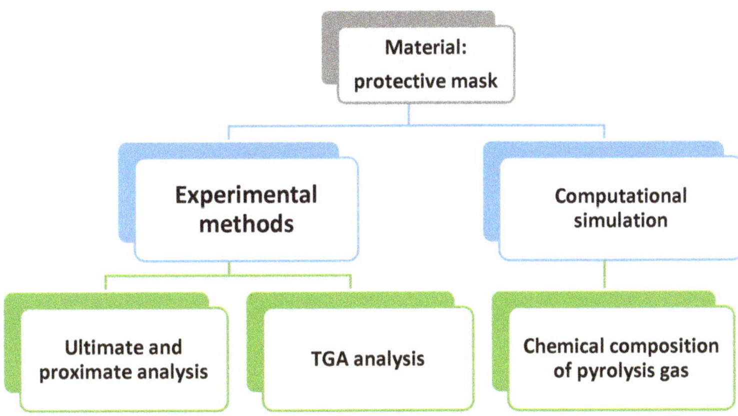

Figure 3. Research scheme.

2.1. Experimental Procedures

2.1.1. Proximate Analysis

Proximate analysis, e.g., the total moisture and ash content, were determined in the studied material. The total moisture content was determined by drying a sample in a laboratory dryer to a constant weight at 105 ± 3 °C. The ash content of the samples was determined by sample incineration. The investigated sample was put in a muffle furnace and heated in an air atmosphere at 600 ± 10 °C. This temperature was maintained until the sample reached a constant weight [48].

2.1.2. Ultimate Analysis

The elementary C_t^a, H_t^a, N^a, S_t^a, and O_t^a compositions in the examined material (protective mask) were measured using a Vario MACRO Cube automatic elemental analyser [49]. At 1150 °C, the material was put through an automated quantitative combustion process in an oxygen stream. Then, the combusted gases were transferred to a reduction tube (filled with copper) in which the sulphur and nitrogen oxides were reduced to SO_2 and N_2. The combustion gases were sent to a dynamic separation system after passing through the reduction tube. The absorption columns were desorbed thermally in order in the separation system. A thermal conductivity detector was used to identify gases including N_2, CO_2, H_2O, and SO_2 (TCD). An NDIR detector was used to examine low SO_2 concentrations [47].

The oxygen concentration was measured quantitatively using sample pyrolysis at 1120–1150 °C in an H_2O-, CO_2-, and O_2-free reductive environment (95% N_2 and 5% H_2). The sample was put in a pyrolysis tube containing elemental carbon, and carbon dioxide was produced as a result of the interaction between the oxygen in the sample and the carbon in the filler (Boudouard equilibrium). Granulated NaOH absorbed the acidic compounds obtained during the pyrolysis process, such as H_2S, HCN, and HCl, while a dehumidifier absorbed the generated water. The other gases did not need to be separated because the NDIR detector was only sensitive to CO. The inert pyrolysis products (N_2 and CH_4) might be immediately transported into a carbon monoxide measuring device.

2.1.3. Chlorine Determination

The chlorine content in the studied sample was determined by a potentiometric method [50]. The approach involves thoroughly burning a sample in an oxygen bomb calorimeter with an Eschka mixture (MgO and Na_2CO_3, ratio 2:1). The generated chlorides (combustion products) were extracted in nitric acid (V) and titrated for chloride ions with 0.1 mol/L $AgNO_3$. The potentiometric titration method used a silver sulphide electrode with a potential that was dependent on that of a reference electrode. The concentration of chloride ions is measured by the potential difference between the electrodes.

3. Results and Discussion

The results of the above-mentioned studies are summarised in Table 2.

Table 2. Proximate and ultimate analysis of plastic waste of protective masks.

Material	Proximate Analysis (wt% Dry Basis)			Elemental Analysis (wt% Dry Basis)					
	Ash	Moisture	Voltaire Matter	C	H	N	S	O	Cl
Plastic waste of protective mask	4.76	1.23	95.24	76.4	11.65	1.12	0.16	4.65	0.03

3.1. TGA Analysis

The tested material (protective mask) was homogenised and then pulverised with a cryogenic mill for thermogravimetric analysis. A Luxz 409 PG thermogravimetric analyser was used in conjunction with a Netzsch QMS 403D Aëolos mass spectrometer. The experiments were conducted in an argon environment with a 25 mL/min flow rate. Measurement began at a temperature of 40 °C. During the research, water jacket testing was used to stabilise the equilibrium. The pyrolysis operations reached a maximum temperature of 700 °C. The sample was 5.0 0.1 mg in weight. The samples were put in 6 mm diameter Al_2O_3 crucibles. The mass loss (TG) and maximum rate of mass loss (DTG) during the reaction, as well as the beginning and final temperatures at each step of the pyrolysis processes, could all be calculated using the data acquired (Figure S1 in Supplementary Material). To illustrate the differences between the various materials, in Figure 3, in addition to the protective mask test sample, the results of thermogravimetric analysis for pure polypropylene and pure polyethylene terephthalate are shown, as well as three biomass samples [49].

3.2. Computational Simulation

Calculations were made for the pyrolysis process carried out in a pilot continuous pyrolysis reactor, described in detail in [47], using licensed Ansys CHEMKIN-PRO software. Surgical masks with the elemental composition presented in Table 2, as well as PET and PP, the elemental composition of which were taken from studies of [51] and [49] were analysed. A detailed mechanism for the thermal conversion of solid fuels was implemented for the calculations, based on the Arrhenius equation for the rate constant of a reaction [52], developed by the CRECK Modeling Group, including 169 compounds and 4656 chemical reactions [53]. The mechanism was based on the detailed mechanism of HCl and Cl_2 high-temperature chemistry reported in [54]. The chemical mechanism adopted for calculations has been used many times by Ranzi [55,56], Faravelli [57], as well as by the authors of this article to model the thermal conversion of fuels and waste [47,58,59]. The calculations assume that pyrolysis occurs in a reactor with perfect reactant mixing, i.e., a perfectly stirred reactor (PSR) [60]. A diagram of the performed calculations with the input data and the conditions of the pyrolysis process is shown in Figure 4.

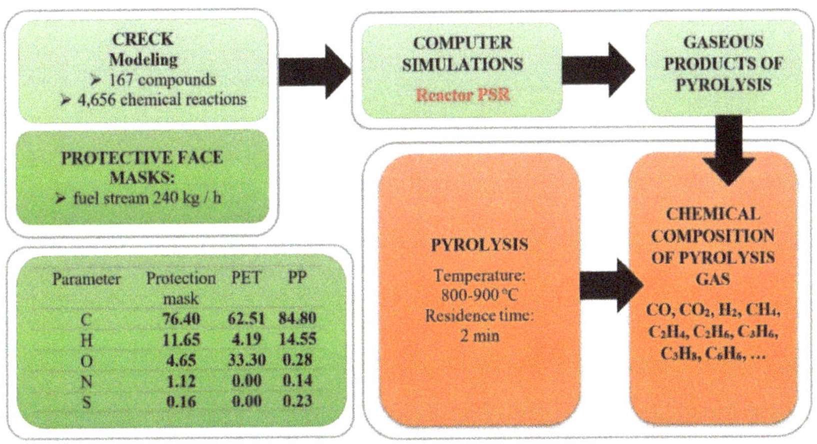

Figure 4. Scheme of calculation procedure.

As part of the computer simulations, the chemical composition of the gaseous pyrolysis products of the above-mentioned waste was determined, on the basis of which the calorific value of the pyrolysis gas was calculated. The following compounds were analysed in detail: CO, CO_2, H_2, CH_4, C_3H_8, C_4H_4, C_6H_6, C_2H_2, C_2H_4, C_2H_6, and C_3H_6. The obtained results were compared with the experimental data obtained from RDF pyrolysis with a high content of plastic waste (over 60%).

3.3. TGA Analysis

Thermogravimetric analysis (TGA) quickly shows the subsequent phases of the thermal conversion process of the studied sample. The mass loss of the examined material during heating is explained by this analysis. The TGA function's second derivative, which is mass change as a function of temperature (DTG), is used to determine the temperature at which the decomposition rate (in the considered example) is highest. The type of studied material is determined by the temperature and the shape of the DTG curve (e.g., one small and sharp peak or a wide peak).

The outcomes of the obtained repetitions are comparable, as can be shown (Figure 5.).

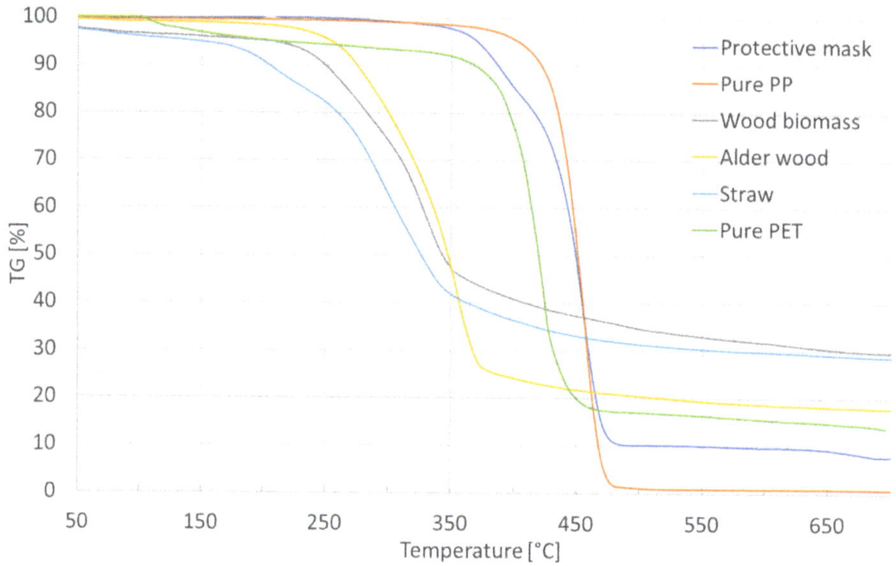

(a)

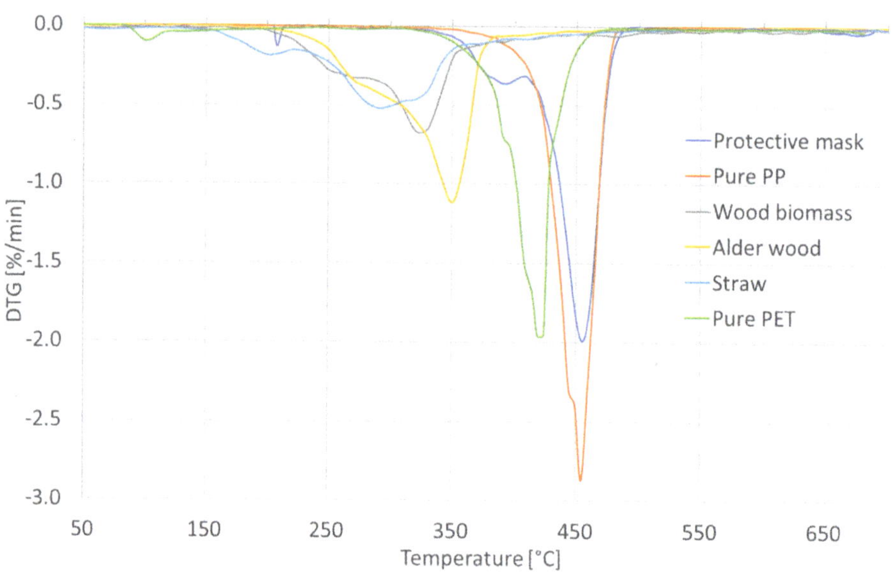

(b)

Figure 5. Comparison of repeated TG curves (**a**) and the TG and DTG curve (**b**).

As a result, further research may be conducted.

To optimise the number of gaseous products generated during the pyrolysis process, thermogravimetric analysis was carried out at a final temperature of 700 °C for the TGA analysis and 900 °C for the TGA-MS study. The face mask might degrade in four stages, according to TGA data (Figure 4).

The temperature ranges for the four stages were 50–200 °C, 201–392 °C, 393–500 °C, and 501–700 °C. The first stage represents minor degradation regions due to the occurrence of moisture evaporation (~2 wt%). The second stage indicates the sample's first substantial decomposition zone, with a mass loss of 16.11 wt% due to organic material degradation below 392 °C, caused by the release of pyrogenetic water from the research material made from polypropylene and a melt blown filter (PP). The third stage represents the second major decomposition region of the investigated sample, with a mass loss of 72.55 wt%, owing to the decomposition of organic materials below 455 °C, which corresponds to polypropylene thermoplastic material [61]. The last decomposition stage represents minor degradation regions due to char devolatilisation/decomposition (~2 wt%). The decomposition of the mineral portion present in the sample, magnesite (MgCO3), which is used as a filler in polymer processing, is connected with the DTG peak detected at 677.6 °C in this location. The DTG curve (Figure 6) shows that the components included in the material from which the investigated face mask were made decompose into two main components (two major degradation peaks), but both corresponded to PP pyrolysis products. During the decomposition of the polypropylene portion at 403.8 °C, one of the most significant emissions of gaseous chemicals from the examined sample is detected—pyrogenetic water (*m*/*z* 18) and carbon dioxide (*m*/*z* 44). The gaseous fraction, rich in alkanes ($C_nH_{2n+1}{}^+$), alkenes ($C_nH_{2n-1}{}^+$), is observed at a higher temperature of 455.3 °C during the pyrolysis process. They correspond to the mass ions at *m*/*z* 43, 55, 56, 57, 69, and 70, respectively, for $C_3H_7{}^+$, $C_4H_7{}^+$, $C_4H_8{}^+$, $C_4H_9{}^+$, $C_5H_9{}^+$, and $C_5H_{10}{}^+$. The mass ion *m*/*z* 53 corresponds to $C_3H_3N^+$, which may come from the degradation of a pigment added to the plastic material of one side of the mask. The results from mass spectrometry are shown in Figures 6 and 7.

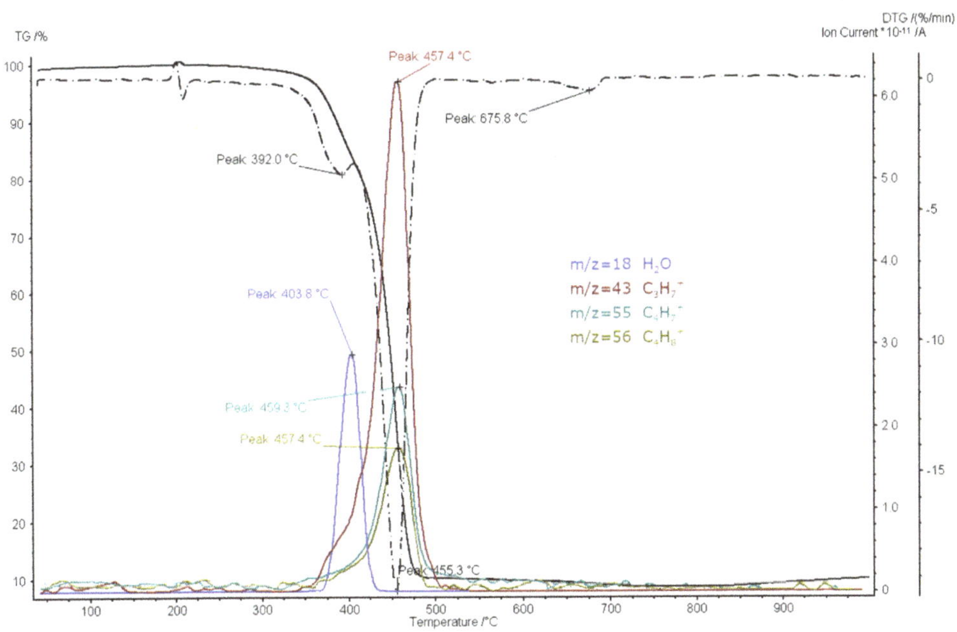

Figure 6. TG-DTG curves with curves of MS ion currents of volatile fraction (*m*/*z* 18, 43, 55, 56).

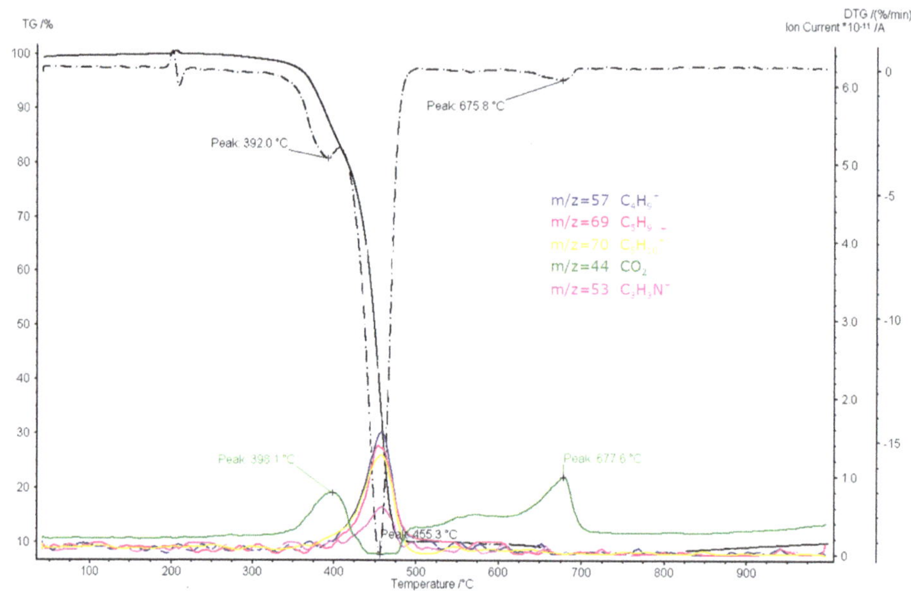

Figure 7. TG-DTG curves with curves of MS ion currents of volatile fraction (*m*/*z* 44, 53, 57, 69, 70).

3.4. Results of Computational Simulation

As part of the computer simulations, the effect of temperature was investigated on the proportion of primary pyrolysis gaseous products determining the calorific value of the pyrolysis gas—namely, CO, H_2, CH_4, C_3H_8, C_4H_4, C_6H_6, C_2H_2, C_2H_4, C_2H_6, and C_3H_6 (Figure 8).

The calculations show that the concentration of all the analysed compounds strongly depends on the process temperature. The largest share is observed for CO (up to 65%) in the entire temperature range but only for polyethylene terephthalate (PET). On the other hand, for PP and protective masks, a high proportion of hydrogen, even 50%, is visible, but only in the lower temperature range, i.e., from 400 °C to 500 °C. At the temperature of 700 °C, there is a clear decrease in the H_2 concentration to 15% for the masks and 22% for PP. At the same time, an increase in the hydrogen content from PET pyrolysis is observed; however, it does not exceed 5%. The opposite situation is observed for CH_4, the share of which increases with the rise in the process temperature, reaching a concentration of approx. 35% for both masks and PP at 700 °C, with a simultaneous low concentration (less than 5%) for PET. Noteworthy is the high proportion of C_6H_6 benzene in the pyrolysis gas for all the analysed wastes, at the level of approx. 20% for PP and masks and 17% for PET. From the energy point of view, the significant content of C_2H_4 is also significant, which grows with the temperature to approx. 9% at 800 °C for the masks and PP, and C_6H_6, reaching even 7% at the temperature of 650 °C for polypropylene. For the remaining compounds—namely, C_3H_8, C_4H_4, and C_2H_2 (except for PET), with increasing temperature, a sharp decrease in the concentration is observed for all the waste, and it does not exceed a 5% share in the pyrolysis gas.

The results of the model were compared with those of experimental research for RDF pyrolysis at 900 °C, presented in [47] (Figure 9).

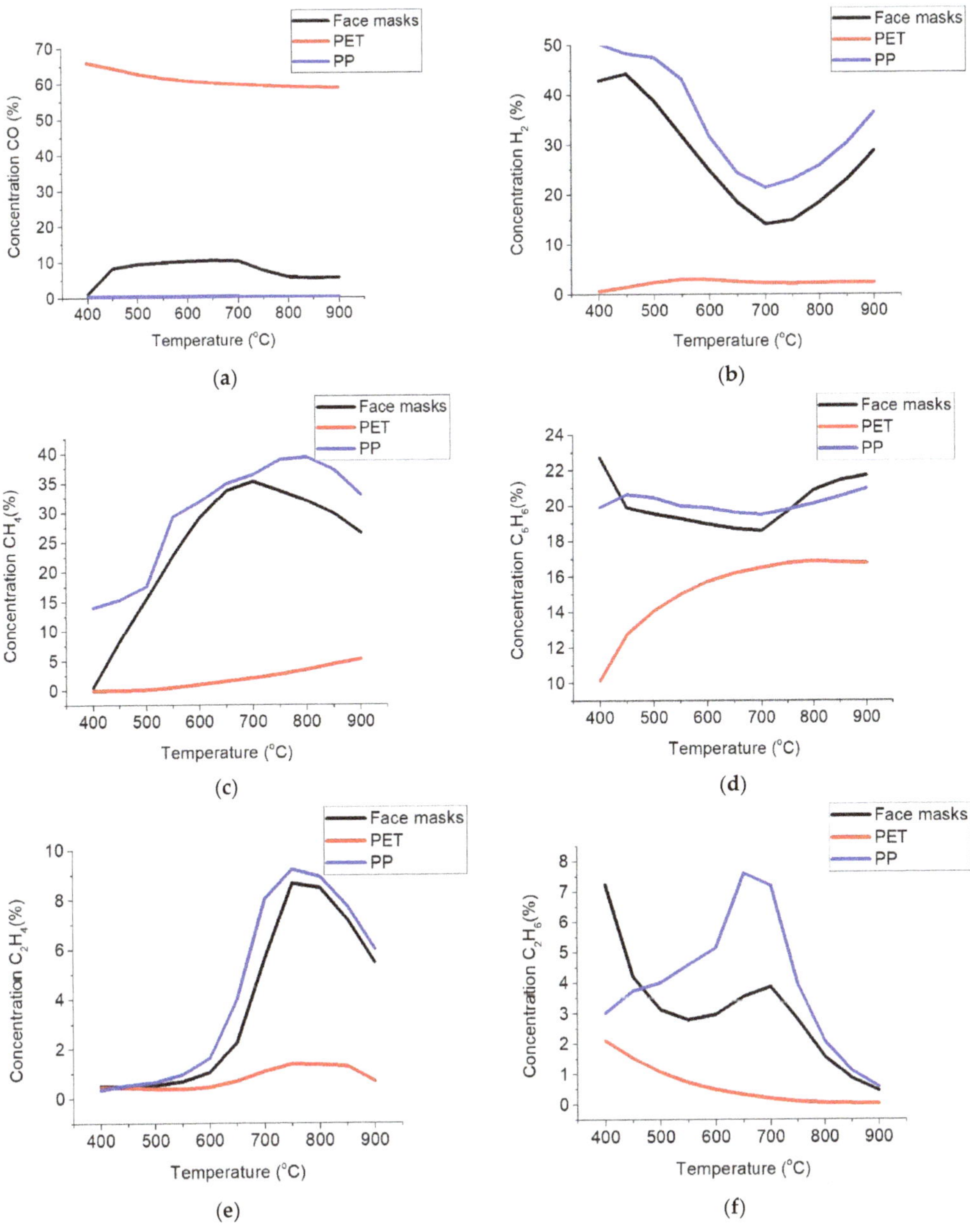

Figure 8. *Cont.*

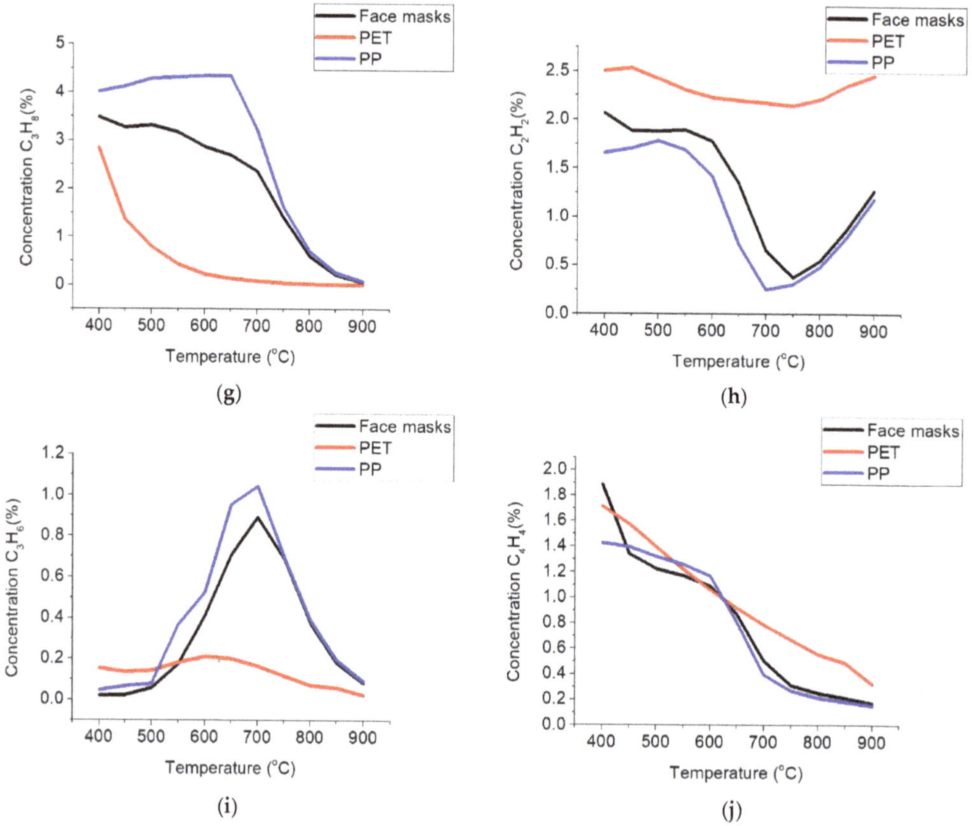

Figure 8. Influence of temperature on formation of (**a**) CO, (**b**) H_2, (**c**) CH_4, (**d**) C_6H_6, (**e**) C_2H_4, (**f**) C_2H_6, (**g**) C_3H_8, (**h**) C_2H_2, (**i**) C_3H_6, (**j**) C_4H_4.

Comparing the results obtained by modelling with the results from the experiment, the potential resulting from the pyrolysis of plastic waste, in particular polypropylene, the main component of protective masks, is visible. The high proportion of flammable compounds such as H_2, CH_4, and C_6H_6 translates into the high calorific value of the pyrolysis gas, amounting to 49.73 MJ/m^3 for PP pyrolysis gas, and 47.74% for mask pyrolysis (Figure 10). It should be mentioned that the presence of heavy organic molecules and their derivatives in the wet pyrolysis gas, which can account for up to 30% of the gas volume, has a substantial impact on the high calorific value. As a result of the condensation of these chemicals, pyrolysis oil is produced. According to preliminary research undertaken by the article's authors, direct management of wet pyrolysis gas (i.e., a mixture of pyrolysis gas and gaseous oil) in high-temperature heating chambers with temperatures exceeding 1300 °C is possible. The disposal of the problematic liquid portion, i.e., pyrolysis oil, is favoured at such a high temperature, resulting in considerable economic gains. The proposed method is described in great detail in [62].

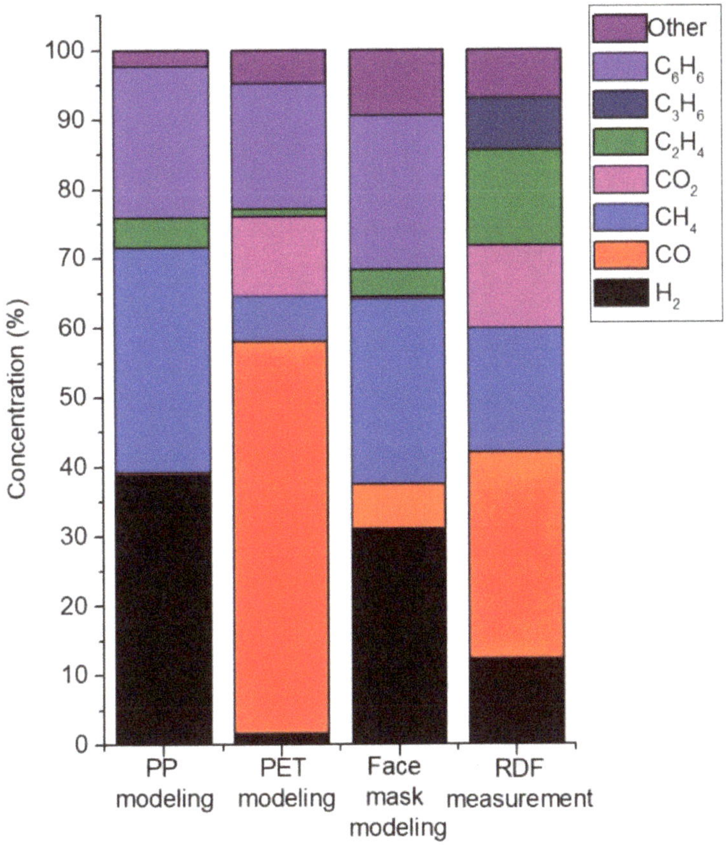

Figure 9. Comparison of selected gaseous pyrolysis products.

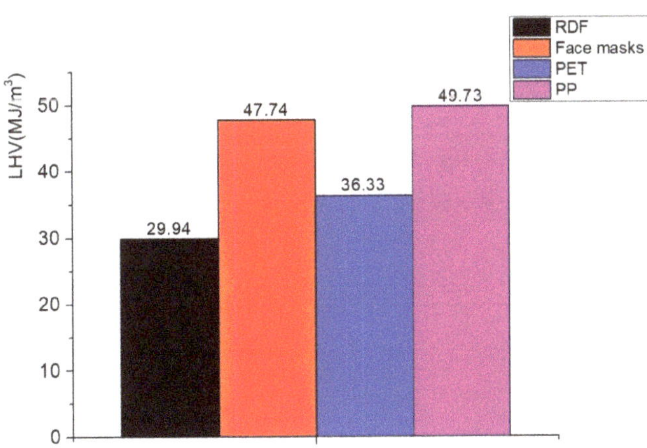

Figure 10. Calorific value of gas from pyrolysis of selected waste.

4. Conclusions

The following conclusions and statements were formulated based on the conducted research and the obtained results:

1. The pyrolysis of plastic waste generated during the COVID-19 pandemic is an effective and environmentally safe solution with great application potential.
2. Thermal conversion of waste by pyrolysis is characterized by a high yield of high-calorific pyrolysis gas. This gas can potentially be used as a substitute for natural gas for energy production.
3. It is possible to forecast the chemical composition of gaseous pyrolysis products of plastic waste and on this basis to estimate the calorific value of the pyrolysis gas produced in the process.
4. The pyrolysis of polypropylene, the main component of protective masks, allows one to obtain high-calorific pyrolysis gas (47.7 MJ/m^3), the main components of which are H_2, CH_4, and C_6H_6.
5. The results of the conducted research and computer simulations show the enormous energy value of plastic waste, especially those generated during the COVID-19 pandemic. Increasing the share of the above-mentioned waste in the municipal waste stream from which the so-called overflow and RDF fraction will increase the calorific value of the pyrolysis gas.
6. Effective disposal of the new category of waste is both a technological- and material-related challenge. Solving this challenge and creating technological techniques and know-how will enable the safe treatment of municipal waste in the COVID-19 era, and will also point the way to effective thermal conversion products management.

Supplementary Materials: The following supporting information can be downloaded at: https://www.mdpi.com/article/10.3390/en15072629/s1. Figure S1 clarified the data of mass loss (TG) and maximum rate of mass loss (DTG) during the reaction, as well as the beginning and final temperatures at each step of the pyrolysis processes.

Author Contributions: Conceptualisation, M.S. (Magdalena Skrzyniarz), M.Z., J.I., A.S. and P.N.; methodology, M.S. (Magdalena Skrzyniarz), M.S. (Marcin Sajdak), M.Z. and J.I.; software, M.Z.; validation, M.Z.; formal analysis, M.S. (Marcin Sajdak) and M.Z.; resources, M.S. (Magdalena Skrzyniarz) and M.Z.; data curation, M.S. (Magdalena Skrzyniarz); writing—original draft preparation, M.S. (Magdalena Skrzyniarz), M.S. (Marcin Sajdak), M.Z., J.I., A.S., P.N. and S.M.; writing—review and editing M.S. (Magdalena Skrzyniarz), M.S. (Marcin Sajdak), M.Z., J.I., A.S., P.N. and S.M.; supervision, A.B.-P.; project administration, A.B.-P. All authors have read and agreed to the published version of the manuscript.

Funding: This research received no external funding.

Institutional Review Board Statement: Not applicable.

Informed Consent Statement: Not applicable.

Data Availability Statement: Data are contained within the article.

Conflicts of Interest: The authors declare no conflict of interest.

References

1. Rahimi, N.R.; Fouladi-Fard, R.; Aali, R.; Shahryari, A.; Rezaali, M.; Ghafouri, Y.; Ghalhari, M.R.; Asadi-Ghalhari, M.; Farzinnia, B.; Conti Gea, O.; et al. Bidirectional association between COVID-19 and the environment: A systematic review. *Environ. Res.* **2021**, *194*, 110692. [CrossRef] [PubMed]
2. Penteado, C.S.G.; de Castro, M.A.S. COVID-19 effects on municipal solid waste management: What can effectively be done in the Brazilian scenario? *Resour. Conserv. Recycl.* **2021**, *164*, 105152. [CrossRef] [PubMed]
3. Nanda, S.; Berruti, F. A technical review of bioenergy and resource recovery from municipal solid waste. *J. Hazard. Mater.* **2021**, *403*, 123970. [CrossRef] [PubMed]
4. Chand Malav, L.; Yadav, K.K.; Gupta, N.; Kumar, S.; Sharma, G.K.; Krishnan, S.; Rezania, S.; Kamyab, H.; Pham, Q.B.; Yadav, S.; et al. A review on municipal solid waste as a renewable source for waste-to-energy project in India: Current practices, challenges, and future opportunities. *J. Clean. Prod.* **2020**, *277*, 123227. [CrossRef]
5. Anastopoulos, I.; Pashalidis, I. Single-use surgical face masks, as a potential source of microplastics: Do they act as pollutant carriers? *J. Mol. Liq.* **2021**, *326*, 115247. [CrossRef]

6. You, S.; Sonne, C.; Ok, Y.S. COVID-19: Resource recovery from plastic waste against plastic pollution. *Cogent Environ. Sci.* **2020**, *6*, 1801220. [CrossRef]
7. Ikiz, E.; Maclaren, V.W.; Alfred, E.; Sivanesan, S. Impact of COVID-19 on household waste flows, diversion and reuse: The case of multi-residential buildings in Toronto, Canada. *Resour. Conserv. Recycl.* **2021**, *164*, 105111. [CrossRef]
8. Dharmaraj, S.; Ashokkumar, V.; Pandiyan, R.; Halimatul Munawaroh, H.S.; Chew, K.W.; Chen, W.H.; Ngamcharussrivichai, C. Pyrolysis: An effective technique for degradation of COVID-19 medical wastes. *Chemosphere* **2021**, *275*, 130092. [CrossRef]
9. Sharma, H.B.; Vanapalli, K.R.; Cheela, V.S.; Ranjan, V.P.; Jaglan, A.K.; Dubey, B.; Goel, S.; Bhattacharya, J. Challenges, opportunities, and innovations for effective solid waste management during and post COVID-19 pandemic. *Resour. Conserv. Recycl.* **2020**, *162*, 105052. [CrossRef] [PubMed]
10. Purnomo, C.W.; Kurniawan, W.; Aziz, M. Technological Review on Thermochemical Conversion of COVID-19-related Medical Wastes. *Resour. Conserv. Recycl.* **2021**, *167*, 105429. [CrossRef] [PubMed]
11. Plastics Europe Market Research Group (PEMRG) and Conversio Market & Strategy GmbH. Plastics—The Fact. 2021. Available online: https://plasticseurope.org/wp-content/uploads/2021/12/Plastics-the-Facts-2021-web-final.pdf (accessed on 3 March 2022).
12. Kulkarni, B.N.; Anantharama, V. Repercussions of COVID-19 pandemic on municipal solid waste management: Challenges and opportunities. *Sci. Total Environ.* **2020**, *743*, 140693. [CrossRef]
13. Elsner, W.; Wysocki, M.; Niegodajew, P.; Borecki, R. Experimental and economic study of small-scale CHP installation equipped with downdraft gasifier and internal combustion engine. *Appl. Energy* **2017**, *202*, 213–227. [CrossRef]
14. Park, C.; Choi, H.; Andrew Lin, K.Y.; Kwon, E.E.; Lee, J. COVID-19 mask waste to energy via thermochemical pathway: Effect of Co-Feeding food waste. *Energy* **2021**, *230*, 120876. [CrossRef] [PubMed]
15. Ragazzi, M.; Rada, E.C.; Schiavon, M. Municipal solid waste management during the SARS-CoV-2 outbreak and lockdown ease: Lessons from Italy. *Sci. Total Environ.* **2020**, *745*, 141159. [CrossRef]
16. Zhao, X.; You, F. Waste respirator processing system for public health protection and climate change mitigation under COVID-19 pandemic: Novel process design and energy, environmental, and techno-economic perspectives. *Appl. Energy* **2021**, *283*, 116129. [CrossRef]
17. Skibiński, A. Demograficzne aspekty zarządzania gospodarką stałymi odpadami komunalnymi w dobie COVID-19. *Przemysł Chem.* **2021**, *1*, 94–96. [CrossRef]
18. Variny, M.; Varga, A.; Rimár, M.; Janošovský, J.; Kizek, J.; Lukáč, L.; Jablonský, G.; Mierka, O. Advances in biomass co-combustion with fossil fuels in the European context: A review. *Processes* **2021**, *9*, 100. [CrossRef]
19. Mohammad, A.; Goli, V.S.N.S.; Singh, D.N. Discussion on 'Challenges, opportunities, and innovations for effective solid waste management during and post COVID-19 pandemic, by Sharma et al. (2020)'. *Resour. Conserv. Recycl.* **2021**, *164*, 105175. [CrossRef]
20. Available online: https://raportsdg.stat.gov.pl/2020/index.html (accessed on 3 March 2022).
21. Solarz, J.; Waliszewski, K. Pandrmia czy wojna pokoleń? *J. Financ. Financ. Law* **2020**, *2*, 99–114.
22. Zhou, C.; Yang, G.; Ma, S.; Liu, Y.; Zhao, Z. The impact of the COVID-19 pandemic on waste-to-energy and waste-to-material industry in China. *Renew. Sustain. Energy Rev.* **2021**, *139*, 110693. [CrossRef]
23. Dharmaraj, S.; Ashokkumar, V.; Hariharan, S.; Manibharathi, A.; Show, P.L.; Tung, C.C.; Ngamcharussrivichai, C. The COVID-19 pandemic face mask waste: A blooming threat to the marine environment. *Chemosphere* **2021**, *272*, 129601. [CrossRef]
24. Shams, M.; Alam, I.; Mahbub, M.S. Plastic Pollution During COVID-19: Plastic Waste Directives and Its Long-term Impact on The Environment. *Environ. Adv.* **2021**, *5*, 100119. [CrossRef] [PubMed]
25. Haque, M.S.; Sharif, S.; Masnoon, A.; Rashid, E. SARS-CoV-2 pandemic-induced PPE and single-use plastic waste generation scenario. *Waste Manag. Res.* **2021**, *39*, 3–17. [CrossRef] [PubMed]
26. Benson, N.U.; Fred-Ahmadu, O.H.; Bassey, D.E.; Atayero, A.A. COVID-19 Pandemic and Emerging Plastic-based Personal Protective Equipment Waste Pollution and Management in Africa. *J. Environ. Chem. Eng.* **2021**, *9*, 105222. [CrossRef] [PubMed]
27. Aragaw, T.A.; Mekonnen, B.A. Current plastics pollution threats due to COVID-19 and its possible mitigation techniques: A waste-to-energy conversion via Pyrolysis. *Environ. Syst. Res.* **2021**, *10*, 8. [CrossRef] [PubMed]
28. Wang, J.; Shen, J.; Ye, D.; Yan, X.; Zhang, Y.; Yang, W.; Li, X.; Wang, J.; Zhang, L.; Pan, L. Disinfection technology of hospital wastes and wastewater: Suggestions for disinfection strategy during coronavirus Disease 2019 (COVID-19) pandemic in China. *Environ. Pollut.* **2020**, *262*, 114665. [CrossRef] [PubMed]
29. Ramteke, S.; Sahu, B.L. Novel coronavirus disease 2019 (COVID-19) pandemic: Considerations for the biomedical waste sector in India. *Case Stud. Chem. Environ. Eng.* **2020**, *2019*, 100029. [CrossRef]
30. Zhang, Y.; Ji, G.; Chen, C.; Wang, Y.; Wang, W.; Li, A. Liquid oils produced from pyrolysis of plastic wastes with heat carrier in rotary kiln. *Fuel Process. Technol.* **2020**, *206*, 106455. [CrossRef]
31. Veksha, A.; Yin, K.; Moo, J.G.S.; Oh, W.D.; Ahamed, A.; Chen, W.Q.; Weerachanchai, P.; Giannis, A.; Lisak, G. Processing of flexible plastic packaging waste into pyrolysis oil and multi-walled carbon nanotubes for electrocatalytic oxygen reduction. *J. Hazard. Mater.* **2020**, *387*, 121256. [CrossRef] [PubMed]
32. Mishra, R.K.; Iyer, J.S.; Mohanty, K. Conversion of waste biomass and waste nitrile gloves into renewable fuel. *Waste Manag.* **2019**, *89*, 397–407. [CrossRef]
33. Kumar Mishra, R.; Mohanty, K. Co-pyrolysis of waste biomass and waste plastics (polystyrene and waste nitrile gloves) into renewable fuel and value-added chemicals. *Carbon Resour. Convers.* **2020**, *3*, 145–155. [CrossRef]

34. Singh, R.K.; Ruj, B.; Sadhukhan, A.K.; Gupta, P. Thermal degradation of waste plastics under non-sweeping atmosphere: Part 1: Effect of temperature, product optimization, and degradation mechanism. *J. Environ. Manag.* **2019**, *239*, 395–406. [CrossRef]
35. Kataki, S.; Chatterjee, S.; Vairale, M.G.; Sharma, S.; Dwivedi, S.K. Concerns and strategies for wastewater treatment during COVID-19 pandemic to stop plausible transmission. *Resour. Conserv. Recycl.* **2021**, *164*, 105156. [CrossRef] [PubMed]
36. Palmieri, V.; De Maio, F.; De Spirito, M.; Papi, M. Face masks and nanotechnology: Keep the blue side up. *Nano Today* **2021**, *37*, 101077. [CrossRef]
37. Mohajerani, A.; Burnett, L.; Smith, J.V.; Markovski, S.; Rodwell, G.; Rahman, M.T.; Kurmus, H.; Mirzababaei, M.; Arulrajah, A.; Horpibulsuk, S.; et al. Recycling waste rubber tyres in construction materials and associated environmental considerations: A review. *Resour. Conserv. Recycl.* **2020**, *155*, 104679. [CrossRef]
38. Ippolito, M.; Vitale, F.; Accurso, G.; Iozzo, P.; Gregoretti, C.; Giarratano, A.; Cortegiani, A. Medical masks and Respirators for the Protection of Healthcare Workers from SARS-CoV-2 and other viruses. *Pulmonology* **2020**, *26*, 204–212. [CrossRef]
39. Rowan, N.J.; Laffey, J.G. Unlocking the surge in demand for personal and protective equipment (PPE) and improvised face coverings arising from coronavirus disease (COVID-19) pandemic—Implications for efficacy, re-use and sustainable waste management. *Sci. Total Environ.* **2021**, *752*, 142259. [CrossRef]
40. Jung, S.; Lee, S.; Dou, X.; Kwon, E.E. Valorization of disposable COVID-19 mask through the thermo-chemical process. *Chem. Eng. J.* **2021**, *405*, 126658. [CrossRef]
41. Anuar Sharuddin, S.D.; Abnisa, F.; Wan Daud, W.M.A.; Aroua, M.K. A review on pyrolysis of plastic wastes. *Energy Convers. Manag.* **2016**, *115*, 308–326. [CrossRef]
42. Das, P.; Tiwari, P. The effect of slow pyrolysis on the conversion of packaging waste plastics (PE and PP) into fuel. *Waste Manag.* **2018**, *79*, 615–624. [CrossRef] [PubMed]
43. Variny, M.; Jediná, D.; Rimár, M.; Kizek, J.; Kšiňanová, M. Cutting Oxygen Production-Related Greenhouse Gas Emissions by Improved Compression Heat Management in a Cryogenic Air Separation Unit. *Int. J. Environ. Res. Public Health* **2021**, *18*, 10370. [CrossRef] [PubMed]
44. Liu, H.C.; You, J.X.; Lu, C.; Chen, Y.Z. Evaluating health-care waste treatment technologies using a hybrid multi-criteria decision making model. *Renew. Sustain. Energy Rev.* **2015**, *41*, 932–942. [CrossRef]
45. Kumar Jha, K.; Kannan, T.T.M. Recycling of plastic waste into fuel by pyrolysis—A review. *Mater. Today Proc.* **2020**, *37*, 3718–37203. [CrossRef]
46. Zaman, C.Z.; Pal, K.; Yehye, W.A.; Sagadevan, S.; Shah, S.T.; Adebisi, G.A.; Marliana, E.; Rafique, R.F.; Johan, R. *Pyrolysis: A Sustainable Way to Generate Energy from Waste, Pyrolysis*; IntechOpen: Rijeka, Croatia, 2017; Tom 1; pp. 3–36.
47. Rajca, P.; Poskart, A.; Chrubasik, M.; Sajdak, M.; Zajemska, M.; Skibiński, A.; Korombel, A. Technological and economic aspect of Refuse Derived Fuel pyrolysis. *Renew. Energy* **2020**, *161*, 482–494. [CrossRef]
48. Jagustyn, B.; Kmieć, M.; Smędowski, Ł.; Sajdak, M. The content and emission factors of heavy metals in biomass used for energy purposes in the context of the requirements of international standards. *J. Energy Inst.* **2017**, *90*, 704–714. [CrossRef]
49. Sajdak, M.; Muzyka, R. Use of plastic waste as a fuel in the co-pyrolysis of biomass. Part I: The effect of the addition of plastic waste on the process and products. *J. Anal. Appl. Pyrolysis* **2014**, *107*, 267–275. [CrossRef]
50. Mazurek, I.; Skawińska, A.; Sajdak, M. Analysis of chlorine forms in hard coal and the impact of leaching conditions on chlorine removal. *J. Energy Inst.* **2021**, *94*, 337–351. [CrossRef]
51. Gałko, G.; Rejdak, M.; Tercki, D.; Bogacka, M.; Sajdak, M. Evaluation of the applicability of polymeric materials to BTEX and fine product transformation by catalytic and non-catalytic pyrolysis as a part of the closed loop material economy. *J. Anal. Appl. Pyrolysis* **2021**, *154*, 105017. [CrossRef]
52. Wilk, M.; Magdziarz, A.; Gajek, M.; Zajemska, M.; Jayaraman, K.; Gokalp, I. Combustion and kinetic parameters estimation of torrefied pine, acacia and Miscanthus giganteus using experimental and modelling techniques. *Bioresour. Technol.* **2017**, *243*, 304–314. [CrossRef] [PubMed]
53. Mlonka-Mędrala, A.; Evangelopoulos, P.; Sieradzka, M.; Zajemska, M.; Magdziarz, A. Pyrolysis of agricultural waste biomass towards production of gas fuel and high-quality char: Experimental and numerical investigations. *Fuel* **2021**, *296*, 120611. [CrossRef]
54. Pelucchi, M.; Frassoldati, A.; Faravelli, T.; Ruscic, B.; Glarborg, P. High-temperature chemistry of HCl and Cl2. *Combust. Flame* **2015**, *162*, 2693–2704. [CrossRef]
55. Ranzi, E.; Cuoci, A.; Faravelli, T.; Frassoldati, A.; Migliavacca, G.; Pierucci, S.; Sommariva, S. Chemical Kinetics of Biomass Pyrolysis. *Energy Fuels* **2008**, *22*, 4292–4300. [CrossRef]
56. Ranzi, E.; Pierucci, S.; Aliprandi, P.C.; Stringa, S. Comprehensive and detailed kinetic model of a traveling grate combustor of biomass. *Energy Fuels* **2011**, *25*, 4195–4205. [CrossRef]
57. Faravelli, T.; Frassoldati, A.; Migliavacca, G.; Ranzi, E. Detailed kinetic modeling of the thermal degradation of lignins. *Biomass Bioenergy* **2010**, *34*, 290–301. [CrossRef]
58. Poskart, A.; Skrzyniarz, M.; Sajdak, M.; Zajemska, M.; Skibiński, A. Management of lignocellulosic waste towards energy recovery by pyrolysis in the framework of circular economy strategy. *Energies* **2021**, *14*, 5864. [CrossRef]
59. Zajemska, M.; Rajca, P.; Szwaja, S.; Morel, S. The chemical mechanism of the HCL formation in the pyrolysis process of selected wastes. *Przem. Chem.* **2019**, *98*, 907–910. [CrossRef]

60. Sieradzka, M.; Rajca, P.; Zajemska, M.; Mlonka-Mędrala, A.; Magdziarz, A. Prediction of gaseous products from refuse derived fuel pyrolysis using chemical modelling software—Ansys Chemkin-Pro. *J. Clean. Prod.* **2020**, *248*, 119277. [CrossRef]
61. Yousef, S.; Eimontas, J.; Striūgas, N.; Abdelnaby, M.A. Pyrolysis kinetic behaviour and TG-FTIR-GC–MS analysis of Coronavirus Face Masks. *J. Anal. Appl. Pyrolysis* **2021**, *156*, 105118. [CrossRef]
62. Szwaja, S.; Zajemska, M.; Szwaja, M.; Maroszek, A. Integration of waste biomass thermal processing technology with a metallurgical furnace to improve its efficiency and economic benefit. *Clean Technol. Environ. Policy* **2021**. [CrossRef]

energies

MDPI

Article

Evaluation of Physical and Chemical Properties of Residue from Gasification of Biomass Wastes

Małgorzata Sieradzka [1,*], Agata Mlonka-Mędrala [2], Izabela Kalemba-Rec [1], Markus Reinmöller [3], Felix Küster [4], Wojciech Kalawa [2] and Aneta Magdziarz [1]

1 AGH University of Science and Technology, Faculty of Metals Engineering and Industrial Computer Science, Mickiewicza 30 Av., 30-059 Krakow, Poland; kalemba@agh.edu.pl (I.K.-R.); amagdzia@agh.edu.pl (A.M.)
2 AGH University of Science and Technology, Faculty of Energy and Fuels, Mickiewicza 30 Av., 30-059 Krakow, Poland; amlonka@agh.edu.pl (A.M.-M.); kalawa@agh.edu.pl (W.K.)
3 DTU Engineering Technology, Technical University of Denmark, Lautrupvang 15, 2750 Ballerup, Denmark; markre@dtu.dk
4 Institute of Energy Process Engineering and Chemical Engineering (IEC), TU Bergakademie Freiberg, Fuchsmuehlenweg 9, 09599 Freiberg, Germany; felix.kuester@iec.tu-freiberg.de
* Correspondence: msieradz@agh.edu.pl

Abstract: Thermochemical conversion of biomass waste is a high potential option for increasing usage of renewable energy sources and transferring wastes into the circular economy. This work focuses on the evaluation of the energetic and adsorption properties of solid residue (char) of the gasification process. Gasification experiments of biomass wastes (wheat straw, hay and pine sawdust) were carried out in a vertical fixed bed reactor, under a CO_2 atmosphere and at various temperatures (800, 900 and 1000 °C). The analysis of the energy properties of the obtained chars included elemental and thermogravimetric (TGA) analysis. TGA results indicated that the chars have properties similar to those of coal; subjected data were used to calculate key combustion parameters. As part of the analysis of adsorption properties, BET, SEM, FTIR and dynamic methanol vapor sorption tests were conducted. The specific surface area has risen from 0.42–1.91 m^2/g (biomass) to 419–891 m^2/g (char). FTIR spectroscopy confirmed the influence of gasification on the decomposition of characteristic chemical compounds for biomass. Methanol sorption has revealed for the 900 °C chars of pine sawdust the highest sorption capacity and its mass change was 24.15% at P/P_0 = 90%. Selected chars might be an appropriate material for volatile organic compounds sorption.

Keywords: CO_2-gasification; biomass wastes; char adsorption; active carbon; BET specific surface area

Citation: Sieradzka, M.; Mlonka-Mędrala, A.; Kalemba-Rec, I.; Reinmöller, M.; Küster, F.; Kalawa, W.; Magdziarz, A. Evaluation of Physical and Chemical Properties of Residue from Gasification of Biomass Wastes. *Energies* **2022**, *15*, 3539. https://doi.org/10.3390/en15103539

Academic Editor: Alok Kumar Patel

Received: 12 April 2022
Accepted: 9 May 2022
Published: 12 May 2022

1. Introduction

Modern challenges are related to the limited amount of fossil fuels, greenhouse gas emissions directly related to the consumption of fossil fuels in various forms [1], as well as the improvement of waste management policies towards the circular economy [2]. For this reason, it is necessary to take quick, decisive and wise steps to use renewable energy (RE) sources, such as biomass, to generate clean energy, as well as improve waste management by using different types of waste, such as RE, and minimizing the waste resulting from their production. Biomass as a renewable energy source has enormous potential. It can be converted by thermochemical processes such as pyrolysis [3,4], hydrothermal carbonization [5,6] or gasification [7], which allows the production of clean energy resources. This study is focused on the gasification process, where the main product is the synthesis gas (syngas) [8,9]. The applied process temperature depends on the material to be gasified and the targeted gas composition. Gasification can be carried out in a temperature range of 600 to 1600 °C in the case of municipal solid waste (MSW) [10,11], in the case of biomass the process temperature range is lower and its ranges vary from 400 to 1100 °C [12,13]. Some of the most commonly used gasifying agents are steam, carbon dioxide and air [14]. This process leads to the formation of gaseous (syngas), liquid (tar) and solid (char) products at lower/moderate gasification temperatures. However, any remaining solids from this process are normally treated as waste [15], leading to a decrease in process efficiency and an increase in waste production. Current studies indicate that gasification chars have a huge potential, both energetic and adsorptive, as a precursor to the formation of activated carbon (AC) [15,16].

AC is characterized by a high surface area, which indicates its great adsorption capacity. Generally, the production and regeneration costs of AC are high, and because of that it is important to find a cheaper way to produce this key solid carrier [17]. The key properties of AC are the small ash content, the ability to maintain physiological characteristics, hardness, and a high carbon content [18]. Figure 1 presents the main steps of AC synthesis; they include preliminary, carbonization, and activation steps. Within the preliminary step, removal of impurities from raw material is provided by washing, then it is dried to remove moisture [19], to obtain similar particle sizes, the raw material is milled and sieved [20], and in the last stage of this step it might be deashed for demineralization by acidic or basic solutions [21]. Volatile matter is removed from raw material during the carbonization step, leading to an increase in carbon concentration in the material produced [22]. After the second stage, a porous structure is formed; it can be developed through physical and chemical activation [17].

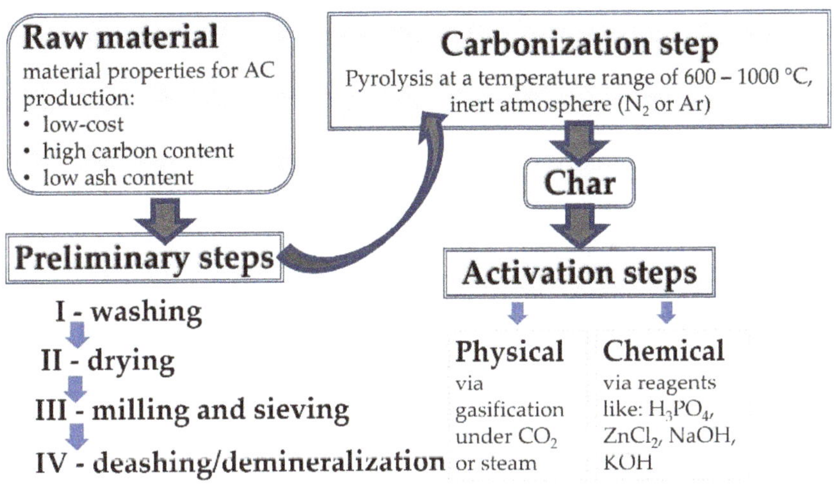

Figure 1. Pathway of activated carbon formation.

The production of activated carbon via physical activation depends on the type of material, the high temperature, the atmosphere with oxidizing gases such as steam, CO_2 or air, and leads to the production of expanded pores structure through the carbon-consuming Reactions (1)–(3) [17]:

$$C + O_2 \rightarrow CO_2 \quad (1)$$

$$C + CO_2 \rightarrow 2CO \quad (2)$$

$$C + H_2O \rightarrow CO_2 + 2H_2 \quad (3)$$

Activated carbon produced by chemical activation includes biomass impregnation via a dehydrating chemical agent, like phosphoric acid, sulphuric acid or zinc chloride in an inert atmosphere by heat treatment. This leads to the production of activated carbon with a large pore structure, which is a desired phenomenon [20].

In a number of studies, the similarities between gasification chars and activated carbon are indicated: Benedetti et al. [15] have pointed out an analogy of gasification char properties to activated carbon (AC) and reviewed its successful applications in the field of adsorption. They used a few chars obtained from gasification plants in the process of conversion of wood chips and pellets after gasification under air at various temperatures (650–900 °C). The main conclusion of this study was that chars are very similar to AC, however, a detailed analysis of each sample is necessary to choose the most suitable application. Benedetti et al. [23] have focused on the gasification char CO_2 adsorption/desorption capacity. The final results indicated that chars obtained from gasification plants offer significant potential in carbon capture storage (CCS) technology. The ability of gasification chars to remove cationic, crystal violet, and anionic congo red chars obtained from industrial gasification of municipal solid waste (MSW) was investigated by Jung et al. [24]. This study indicates that these types of chars have high potential for applications of adsorption in wastewater treatment. AC obtained from gasification chars has the ability to remove H_2S (hydrogen sulphide). This is a common pollutant in biogas produced during anaerobic digestion and syngas obtained from gasification. This capability of gasification chars was studied in other work [25].

The gasification process is an effective waste conversion method. It leads to the production of valuable syngas and solid residues. Syngas analysis has been widely investigated in other works [14,26]. In the presented study, the main focus was on the analysis of solid residue, which is considered as a waste. The two main aims of this study were proposed. First, the energetic properties of the chars obtained from the gasification process were investigated. Second, as other studies conducted, each char mimics activated carbon properties, and individual analysis is required to choose the most adequate application in the adsorption area. Therefore, in this study, the chars formed during the gasification process of waste biomass were examined in terms of methanol adsorption. Methanol is one of the typical volatile organic compounds (VOCs), leading to serious health issues [27]. Another possible utilization of the produced solid residue is in a working pair with methanol in heat-driven sorption cooling devices.

2. Materials and Methods

2.1. Materials

In this study, two types of biomass wastes were selected (agriculture: wheat straw (ABW) and hay (ABH); wood: pine sawdust (WBP)). The materials used in the experiments were dried first and after that independently ground by a knife mill to achieve a similar particle size.

2.2. Test Rig and Experimental Procedure

Gasification experiments were conducted under CO_2 atmosphere at three temperatures of 800, 900, and 1000 °C. During the process, formed syngas was collected and analysed. This paper is focused on solid residue and its properties, and because of that, syngas analysis was omitted. The test rig overview scheme is presented in Figure 2. The gasifier consists of a fixed-bed vertical quartz reactor heated by a furnace to the target temperature. The maximum furnace power was 1.8 kW and was equipped with a (NiCr-NiAl) type thermocouple, which allowed the process temperature to be controlled in real time. The process atmosphere was injected from an external cylinder with a high purity of CO_2, while the flow of the gasification agent was controlled by a rotameter.

The experimental procedure included the following steps: weighing the sample, weighing the quartz wool, placing the quartz wool and the sample in the reactor, assembling the reactor in the oven, providing the gasification process, cooling the oven, disassembling the reactor and weighing obtained chars. The input weight of the sample was constantly 8 g. Quartz wool was placed at the bottom of the sample to prevent the sample from falling to the bottom of the reactor and to ensure that it was located in the centre of the oven heating zone. Prior the gasification process, the reactor was flushed with pure CO_2 at ambient temperature, with a flow rate of 80 mL/min, for 15 min, to remove all air. Next, the sample was heated for 15 min, and once the set temperature was achieved the gasification process was carried out for the next 15 min. Once the gasification process was finished, the reactor with the formed char was cooled to ambient temperature in a CO_2 atmosphere.

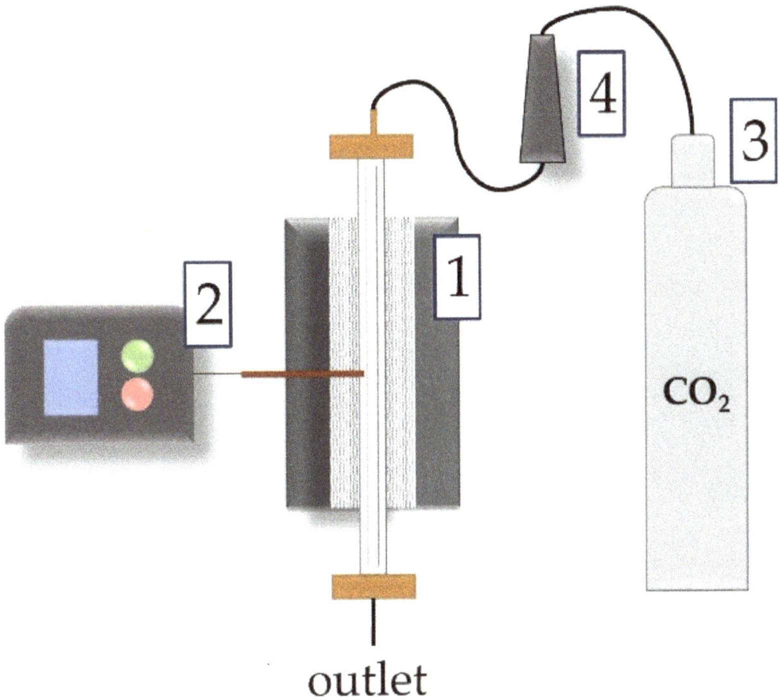

Figure 2. Scheme of the test rig: 1—fixed-bed vertical reactor, 2—temperature controller, 3—gas cylinder, 4—rotameter.

2.3. Analysis of Raw Materials and Chars

To determine raw material properties, proximate and elemental analyses of all samples were performed. Moisture (M), ash (A), and volatile matter (VM) content were established according to the following European standards: PN-EN ISO 18134-1:2015-11, PN-EN ISO 18122:2016-01, and PN-EN ISO 18123:2016-01, respectively. The elemental composition was investigated using a Truspec CHNS 628 Leco analyzer determining the contents of carbon (C), hydrogen (H) and nitrogen (N). The same analyzer was used to establish the elemental composition of the obtained chars. This apparatus has a standard deviation for carbon and nitrogen at a level of ±0.5% and for hydrogen at a level of ±1.0%. This device uses the Dumas method to establish the concentration of the elements mentioned above. It involves the combustion of the sample in pure oxygen at a temperature of 950 °C.

To indicate more valuable data on the energetic properties of materials, the higher heating value (HHV) in MJ/kg, was calculated based on the following equation [28]:

$$HHV = 5.22C^2 - 319C - 1647H + 38.6C \cdot H + 133N + 21028 \quad (4)$$

2.4. Thermogravimetric Analysis

A Mettler Toledo TGA/DSC 1 Star System thermogravimetric analyzer (TGA) was used to study the combustion process of the raw materials and the obtained chars. The analyses took place in air atmosphere and started from the ambient temperature up to 800 °C with a heating rate of 10 K/min. The mass sample was around 5 mg and placed in the platinum crucible.

2.5. Specific Surface Analysis (BET)

The Brunauer–Emmett–Teller (BET) method was used to determine specific surface areas in the raw biomass and the chars obtained. A nitrogen adsorption analyzer ASAP 2010 system (Micromeritics, Norcross, GA, USA) was used. Before analyses, all samples were pretreated with degasses to eliminate faults caused by other gases during the analysis. The studied samples were degassed for 24 h at a temperature of 200 °C in a case of raw biomass and at a temperature of 250 °C in a case of chars.

2.6. Fourier-Transformed Infrared Spectroscopy (FTIR)

The surface functional groups of the studied samples were found by Fourier-transformed infrared spectroscopy (FTIR). This apparatus analyzes the chemical groups of the biomass/char based on the characteristic vibrations of the chemical functional groups. The Bruker Alpha II system was applied in the range of collected spectra from 400 to 4000 cm^{-1}.

2.7. Scanning Electron Microscopy (SEM)

For the investigation of raw materials and chars obtained at a gasification temperature of 1000 °C, the morphology and surface topography were analyzed using a Nova NanoSEM 450 scanning electron microscopy (SEM). Analysis was carried out under high vacuum with a secondary electron detector with a voltage of 10 kV. In these analyses, chosen chars obtained at the highest process temperature are used to compare the modification of the raw material subjected to gasification.

2.8. Dynamic Methanol Vapor Sorption Tests

The Dynamic Gravimetric Vapor Sorption System DVS Vacuum was used for the investigation of the methanol adsorption properties of the studied samples. Methanol was selected as it is a nonhazardous and very common adsorbate used in sorption cooling devices in a working pair with activated carbon [27,29–31]. The mass of the sample was measured with high sensitivity (0.1 μg) by the apparatus during the adsorption and desorption processes. The stability of the temperature was equal to ±0.02 K, at 25 °C, where the humidity condition, with respect to the target value, was generated in the range of ±0.1% [32,33]. The DVS Vacuum allowed performing experiments of static and dynamic sorption with measurements of isotherms and isobars of adsorption–desorption over a wide range of temperatures.

In the presented study, methanol was used as an adsorbate. About 20 mg of produced char obtained at 900 °C was placed in a crucible and dried at 100 °C for approximately 60 min to achieve sample degassing. After a 60 min stabilization stage at given process temperature, a series of 18 experimental stages started. Each stage lasted 20 min and had a different setting of a relative pressure P/P_0, starting from 10% to 90%. Based on the obtained results, adsorption and desorption isotherms were calculated. The methanol flow rate was set constant and equal to 15 sccm (standard cubic centimeters per minute). The experiment was carried out at adsorption temperature of 30 °C, and assumed desorption temperature of 60 °C [33].

3. Results and Discussion

3.1. Char as a Material with Energy Potential

3.1.1. Characterization of Raw Biomass and Chars

The main results of the elemental and proximate analysis of raw biomasses are presented in Table 1. The carbon content of the studied biomasses obtained the lowest value for ABH (39.6%) and the highest value for wood biomass WBP (48.6%). Its concentration after the gasification process increased significantly. The greatest change was observed for the ABH sample, which reached 92.1% for the char obtained at temperature 1000 °C. For other char samples, an increase in the process temperature results in a decrease in the C content from 76.8% to 71.7% for ABW and from 60.7% to 53.1% for WBP. Nitrogen content did not exceed 2%. In the case of hydrogen, a decrease in its content was noted when comparing raw material with chars; this is related to the release of hydrocarbons that form syngas during gasification. When analyzing the results for the chars, the highest decrease occurred at a temperature of 1000 °C and it was 0.9% for ABW_1000 and 0.8% for ABH_1000 and WBP_1000.

Table 1. Raw biomass characterization.

	Raw Biomass		
	ABW	**ABH**	**WBP**
	Proximate Analysis (%)		
VM	74.06	59.94	79.89
A	3.86	21.83	0.19
M	4.99	7.71	6.76
FC	17.08	10.52	13.16
	Elemental Analysis (%) *		
C	44.65	39.59	48.56
H	6.47	5.76	6.57
N	0.68	1.16	0.07
O **	39.35	23.95	37.86

* moisture and ash-free. ** calculated by difference.

The results of the elemental analysis were used to calculate the hydrogen (H/C ratio) and oxygen (O/C ratio) indexes in the studied samples (Figure 3). Biomass is characterized by a high content of oxygen compared to coal, which influences its heating value. It is related to the fact that the strength of carbon–hydrogen and carbon–oxygen bonds is lower than that of carbon-carbon bonds [34]. The ratio of H/C and O/C in raw biomass was averaged at 1.7 and 0.7, respectively, indicating that the raw material has a low aromaticity and a high aliphatic content [35]. The results for the hydrogen and oxygen indexes of the chars decreased significantly, which indicates that it is more similar to the coal, where the H/C of the coal is in the range of 0.4–1 and the O/C is up to 0.2 [35]. Based on the results presented in Figure 3, it can be observed that the H/C and O/C values for the chars did not exceed the value of 0.2.

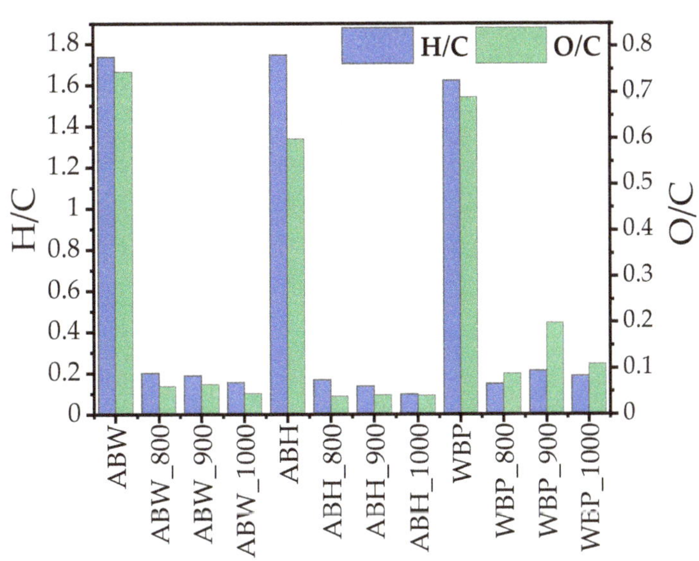

Figure 3. Hydrogen (H/C ratio) and oxygen (O/C ratio) indexes of the studied samples.

The influence of gasification temperature on char HHV and raw biomass HHV, as a reference, is presented in Table 2. Raw materials obtained HHV at a level of 17.8, 16 and 19.4 MJ/kg for ABW, ABH and WBP, respectively. For comparison, this value in other studies was 16.6 MJ/kg for mixed sawdust reported in [36], 19.9 MJ/kg for pine sawdust and 19.1 MJ/kg for chestnut sawdust reported in [37]. The HHV of millet husk char obtained from gasification was 22.9 MJ/kg [38]. In this study, a significant increase in HHV was observed for chars, especially for ABW and ABH samples. With an increase in the process temperature, there was a slight decrease in the HHV as a result of the higher release of carbon content towards syngas production. Additionally, the value of HHV for other carbonized materials and various types of carbon reported in [39] was 31.8 MJ/kg for anthracite coal, 18.5 MJ/kg for briquette, 35 MJ/kg for coal, 31.3 MJ/kg for lignite char and 34.4 MJ/kg for charcoal. When the results presented in other studies and the data obtained are analyzed, it can be concluded that chars have values similar to coal. The best results were obtained for the chars produced at a temperature of 800 °C.

Table 2. Higher heating value of studied samples.

Sample	HHV, MJ/kg
ABW	17.78
ABW_900	27.77
ABW_1000	26.25
ABH	16.05
ABH_800	38.12
ABH_900	37.46
ABH_1000	37.47
WBP	19.35
WBP_800	21.52
WBP_900	19.80
WBP_1000	19.33

Table 3 presents the yield of the char formation during gasification. An increase in process temperature leads to a decrease in the amount of solid residue. The highest values were obtained for the ABH sample, which contains the highest ash content (21.8%), where the WBP yield was the lowest, which corresponds to the lowest ash content (0.2%). In the case of ABH, the char yield decreased from 25.7% for 800 °C to 21.8% for 1000 °C. For WBP samples, for the same parameters, the char yields decreased from 15.3% to 13.1%. This phenomenon is related to an increase in char consumption during the Boudouard reaction due to the shift of the chemical equilibrium toward CO at higher temperatures [40].

Table 3. The yield of char formation during gasification.

Sample	Char Yield, %
ABW_800	21.0
ABW_900	17.4
ABW_1000	14.9
ABH_800	25.7
ABH_900	22.6
ABH_1000	21.8
WBP_800	15.3
WBP_900	13.2
WBP_1000	13.1

3.1.2. Thermogravimetric Analysis of Combustion Properties of Raw Biomasses and Chars

The comparison of the mass loss (TG curves) and the differential mass loss (DTG curves) of raw biomass and its chars during the combustion process is presented in Figure 4.

During the analysis, the combustion reaction occurred in different temperature zones for raw biomass samples than for chars obtained from it. There can be distinguished two stages of combustion in the case of raw biomass [41]: the first stage is related to the combustion of volatile matter and appeared in temperature zones of 160–330 °C for ABW, 135–330 °C for ABH and 160–360 °C for WBP, which can be observed based on DTG curves. The second stage appeared in temperature ranges 330–457 °C, 330–484 °C, 260–490 °C for ABW, ABH and WBP, respectively. The fact that the combustion process of the chars is missing the first stage is related to the fact that volatile matter was released during gasification. Its second stage appeared in approximate temperature ranges of 244–494 °C, 255–486 °C and 340–620 °C for ABW, ABH and WBP, correspondingly. When comparing the second stages of combustion of raw materials and chars, we can see a higher intensity at DTG peaks in this area for chars than for raw biomass. It is related to the fact that char contains a higher content of carbon, resulting in high-intensity combustion of carbon compounds at this stage. During combustion of raw material, volatiles start to evolve at a lower temperature and then burn in an area of higher temperature by reaching ignition temperature. Generated heat during volatile combustion influences the start of the combustion process of fixed carbon [42,43]. As the volatile matter was removed from the chars, it can be observed that only one peak is detected at the DTG curve, where the mass loss at the TG curve is moved to higher temperature areas.

On the basis of the TG and DTG results, the characteristic parameters like ignition temperature (T_i, °C), burnout temperature (T_b, °C), maximum mass loss rate ($(dW/dT)_{max}$, wt. %/min), average mass loss rate ($(dW/dT)_{av}$, wt. %/min) the temperature of the highest peak of DTG (T_{DTG}, °C), time corresponding to the parameters mentioned above, such as the ignition time (t_i, min), burnout time (t_b, min), time range = 0.5 ($\Delta t_{1/2}$, min) and the maximum peak time (t_p, min) were obtained and used in the calculation of key combustion parameters for all studied samples [44,45].

As described earlier, based on the characteristic parameters of the TG and DTG curves, the parameters were obtained based on the 'Intersection method' described in refs. [46,47]. The data was used to determine the key combustion parameters according to the following equations:

$$D_i = \frac{(dW/dT)_{max}}{t_p \cdot t_i} \tag{5}$$

$$D_f = \frac{(dW/dT)_{max}}{\Delta t_{1/2} \cdot t_p \cdot t_b} \tag{6}$$

$$S = \frac{(dW/dT)_{max} \cdot (dW/dT)_{av}}{T_i^2 \cdot T_b} \tag{7}$$

$$H_f = T_{DTG} \ln\left(\frac{\Delta t_{1/2}}{(dW/dT)_{av}}\right) \tag{8}$$

The group of key parameters includes the following indexes: ignition (D_i), burnout (D_f), comprehensive combustion characteristic (S), rate and intensity of combustion processes (H_f), which are described by the corresponding Equations (5)–(8) and its calculated values are presented in Table 4.

The ignition index D_i describes the ability to release volatile compounds from fuel, which also determines the stability of combustion [44,48]. With an increase in its value, this ability increases with increasing stability of combustion. Burnout performance, combustion intensity and reactivity are described by comprehensive combustion index S, where greater values characterise better combustion behaviour [48]. The rate and intensity of combustion is described by the H_f index, which additionally reflects the stability of the flame after ignition. A decrease in its value indicates a better combustion characteristic [49].

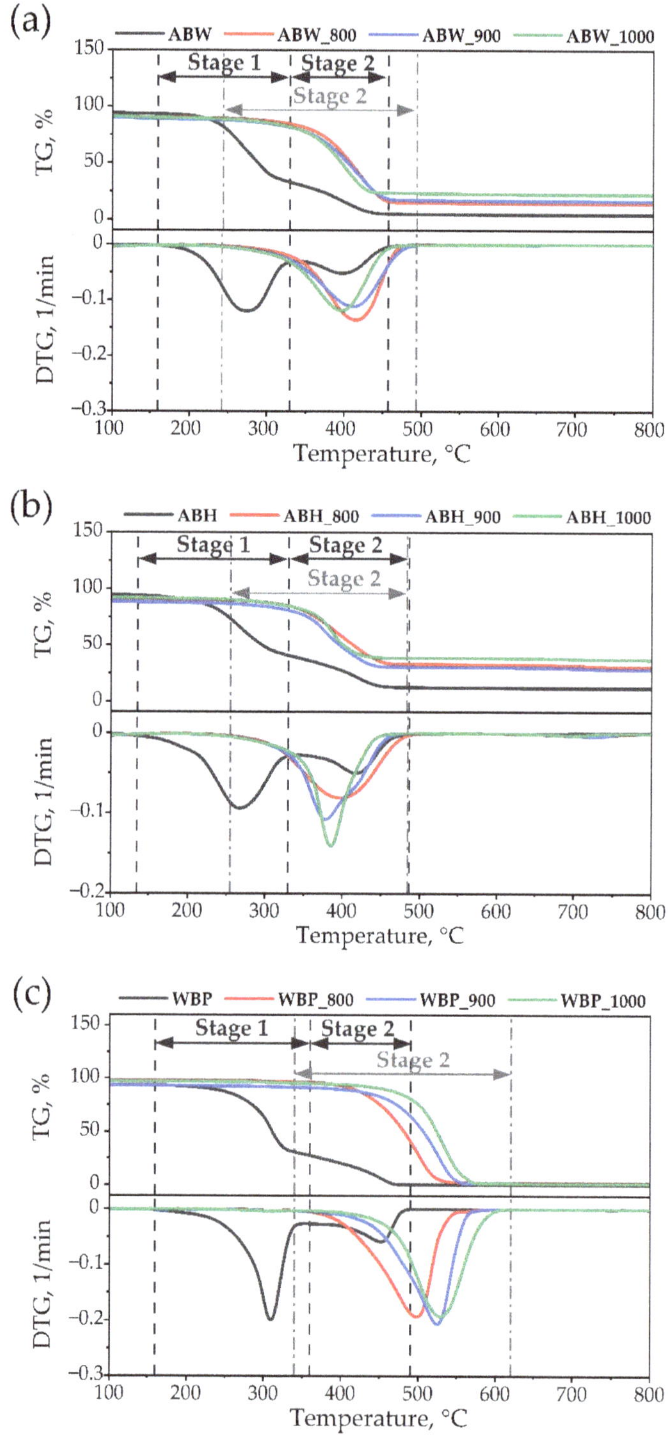

Figure 4. TG and DTG curves of (**a**) ABW, (**b**) ABH and (**c**) WBP.

Table 4. Key combustion parameters.

Sample	T_{ig}, °C	T_b, °C	D_i, wt. %/min^3	D_f, wt. %/min^4	S, min^{-2} °C^{-3}	H_f, °C
ABW	233.1	398.2	5.82×10^{-3}	1.55×10^{-4}	3.27×10^{-7}	966
ABH	222.6	417.7	4.84×10^{-3}	9.00×10^{-5}	2.48×10^{-7}	959
WBP	262.4	452.3	8.03×10^{-3}	1.33×10^{-4}	3.94×10^{-7}	1126
ABW_800	354.1	415.7	3.40×10^{-3}	6.29×10^{-5}	1.38×10^{-7}	1657
ABW_900	345.6	411.8	2.87×10^{-3}	5.37×10^{-5}	1.17×10^{-7}	1640
ABW_1000	339.8	397.1	3.20×10^{-3}	6.17×10^{-5}	1.24×10^{-7}	1606
ABH_800	335.7	397.7	2.23×10^{-3}	4.35×10^{-5}	0.79×10^{-7}	1644
ABH_900	338.6	378.5	3.03×10^{-3}	6.09×10^{-5}	1.10×10^{-7}	1559
ABH_1000	352.5	385.5	3.78×10^{-3}	7.48×10^{-5}	1.14×10^{-7}	1652
WBP_800	420.9	498.4	3.59×10^{-3}	5.57×10^{-5}	1.32×10^{-7}	2001
WBP_900	450.6	525.0	3.46×10^{-3}	5.13×10^{-5}	1.18×10^{-7}	2135
WBP_1000	486.9	529.1	3.01×10^{-3}	4.65×10^{-5}	0.93×10^{-7}	2161

The ignition process is reflected in the T_i parameter and the D_i index. This stage of combustion is boosted by the large amount of volatile matter in the sample, and results in a tendency to burn quickly [50]. As the VM are removed from the chars after gasification, it is expected that the ignition temperature increases for carbonized samples in comparison to the raw biomass. The range of T_i for raw biomass is 222–262 °C, while for chars it is 335–486 °C [51]. This trend indicates that char samples have lower reactivity, which is characteristic for coal [52]. The relationship between VM concentration and ignition performance is noticeable through results of D_i, where the value of this index is the highest for WBP (8.03×10^{-3}), which consists of almost 80% VM in raw form. Another factor affecting ignition temperature is the ash content, which decreases its values with increasing content. The concentration of ABH ash is 21.8%, where its T_i is the lowest of all the samples analysed (222.6 °C).

The burnout process is characterized by the D_f index, which is influenced not only by the temperature of T_b but also by the temperature of T_i. The lower the T_b temperature, the faster the burnout process is as the reduction of unburnt compounds is completed. A higher burnout temperature indicates that the process must be extended, which requires a higher temperature and a longer residence time [52]. This phenomenon is apparent when we analyse the T_b and D_f values for raw biomass and chars. In case of samples ABW and WBP, burnout temperature for its chars increased because of the difficulties in burning the carbonized samples, and the parallel burnout index value also increased. We observe an inverse relationship for the ABH sample, which contains the highest concentration of ash among all studied samples, which shortens the combustion process due to the lower content of combustible parts in the material. The better combustion process was determined by the S index; the increase in the value indicates rapid combustion with the release of a large amount of heat in a short range of time, which is characteristic for biomass. In the case of coal, this parameter is lower. The highest decrease in the S index is reported for the WBP sample where in raw form S is equal to 3.94×10^{-7} min^{-2} °C^{-3}, where in the case of WBP_1000 it is 0.93×10^{-7} min^{-2} °C^{-3}. Lastly, the H_f index shows an upward trend for chars. The range of the Hf index of raw biomass is 959–1126 °C, where for chars it is 1559–2161 °C. Mureddu et al. [52] have evaluated the combustion performance of different kinds of coals and biomasses. On the basis of these results, we can conclude that after the gasification process, the samples of chars are close to the combustible properties of coal.

3.2. Adsorption Properties of Chars

3.2.1. Surface Area (BET)

The surface area of the chars compared to the raw biomass samples measured by N_2 adsorption is shown in Table 5. Additionally, Table 5 indicates a literature review of BET surface area (S_{BET}) of activated chars of different biomass samples via physical activation under different atmospheres and two commercial activated carbons. In the case of the studied raw biomass, S_{BET} has been reported as small values, where chars have exhibited a significant increase. The surface area varies from 469.6 to 891.5 m^2/g, from 256 to 419 m^2/g and from 520.9 to 764.8 m^2/g for chars obtained from ABW, ABH and WBP, respectively. These data indicate that ABH chars showed the lowest S_{BET} due to the highest mineral matter content (21.8%) [15]. The largest surface area was obtained for ABW chars. The increase of process temperature from 800 °C to 900 °C results in a higher surface area. This phenomenon is determined by the release of volatile matter contained in raw biomass. In the case of results for temperatures of 900 and 1000 °C, a decrease of S_{BET} for agriculture biomass was observed; for wood biomass the increase was not significant. Extensive heating of the samples provided for breaking and cracking of the pore structure [20] resulting in a reduction of the S_{BET} value from 891 to 624 m^2/g for ABW and from 419 to 296 m^2/g for ABH, in the case where the WBP value increases only 16 m^2/g. For comparison, the value of the surface area of the BET of biomass solid residue in the literature ranges from 167 to 1355 m^2/g, whereas the activated carbon BET ranges from 984 to 1002 m^2/g.

Table 5. BET analysis of raw biomass and chars.

Experiment		Literature Review			
Sample Name	**S_{BET} Surface Area, m^2/g**	**Raw Sample Name**	**Gasification Conditions**	**BET Surface Area, m^2/g**	**Ref.**
ABW	1.91	Chars			
ABH	1.25	Corn straw	99.9% CO_2, 800 °C	444	[53]
WBP	0.42	Poplar	90% CO_2, 1150 °C	586	[54]
ABW_800	469.6	Corn stover	50% CO_2, 1150 °C	333	[54]
ABW_900	891.5	Walnut shells	99.9% CO_2, 900 °C	765	[55]
ABW_1000	623.5	Pelletized pine sawdust	O_2/steam, 800 °C	235–268	[56]
ABH_800	256.0	Pinewood	Steam, 800, 850 and 900 °C	364–889	[57]
ABH_900	419.1	Olive stone	99.9% CO_2, 800 °C	1355	[58]
ABH_1000	269.2				
WBP_800	520.9	Activated carbon			
WBP_900	748.9	Activated carbon		984	[23]
WBP_1000	764.8	Activated carbon		1002	[23]

3.2.2. Morphology of Raw Biomass and Chars

The surface morphology and structure of the studied samples is presented in Figure 5. In this analysis, the raw samples were chosen as a reference and the chars obtained at a temperature of 1000 °C were selected, due to the fact that the chars obtained at this temperature represent various surface areas according to the BET analysis. In the case of raw agricultural biomasses, the sample structure is more homogenous and fibrous, closed without noticeable pores. After physical activation by CO_2-gasification, the surface revealed an expanded and heterogeneous structure reached with a combination of micropores and mesopores. The ABW_1000 sample consists of sponge construction with irregular shapes and sizes, the surface area for these samples was 623 m^2/g, which indicates that it is still a high value. In case of ABH_1000, S_{BET} obtained the lowest value (269 m^2/g), its morphology is characterized by small holes as well as irregular shapes. Sample WBP_1000 enriched its structure in ordered and regular pores shape and sizes. The surface area of this sample obtained the highest value under 1000 °C. All char samples indicated similarities with the shapes to the parent materials. The release of volatile matter during gasification did not involve damage to the structure of the char particles due to the natural porosity of the biomass [59]. On the other hand, rapid release of volatile compounds at high temperature leads to the softening of the solid matrix or a coverage of the pores by fractional melt formation [60], which then causes clogging of the pores. High-temperature releases additional volatiles, which cannot be exhausted because of closed pores. Overpressure is generated in the samples, resulting in the formation of bubbles, cracks and finally fragmentation [61]. Kamble et al. reported similar behavior of biomass char structure in [62].

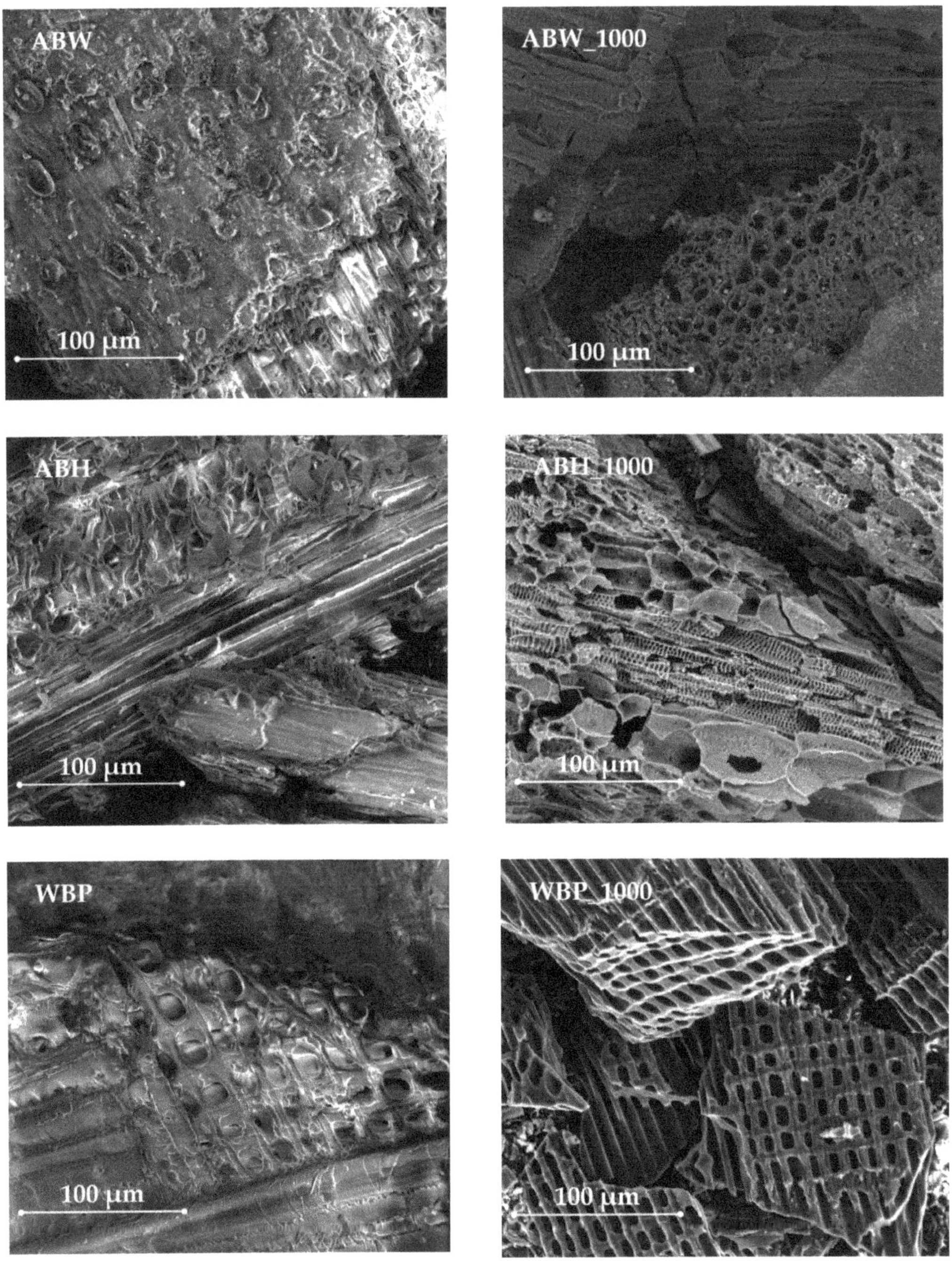

Figure 5. SEM images of the raw biomasses (ABW, ABH, and WBP) and their chars (ABW_1000, ABH_1000, and WBP_1000).

3.2.3. Surface Functional Groups

The FTIR spectra for raw biomass and chars obtained at 900 °C are presented in Figure 6. The results indicate a difference in the FTIR spectrum, where the main peaks appearing in the raw sample after the gasification process did not occur in the char analysis. This phenomenon is related to the decomposition of chemical compounds caused by gasification, and it can be seen by flattening out in the transmittance bands. The first main peak in raw material occurred between 3200 and 3600 cm^{-1}, which corresponded to the hydrogen-bonded O-H stretch, where spectra from approximately 2920 and 2850 cm^{-1} indicated the symmetric/asymmetric stretch of methylene CH. The alkenyl stretch C=C was recognized with a wavenumber of approximately 1640 cm^{-1} [24]. Cellulose bonds C=O appeared at 1030 cm^{-1}; in the case of char ABH_900, we can see that this bond was not completely decomposed. According to the study by Sun et al. [63], samples with functional groups counting oxygen indicated high reactivity for graphene oxide adsorption. The thermal degradation of the chars and the reduction of the peaks compared to the raw materials indicate that an aromatic structure was formed during gasification [64,65].

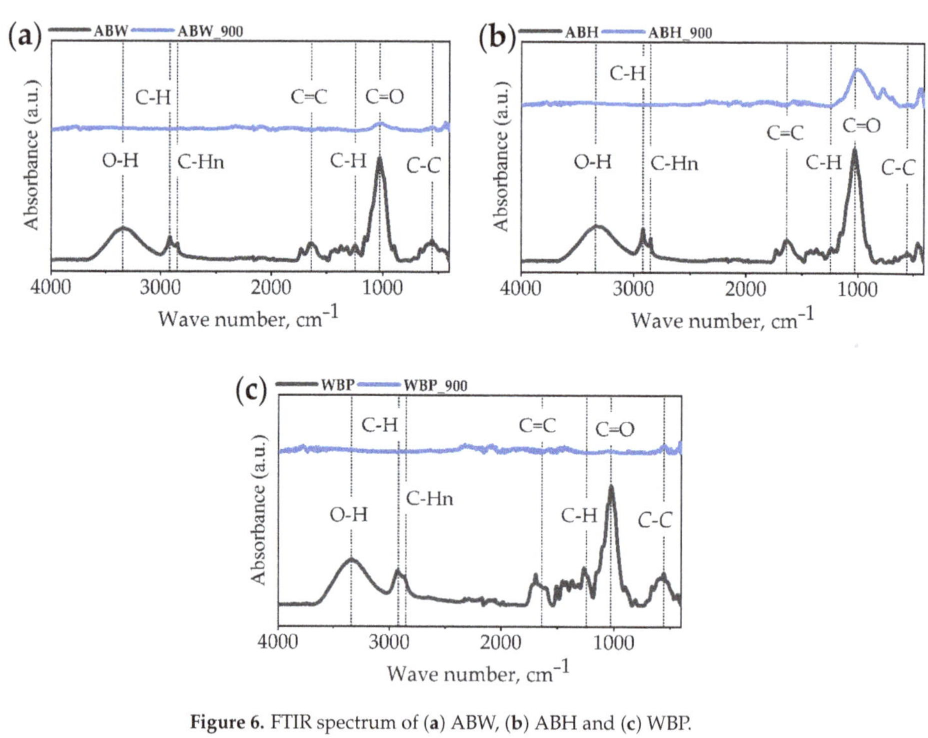

Figure 6. FTIR spectrum of (**a**) ABW, (**b**) ABH and (**c**) WBP.

3.2.4. Dynamic Methanol Vapor Sorption Tests

Methanol adsorption was analyzed to define potential utilization of produced solid residue as VOCs adsorbent or adsorbent in working pair in methanol in sorption cooling devices. The methanol intake was tested in an adsorption temperature 30 °C, and desorption temperature 60 °C, and the sorption isotherms are visualized in Figures 7–9.

The solid residues of ABW_900 in Figure 7 and WBP_900 in Figure 9 were characterized by an adsorption isotherm of type II, and a very small hysteresis of adsorption was observed, which confirms the microporous structure of the analyzed materials. However, a slight H4 type hysteresis might suggest the presence of split pores in the samples. Non-uniform tendency and weakest sorption capacity were observed for sample ABH_900 in Figure 8, which was also characterized by the lowest specific surface area equal to 419.1 m^2/g. However, most probably the results for this sample were not associated with the biomass type, but they were disturbed by high soil contamination of the sample, as its ash content is 21.83%. Whereas, the ash content in hay reported in the literature is in the range of 6–8% [66,67].

Chars produced from wheat straw and wood (ABW and WBP) have exhibited a moderate methanol adsorption capacity, which was similar in both analyzed temperatures. Despite of the quite noticeable BET specific surface area of samples after gasification at 900 °C, the produced chars have shown average sorption properties. The highest weight gain was observed for the process temperature of 60 °C for the WBP_900 sample. For the value of P/P_0 of 90% the mass change was equal to 24.15%. Similar methanol uptakes were noted for commercial activated carbons (e.g., Norit GL 50) of BET not higher than 600 m^2/g. Other commercial, wood- and coal-based activated carbons of specific surface area equal to 1200 m^2/g present the maximal methanol sorption at the level not higher than 40% [68,69]. Unfortunately, the presented sorption properties do not present produced solid residue derived from biomass waste as most suitable candidates for a sorption chiller adsorbent. However, due to noticeable methanol sorption capacity and the origin of the studied material, produced solid residue might be a proper material for VOCs sorption [27] and subscribe in a circular economy concept.

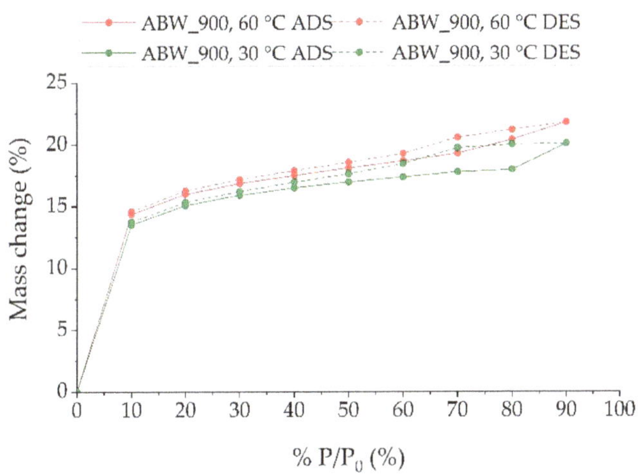

Figure 7. Adsorption and desorption isotherms for the samples ABW_900 at 30 and 60 °C.

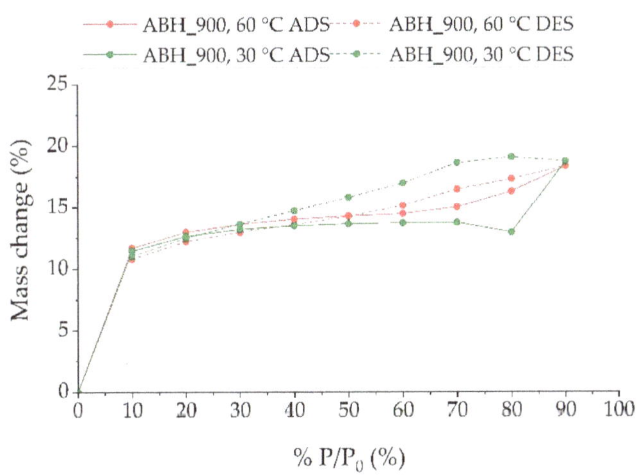

Figure 8. Adsorption and desorption isotherms for the sample ABH_900 at 30 and 60 °C.

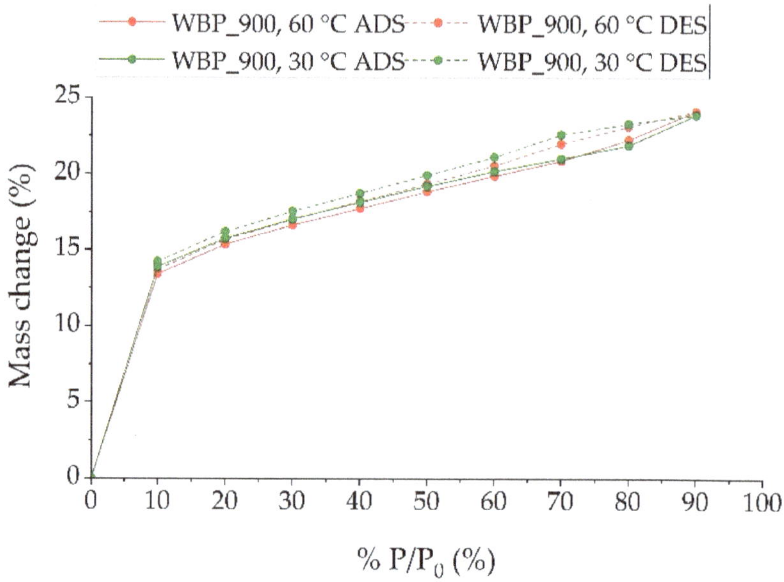

Figure 9. Adsorption and desorption isotherms for the sample WBP_900 at 30 and 60 °C.

4. Conclusions

The main conclusions of this study in agreement with the literature are as follows:

- The gasification process leads to an increase in the content of C in the char, with a parallel decrease in the concentration of H and O. This change affected the hydrogen (H/C ratio) and oxygen (O/C ratio) indexes, which both decreased significantly for the char samples. The above shift indicates that chars become more alike to coal. A similar trend was observed by analyzing the HHV value of the chars compared to the raw biomass;
- The TGA results for the combustion process of raw biomass and chars showed that the char samples lack peaks related to the burning of volatile matter. Additionally, key combustion parameters indicated that the chars have properties comparable to those of coal;
- The provided analysis of the chars obtained from waste biomass has proved similarities of these materials with coals, which confirms the possibility of using them energetically;
- The conducted gasification process leads to the formation of pores and cracks in the morphological structure of the chars, which results in the formation of the heterogeneous structure. Organic compounds were decomposed during the process, which is reflected by flattening out of the FTIR spectra;
- Chars significantly increased their specific surface area in comparison to raw materials. The highest increase is observed for the WBP sample, where the surface area of the raw sample is equal to 0.42 m^2/g, while a value of 764.8 m^2/g for WBP_1000 is determined, indicating that these materials are potential candidates for AC production; and
- Unfortunately, the studied samples are not the most applicable materials for the production of activated carbon (AC) used in working pairs with methanol in sorption chillers. On the other hand, the results indicated that the produced chars can be good candidates for VOCs sorption.

Author Contributions: Conceptualization, M.S., A.M. and A.M.-M.; data curation, M.S.; formal analysis, M.S., A.M., A.M.-M., I.K.-R., W.K. and F.K.; funding acquisition, M.S. and A.M.; investigation, M.S. and A.M.-M.; methodology, M.S., A.M., I.K.-R., W.K. and F.K.; project administration M.S. and A.M.; resources, M.S., A.M.-M., I.K.-R., A.M., F.K., M.R. and W.K.; supervision, A.M. and M.R.; validation, M.S., A.M. and A.M.-M.; writing—original draft, M.S. and A.M.-M.; writing—review & editing, A.M., F.K. and M.R. All authors have read and agreed to the published version of the manuscript.

Funding: Part of this work was supported by the Ministry of Science and Higher Education, Poland (grant AGH UST no 16.16.110.663) and co-financed by the European Union from the European Social Fund under the project: POWR.03.05.00-00-Z307/17-00.

Institutional Review Board Statement: Not applicable.

Informed Consent Statement: Not applicable.

Data Availability Statement: Not applicable.

Conflicts of Interest: The authors declare no conflict of interest.

References

1. Greiner, P.T.; York, R.; McGee, J.A. When are fossil fuels displaced? An exploratory inquiry into the role of nuclear electricity production in the displacement of fossil fuels. *Heliyon* **2022**, *8*, e08795. [CrossRef]
2. Syberg, K.; Nielsen, M.B.; Oturai, N.B.; Clausen, L.P.W.; Ramos, T.M.; Hansen, S.F. Circular economy and reduction of micro(nano)plastics contamination. *J. Hazard. Mater. Adv.* **2022**, *5*, 100044. [CrossRef]
3. Ansari, K.B.; Kamal, B.; Beg, S.; Wakeel Khan, M.A.; Khan, M.S.; Al Mesfer, M.K.; Danish, M. Recent developments in investigating reaction chemistry and transport effects in biomass fast pyrolysis: A review. *Renew. Sustain. Energy Rev.* **2021**, *150*, 111454. [CrossRef]
4. Maniscalco, M.; Infurna, G.; Caputo, G.; Botta, L.; Dintcheva, N.T. Slow pyrolysis as a method for biochar production from carob waste: Process investigation and products' characterization. *Energies* **2021**, *14*, 8457. [CrossRef]
5. Czerwińska, K.; Śliz, M.; Wilk, M. Hydrothermal carbonization process: Fundamentals, main parameter characteristics and possible applications including an effective method of SARS-CoV-2 mitigation in sewage sludge. A review. *Renew. Sustain. Energy Rev.* **2022**, *154*, 111873. [CrossRef]
6. Han, S.; Bai, L.; Chi, M.; Xu, X.; Chen, Z.; Yu, K. Conversion of Waste Corn Straw to Value-Added Fuel via Hydrothermal Carbonization after Acid Washing. *Energies* **2022**, *15*, 1828. [CrossRef]
7. Yu, J.; Guo, Q.; Gong, Y.; Ding, L.; Wang, J.; Yu, G. A review of the effects of alkali and alkaline earth metal species on biomass gasification. *Fuel Process. Technol.* **2021**, *214*, 106723. [CrossRef]
8. Vishwajeet, V.; Pawlak-Kruczek, H.; Baranowski, M.; Czerep, M.; Chorążyczewski, A.; Krochmalny, K.; Ostrycharczyk, M.; Ziółkowski, P.; Madejski, P.; Mączka, T.; et al. Entrained Flow Plasma Gasification of Sewage Sludge–Proof-of-Concept and Fate of Inorganics. *Energies* **2022**, *15*, 1948. [CrossRef]
9. Zhang, W.; Wu, Y.; Huang, S.; Wu, S.; Gao, J. Study on physicochemical characteristics, solidification and utilisation of tannery sludge gasification ash. *J. Environ. Manag.* **2022**, *310*, 114584. [CrossRef]
10. Yayalık, İ.; Koyun, A.; Akgün, M. Gasification of Municipal Solid Wastes in Plasma Arc Medium. *Plasma Chem. Plasma Process.* **2020**, *40*, 1401–1416. [CrossRef]
11. Lv, L.; Zhang, Z.; Li, H. SNG-electricity cogeneration through MSW gasification integrated with a dual chemical looping process. *Chem. Eng. Process.—Process Intensif.* **2019**, *145*, 107665. [CrossRef]
12. Razmi, A.R.; Heydari Afshar, H.; Pourahmadiyan, A.; Torabi, M. Investigation of a combined heat and power (CHP) system based on biomass and compressed air energy storage (CAES). *Sustain. Energy Technol. Assess.* **2021**, *46*, 101253. [CrossRef]
13. Ahmad, A.A.; Zawawi, N.A.; Kasim, F.H.; Inayat, A.; Khasri, A. Assessing the gasification performance of biomass: A review on biomass gasification process conditions, optimization and economic evaluation. *Renew. Sustain. Energy Rev.* **2016**, *53*, 1333–1347. [CrossRef]
14. Hasanzadeh, R.; Azdast, T.; Mojaver, M.; Park, C.B. High-efficiency and low-pollutant waste polystyrene and waste polystyrene foam gasification: Comprehensive comparison analysis, multi-objective optimization and multi-criteria decision analysis. *Fuel* **2022**, *316*, 123362. [CrossRef]
15. Benedetti, V.; Patuzzi, F.; Baratieri, M. Characterization of char from biomass gasification and its similarities with activated carbon in adsorption applications. *Appl. Energy* **2018**, *227*, 92–99. [CrossRef]
16. Nguyen, H.N.; Khuong, D.A.; Tran-Nguyen, P.L.; Tsubota, T. Towards sustainable biomass gasification: Evolution of bagasse char characteristics throughout gasification. *Biomass Bioenergy* **2022**, *158*, 106384. [CrossRef]

17. Naji, S.Z.; Tye, C.T. A review of the synthesis of activated carbon for biodiesel production: Precursor, preparation, and modification. *Energy Convers. Manag. X* **2022**, *13*, 100152. [CrossRef]
18. Narowska, B.E.; Kułażyński, M.; Łukaszewicz, M. Application of activated carbon to obtain biodiesel from vegetable oils. *Catalysts* **2020**, *10*, 1049. [CrossRef]
19. Sadeek, S.A.; Mohammed, E.A.; Shaban, M.; Abou Kana, M.T.H.; Negm, N.A. Synthesis, characterization and catalytic performances of activated carbon-doped transition metals during biofuel production from waste cooking oils. *J. Mol. Liq.* **2020**, *306*, 112749. [CrossRef]
20. Mazlan, M.A.F.; Uemura, Y.; Yusup, S.; Elhassan, F.; Uddin, A.; Hiwada, A.; Demiya, M. Activated Carbon from Rubber Wood Sawdust by Carbon Dioxide Activation. *Procedia Eng.* **2016**, *148*, 530–537. [CrossRef]
21. Phiri, Z.; Everson, R.C.; Neomagus, H.W.J.P.; Wood, B.J. The effect of acid demineralising bituminous coals and de-ashing the respective chars on nitrogen functional forms. *J. Anal. Appl. Pyrolysis* **2017**, *125*, 127–135. [CrossRef]
22. Subramani, T.; Revathi, P.K. Production of Activated Carbon from Agricultural Raw Waste. *IOSR J. Eng.* **2015**, *5*, 54–63.
23. Benedetti, V.; Cordioli, E.; Patuzzi, F.; Baratieri, M. CO_2 Adsorption study on pure and chemically activated chars derived from commercial biomass gasifiers. *J. CO_2 Util.* **2019**, *33*, 46–54. [CrossRef]
24. Jung, H.; Sewu, D.D.; Ohemeng-Boahen, G.; Lee, D.S.; Woo, S.H. Characterization and adsorption performance evaluation of waste char by-product from industrial gasification of solid refuse fuel from municipal solid waste. *Waste Manag.* **2019**, *91*, 33–41. [CrossRef]
25. Marchelli, F.; Cordioli, E.; Patuzzi, F.; Sisani, E.; Barelli, L.; Baratieri, M.; Arato, E.; Bosio, B. Experimental study on H_2S adsorption on gasification char under different operative conditions. *Biomass Bioenergy* **2019**, *126*, 106–116. [CrossRef]
26. Barontini, F.; Frigo, S.; Gabbrielli, R.; Sica, P. Co-gasification of woody biomass with organic and waste matrices in a down-draft gasifier: An experimental and modeling approach. *Energy Convers. Manag.* **2021**, *245*, 114566. [CrossRef]
27. Zhou, C.; Zhou, K.; Li, H.; Xu, X.; Liu, B.; Li, H.; Zeng, Z.; Ma, W.; Li, L. Pressure swing adsorption properties of activated carbon for methanol, acetone and toluene. *Chem. Eng. J.* **2021**, *413*, 127384. [CrossRef]
28. Friedl, A.; Padouvas, E.; Rotter, H.; Varmuza, K. Prediction of heating values of biomass fuel from elemental composition. *Anal. Chim. Acta* **2005**, *544*, 191–198. [CrossRef]
29. Shmroukh, A.N.; Ali, A.H.H.; Ookawara, S. Adsorption working pairs for adsorption cooling chillers: A review based on adsorption capacity and environmental impact. *Renew. Sustain. Energy Rev.* **2015**, *50*, 445–456. [CrossRef]
30. Jerai, F.; Miyazaki, T.; Saha, B.B.; Koyama, S. Overview of adsorption cooling system based on activated carbon—Alcohol pair. *Evergreen* **2015**, *2*, 30–40. [CrossRef]
31. Sur, A.; Das, R.K. Experimental investigation on waste heat driven activated carbon-methanol adsorption cooling system. *J. Braz. Soc. Mech. Sci. Eng.* **2017**, *39*, 2735–2746. [CrossRef]
32. Sztekler, K.; Kalawa, W.; Mika, Ł.; Mlonka-Medrala, A.; Sowa, M.; Nowak, W. Effect of additives on the sorption kinetics of a silica gel bed in adsorption chiller. *Energies* **2021**, *14*, 1083. [CrossRef]
33. Sztekler, K.; Mlonka-Mędrala, A.; Khdary, N.H.; Kalawa, W.; Nowak, W.; Mika, Ł. Possibility of Advanced Modified-Silica-Based Porous Materials Utilisation in Water Adsorption Processes—A Comparative Study. *Energies* **2022**, *15*, 368. [CrossRef]
34. Munir, S.; Daood, S.S.; Nimmo, W.; Cunliffe, A.M.; Gibbs, B.M. Thermal analysis and devolatilization kinetics of cotton stalk, sugar cane bagasse and shea meal under nitrogen and air atmospheres. *Bioresour. Technol.* **2009**, *100*, 1413–1418. [CrossRef] [PubMed]
35. Ren, J.; Cao, J.P.; Zhao, X.Y.; Liu, Y.L. Fundamentals and applications of char in biomass tar reforming. *Fuel Process. Technol.* **2021**, *216*, 106782. [CrossRef]
36. Singh, M.; Salaudeen, S.A.; Gilroyed, B.H.; Dutta, A. Simulation of biomass-plastic co-gasification in a fluidized bed reactor using Aspen plus. *Fuel* **2022**, *319*, 123708. [CrossRef]
37. González-Vázquez, M.P.; García, R.; Gil, M.V.; Pevida, C.; Rubiera, F. Comparison of the gasification performance of multiple biomass types in a bubbling fluidized bed. *Energy Convers. Manag.* **2018**, *176*, 309–323. [CrossRef]
38. Wang, G.; Zhang, J.; Huang, X.; Liang, X.; Ning, X.; Li, R. Co-gasification of petroleum coke-biomass blended char with steam at temperatures of 1173–1373 K. *Appl. Therm. Eng.* **2018**, *137*, 678–688. [CrossRef]
39. Noushabadi, A.S.; Dashti, A.; Ahmadijokani, F.; Hu, J.; Mohammadi, A.H. Estimation of higher heating values (HHVs) of biomass fuels based on ultimate analysis using machine learning techniques and improved equation. *Renew. Energy* **2021**, *179*, 550–562. [CrossRef]
40. Wang, Z.; Burra, K.G.; Li, X.; Zhang, M.; He, X.; Lei, T.; Gupta, A.K. CO_2-assisted gasification of polyethylene terephthalate with focus on syngas evolution and solid yield. *Appl. Energy* **2020**, *276*, 115508. [CrossRef]
41. Zhang, X.; Zhu, S.; Zhu, J.; Lyu, Q.; Wei, K.; Huang, Q.; Li, G.; Xia, H. TG-MS study on co-combustion characteristics and coupling mechanism of coal gasification fly ash and coal gangue by ECSA®. *Fuel* **2022**, *314*, 123086. [CrossRef]
42. Ding, H.; Ouyang, Z.; Wang, W.; Zhang, X.; Zhu, S. Experimental study on the influence of O_2/CO_2 ratios on NO conversion and emission during combustion and gasification of high-temperature coal char. *Fuel* **2022**, *310*, 122311. [CrossRef]

43. Guo, Y.; Guo, F.; Zhou, L.; Guo, Z.; Miao, Z.; Liu, H.; Zhang, X.; Wu, J.; Zhang, Y. Investigation on co-combustion of coal gasification fine slag residual carbon and sawdust char blends: Physiochemical properties, combustion characteristic and kinetic behavior. *Fuel* **2021**, *292*, 120387. [CrossRef]
44. Sieradzka, M.; Gao, N.; Quan, C.; Mlonka-Mędrala, A.; Magdziarz, A. Biomass thermochemical conversion via pyrolysis with integrated CO_2 capture. *Energies* **2020**, *13*, 1050. [CrossRef]
45. Sieradzka, M.; Kirczuk, C.; Kalemba-Rec, I.; Mlonka-Mędrala, A.; Magdziarz, A. Pyrolysis of Biomass Wastes into Carbon Materials. *Energies* **2022**, *15*, 1941. [CrossRef]
46. Ma, B.G.; Li, X.G.; Xu, L.; Wang, K.; Wang, X.G. Investigation on catalyzed combustion of high ash coal by thermogravimetric analysis. *Thermochim. Acta* **2006**, *445*, 19–22. [CrossRef]
47. Lu, J.J.; Chen, W.H. Investigation on the ignition and burnout temperatures of bamboo and sugarcane bagasse by thermogravimetric analysis. *Appl. Energy* **2015**, *160*, 49–57. [CrossRef]
48. Liu, Y.; Cao, X.; Duan, X.; Wang, Y.; Che, D. Thermal analysis on combustion characteristics of predried dyeing sludge. *Appl. Therm. Eng.* **2018**, *140*, 158–165. [CrossRef]
49. Niu, S.L.; Lu, C.M.; Han, K.H.; Zhao, J.L. Thermogravimetric analysis of combustion characteristics and kinetic parameters of pulverized coals in oxy-fuel atmosphere. *J. Therm. Anal. Calorim.* **2009**, *98*, 267–274. [CrossRef]
50. Li, X.G.; Lv, Y.; Ma, B.G.; Jian, S.W.; Tan, H.B. Thermogravimetric investigation on co-combustion characteristics of tobacco residue and high-ash anthracite coal. *Bioresour. Technol.* **2011**, *102*, 9783–9787. [CrossRef]
51. Reinmöller, M.; Sieradzka, M.; Laabs, M.; Schreiner, M.; Mlonka-Mędrala, A.; Kopia, A.; Meyer, B.; Magdziarz, A. Investigation of the thermal behaviour of different biomasses and properties of their low- and high-temperature ashes. *Fuel* **2021**, *301*, 121026. [CrossRef]
52. Mureddu, M.; Dessì, F.; Orsini, A.; Ferrara, F.; Pettinau, A. Air- and oxygen-blown characterization of coal and biomass by thermogravimetric analysis. *Fuel* **2018**, *212*, 626–637. [CrossRef]
53. Xu, M.X.; Wu, Y.C.; Nan, D.H.; Lu, Q.; Yang, Y.P. Effects of gaseous agents on the evolution of char physical and chemical structures during biomass gasification. *Bioresour. Technol.* **2019**, *292*, 121994. [CrossRef] [PubMed]
54. Wu, R.; Beutler, J.; Price, C.; Baxter, L.L. Biomass char particle surface area and porosity dynamics during gasification. *Fuel* **2020**, *264*, 116833. [CrossRef]
55. Jung, S.H.; Oh, S.J.; Choi, G.G.; Kim, J.S. Production and characterization of microporous activated carbons and metallurgical bio-coke from waste shell biomass. *J. Anal. Appl. Pyrolysis* **2014**, *109*, 123–131. [CrossRef]
56. Chen, T.; Zhang, J.; Wang, Z.; Zhao, R.; He, J.; Wu, J.; Qin, J. Oxygen-enriched gasification of lignocellulosic biomass: Syngas analysis, physicochemical characteristics of the carbon-rich material and its utilization as an anode in lithium ion battery. *Energy* **2020**, *212*, 118771. [CrossRef]
57. Alvarez, J.; Lopez, G.; Amutio, M.; Bilbao, J.; Olazar, M. Evolution of biomass char features and their role in the reactivity during steam gasification in a conical spouted bed reactor. *Energy Convers. Manag.* **2019**, *181*, 214–222. [CrossRef]
58. Guerrero-Pérez, M.O.; Valero-Romero, M.J.; Hernández, S.; Nieto, J.M.L.; Rodríguez-Mirasol, J.; Cordero, T. Lignocellulosic-derived mesoporous materials: An answer to manufacturing non-expensive catalysts useful for the biorefinery processes. *Catal. Today* **2012**, *195*, 155–161. [CrossRef]
59. Khasraw, D.; Theint Htet, T.; Yang, X.; Degirmenci, V.; Hage, H.; Meijer, K.; Li, Z. Gasification and structural behaviour of different carbon sources and resultant chars from rapid devolatilization for HIsarna alternative ironmaking process. *Fuel* **2022**, *309*, 122210. [CrossRef]
60. Reinmöller, M.; Schreiner, M.; Guhl, S.; Neuroth, M.; Meyer, B. Formation and transformation of mineral phases in various fuels studied by different ashing methods. *Fuel* **2017**, *202*, 641–649. [CrossRef]
61. Biagini, E.; Simone, M.; Tognotti, L. Characterization of high heating rate chars of biomass fuels. *Proc. Combust. Inst.* **2009**, *32*, 2043–2050. [CrossRef]
62. Kamble, A.D.; Mendhe, V.A.; Chavan, P.D.; Saxena, V.K. Insights of mineral catalytic effects of high ash coal on carbon conversion in fluidized bed Co-gasification through FTIR, XRD, XRF and FE-SEM. *Renew. Energy* **2022**, *183*, 729–751. [CrossRef]
63. Sun, J.; Liu, X.; Duan, S.; Alsaedi, A.; Zhang, F.; Hayat, T.; Li, J. The influential factors towards graphene oxides removal by activated carbons: Activated functional groups vs. BET surface area. *J. Mol. Liq.* **2018**, *271*, 142–150. [CrossRef]
64. Pelaez-Samaniego, M.R.; Perez, J.F.; Ayiania, M.; Garcia-Perez, T. Chars from wood gasification for removing H_2S from biogas. *Biomass Bioenergy* **2020**, *142*, 105754. [CrossRef]
65. Rubio-Clemente, A.; Gutiérrez, J.; Henao, H.; Melo, A.M.; Pérez, J.F.; Chica, E. Adsorption capacity of the biochar obtained from Pinus patula wood micro-gasification for the treatment of polluted water containing malachite green dye. *J. King Saud Univ.—Eng. Sci.* 2021; *in press*. [CrossRef]
66. Zając, G.; Szyszlak-Bargłowicz, J.; Gołębiowski, W.; Szczepanik, M. Chemical characteristics of biomass ashes. *Energies* **2018**, *11*, 2885. [CrossRef]
67. Kraszkiewicz, A.; Kachel-Jakubowska, M.; Lorencowicz, E.; Przywara, A. Influence of Cellulose Content in Plant Biomass on Selected Qualitative Traits of Pellets. *Agric. Agric. Sci. Procedia* **2015**, *7*, 125–130. [CrossRef]

68. Gordeeva, L.; Aristov, Y. Dynamic study of methanol adsorption on activated carbon ACM-35.4 for enhancing the specific cooling power of adsorptive chillers. *Appl. Energy* **2014**, *117*, 127–133. [CrossRef]
69. Wu, J.W.; Madani, S.H.; Biggs, M.J.; Phillip, P.; Lei, C.; Hu, E.J. Characterizations of Activated Carbon-Methanol Adsorption Pair Including the Heat of Adsorptions. *J. Chem. Eng. Data* **2015**, *60*, 1727–1731. [CrossRef]

energies

MDPI

Article

Hydrothermal Carbonization Process of Digestate from Sewage Sludge: Chemical and Physical Properties of Hydrochar in Terms of Energy Application

Małgorzata Wilk [1,*], Marcin Gajek [2], Maciej Śliz [1], Klaudia Czerwińska [1] and Lidia Lombardi [3]

1 Department of Heat Engineering and Environment Protection, AGH University of Science and Technology, Mickiewicza 30 Av., 30-059 Kraków, Poland
2 Faculty of Materials Science and Ceramics, AGH University of Science and Technology, Mickiewicza 30 Av., 30-059 Kraków, Poland
3 Niccolò Cusano University, Via don Carlo Gnocchi 3, 00166 Rome, Italy
* Correspondence: mwilk@agh.edu.pl

Abstract: Hydrochars (HTCD) derived from digestates, namely D1 and D2 (from two plants) of sewage sludge, were examined with respect to their fuel properties. The hydrothermal carbonization (HTC) tests were performed at temperatures of 200 and 220 °C, for 2 and 4 h of residence times, and with 1:10 and 1:8 digestate to water ratios (D/W), causing an increase of ash content (max. 55.8%), and a decrease c.a. 20% of the higher heating value except for a slight increase to 15 kJ/kg at 200 °C and 4 h in hydrochars. Conversely, the combustion profiles of hydrochars moved towards higher temperatures (225–257 °C) and finished earlier at lower temperatures (423–438 °C). The HTCD from D1 and D2 showed very similar properties under the same conditions (200 °C, 4 h, 1:8 D/W) for combustion characteristic temperatures, indices and profiles. The best efficiency was found for HTCD2. In addition, the polluted post-processing liquid phase was treated by a distillation process providing 30% higher pH, 50% lower BOD values, up to 15 times lower COD values, and c.a. three times lower conductivity.

Keywords: hydrothermal carbonization; sewage sludge; hydrochar; digestate; thermal analysis; post-processing water

Citation: Wilk, M.; Gajek, M.; Śliz, M.; Czerwińska, K.; Lombardi, L. Hydrothermal Carbonization Process of Digestate from Sewage Sludge: Chemical and Physical Properties of Hydrochar in Terms of Energy Application. *Energies* **2022**, *15*, 6499. https://doi.org/10.3390/en15186499

Academic Editor: Andrea De Pascale

Received: 13 August 2022
Accepted: 2 September 2022
Published: 6 September 2022

1. Introduction

Every sector that utilizes biomass generates biodegradable waste, which can then be transformed via different methods, is primarily for energy production. One method is anaerobic digestion, considered to be the best biological transformer of organic waste [1], employing the help of microbes, and resulting in the production of biogas [2]. Organic and municipal wastewater waste can be successfully utilized in biogas plants producing biogas as the main product and digestate as a by-product. Anaerobic digestion involves a sequence of reactions such as hydrolysis, acidogenesis and methanogenesis [3]. Digestate is a substance which is rich in nutrients and contains both organic and inorganic compounds as well as micro-organisms. Due to a significant amount of organic matter in the digestate, it can be used as a solid biofuel, a soil amendment or a substrate for activated carbon production. Most of the digestate (c.a. 95%) produced in Europe is used as fertilizer, but some impurities (e.g., biological and heavy metals), and law regulations can limit this utilization, such as European Nitrates Directive (91/676/EEC) [4]. There are a few disadvantages linked with the anaerobic digestion process. The digestates are often significantly diluted (commonly from 3 to 6% dry mass), have a colloidal structure responsible for significant problems with their mechanical dewaterability, and contain inorganic impurities, including alkali and/or heavy metal compounds. They are characterized by a low calorific value of dry matter (organic compounds contained in the substrates for biogas plants have already been mostly converted into biogas in the process of methane fermentation) and they create strong odours. If the digestate cannot be used for field fertilization due to exceeding the permissible content of contaminants, additional methods concerning utilization have to be found.

The digested sewage sludge still contains a large quantity of non-biodegradable organic matter that can be used for additional energy production. The most common usage of digested sewage sludge is drying combined with combustion, but the high moisture content in a feedstock is cost consuming [5]. Effective energy management may be improved by a suitable pretreatment to help solve this problem. In fact, a high amount of moisture in the digestate makes hydrothermal carbonization, treatment or conditioning, the most adequate method. Additionally, a part of the biogas could be used for its energy demand. Moreover, the hydrothermal conditioning of sewage sludge significantly improves its dewaterability, which may help with a reduction of heat consumption in all processes, in accordance with the European key objectives action plan, by focusing on "waste-to-energy" in the circular economy and, thereby, increasing waste prevention, reuse and recycling.

Hydrothermal carbonization (HTC) is a thermochemical process which involves the application of heat and pressure to convert the digestate in the presence of water into a solid product, namely hydrothermally carbonized material or hydrochar. The process is generally carried out within a temperature range of 180–350 °C, and the pressure is maintained above saturation pressure to ensure the liquid state of water. Additionally, the residence time is maintained between one and several hours. Under those rigid conditions, the chemical destruction of the dry mass of feedstock occurs and the following reactions take place: hydrolysis, decarboxylation, and dehydration [6]. This results in three types of products: gases, mainly CO_2, water and some simple organic compounds dissolved in water. A solid product, (30–40% moisture content), can be easily filtered from the reaction solution, and has upgraded carbon-like physical and chemical properties (e.g., hydrophobic, non-biodegradable). The liquid phase, which is often a major product, may be highly toxic, reaching relatively high values of chemical oxygen demand (COD) including up to 45 gO_2/L in the case of sewage sludge used as feedstock and, hence, requires proper management. The efficiency of the process depends chiefly on the physical and chemical structure of feedstock, reaction temperature, reaction time, pressure and pH. The main advantage of this pretreatment method is to increase the dewaterability (drying) of the obtained hydrochar with respect to the feedstock, which is five times less energy-consuming than drying. The researchers attracted a great deal of attention for the HTC process, because it is energy-efficient, simple and low cost. Moreover, the energy required to heat the reacting water is very low in comparison with traditional thermochemical processes. As previously mentioned, the process is specific to biomass with a high moisture content, and mainly includes municipal waste, e.g., sewage sludge [6,7]; food factory waste, e.g., beetroot, citrus, orange, herbal tea, pulp mill waste, olive mill, wine industry or tobacco stalks [8–13]; agricultural waste, e.g., straw, lignocellulosic or algae biomass [14–18], and also digestate from biogas plants [19]. The physical and chemical properties of the hydrochar and its yield depend primarily on the properties of the feedstock and the processing conditions (temperature and residence time), which is why research on their impact is required for every studied feedstock. Moreover, during hydrothermal carbonization, the morphology changes significantly with temperature from macroporous to micro carbonaceous spheres on the surface, mainly caused by dehydration and decarboxylation of polymers followed by aromatization and re-condensation reactions [20], resulting in hydrophilic heterogeneous structures of hydrochar. In general, during the biomass degradation under hydrothermal carbonization, the inorganic elements are released and dissolved in the processing solution; thus, a lower ash content is found in hydrochar than in raw biomass [21]. In the case of sewage sludge, the opposite results have been found [22]. Due to the excess loss of volatile matter and retention of minerals, an increase of ash in hydrochar is observed. The HTC also modifies the porous structure and surface area of sewage sludge, but high temperatures and long residence times are not desirable as they cause limited porosity and a small surface area. However, according to Wang [22], the abundant oxygen-containing functional groups on its surface indicate the potential for good adsorptive properties for heavy metals, and xenobiotic organism retention. Sewage sludge was also studied by Tasca et al. [23], who reviewed the hydrochar properties towards different applications, namely: energy sector, soil amelioration, wastewater pollution remediation, and carbon and greenhouse gases sequestration. In addition, the methods for treatment of the post-processing liquid phase were also analyzed. Furthermore, the fate of heavy metals was also investigated since sewage sludge is considered as an extremely complicated heterogeneous material that contains organic debris, inorganic particles, bacteria, colloid sewage sludge and moisture, as well as undesirable components, such as heavy metals, organic pollutants and pathogens, which are environmentally hazardous [24–27]. The hydrothermal treatment of sewage sludge has been investigated, taking into account the possible application of hydrochars as adsorbents, land fertilizer or renewable fuel [28–30]. However, the contents of heavy metals seriously limit their application. The reported research suggests that changing HTC process parameters (e.g., temperature, time, catalyst, and the addition of other biomass) could influence the total contents and chemical forms of heavy metals [31], mainly with an increase of temperature in the process. The significant impact of feedstock properties has created a wide range of investigations resulting in many documented studies concerning a variety of feedstock, primarily in renewable energy generation. Sharma et al. [32] reviewed the waste biomass and its conversion by using the conventionally heated HTC process, modified by microwave assistance, and confirmed the advantages of hydrothermal carbonization of waste biomass towards renewable energy generation potential. Although there have been many investigations concerning the hydrothermal carbonization of sewage sludge [33,34], fewer studies have considered the hydrothermal carbonization process of its digestate [35–38]. Aragón-Briceño et al. [39] investigated the process temperature on the characteristics of hydrochars including the fate of nitrogen and phosphorus species in energy application. The post-processing liquid phase from the hydrothermal process in terms of the biochemical methane potential was also tested. Both hydrochar and the post-processing liquid phase were found to have some potential for use in additional energy production. The same idea was investigated by Parmar and Ross [40], who studied the HTC of four different digestates from disparate origins: agricultural residue, sewage sludge, residual municipal solid waste, and vegetable, garden, and fruit waste. It appeared that the hydrothermal carbonization process enhanced biogas production, but the solid product, due to a high ash content, was not recommended as a fuel. It was indicated that hydrochar should be studied further in terms of utilization as a soil amender.

Cao et al. [41] investigated the fundamentals of the effect of digestate origin and process conditions. Although many researchers are interested in the hydrothermal treatment of digestate and are enthusiastic about this idea, they underline that not only the conditions of the process, but also the feedstock origin and its properties highly influence the resulting products and their properties. Composition and chemical energy can vary even between the same types of digestate depending on the source [6,9,19,39–43]. Moreover, discussion of the optimal conditions at a lower range of temperatures for producing hydrochar with homogeneous and constant fuel properties derived from digestate of sewage sludge is crucial from an industrial point of view to provide clear guidance for the design, optimization and economy of the HTC process when hydrochar is dedicated to energy application. The use of a higher range of temperature and pressure requires higher capital and operating costs of the HTC installation and more complex procedures concerning technical inspection. Therefore, further research on process conditions and properties of feedstock is still required to discover the overall benefits from hydrothermal processes as a post-treatment step after the anaerobic digestion of sewage sludge.

Accordingly, the main goal of this study was to investigate the hydrothermal carbonization process of digestate under different conditions of the HTC process. The novelty of this work is a comparative analysis of physical and chemical properties of hydrochars and liquid phases derived from the hydrothermal carbonization of digestates from two different sewage treatment plants. Therefore, the fuel characteristics as well as the thermal combustion behaviour (TGA), including combustibility indices and kinetic analyses of solid hydrochars, were investigated and compared to feedstock. Following-up the preliminary, less developed study of the distillation of post-processing water from sewage sludge, previously presented by the authors [6], where it was highlighted it as a very promising approach, further analyses were carried out. The liquid phase was analyzed in terms of its physical properties, and possible disposal, and application in, for example., biogas production or liquid biofertilizer. For these reasons, the following indicators were characterized according to standard analytical methods for Chemical Oxygen Demand, Biochemical Oxygen demand, conductivity, pH and acidity. The results could provide a better understanding of the hydrothermal conversion of the digestate of sewage sludge in order to successfully optimize this pre-treatment process and give a multifaceted description of digestate and its hydrothermally carbonized products.

2. Materials and Methods

2.1. Materials

Two filtrated, dewatered, aerobically stabilized types of sewage sludge were investigated. Both were collected from municipal sewage treatment plants at the same time of the year: in the summer. The first (D1) was taken from Płaszów, a metropolitan area of Kraków, (c.a. 780,000 inhabitants), Poland. The second (D2) was taken from the Silesian industrial district area of Gliwice, (c.a. 184,000 inhabitants), Poland, where wastewater was collected. Both digestates were stored, briefly in the refrigerator at 4 °C prior to the experiments. Samples from both materials were air dried, pulverized, and then stored in sealed containers before further comparative tests. The moisture content in the raw digestates were determined: for D1 it was 76.25%, whereas, in the case of D2, it was 80%. Both materials were diluted with deionized water for hydrothermal carbonization tests.

2.2. Hydrothermal Carbonization Test

The hydrothermal carbonization tests were performed using a stainless steel Zipperclave® Stirred Reactor equipped with a built-in stirrer produced by Parker Autoclave Engineers, Hessinger Dr, Erie, PA, USA. The full procedure was presented by Wilk [44]. Briefly, the diluted material D1 at either 1:8 or 1:11 digestate to water ratio was introduced to the reactor. The isolated reactor was heated up to 200 °C or 220 °C and maintained for a residence time of 4 h. In the case of D2, material was diluted at a 1:8 digestate to water ratio, and two residence times, 2 and 4 h, and a temperature of 200 °C were applied. Finally, the heat was turned off and the main reactor was cooled down by cooling water to room temperature. The solution was evacuated from the reactor and was filtrated through microfiltration paper using the ceramic Buchner funnel vacuum filtration setup. Both products, solid and liquid, were weighed before further testing was undertaken. The solid product was oven dried at 105 °C for 24 h and then pulverized prior to physical and chemical analyses. The liquid product was immediately investigated to avoid any changes in its properties. The hydrothermally carbonized samples were named according to the method used and the feedstock number, namely HTCD1 and HTCD2, which corresponded to digestates D1 and D2, and then were numbered according to the applied conditions.

2.3. Methods for Analysis of Solid Materials

The ultimate analysis was performed using the Elemental Analyser Leco 628, according to PKN-ISO/TS 12902:2007. The proximate analysis including moisture, ash and volatile content were determined under the following standards, PN EN ISO 18134-2:2017, PN EN ISO 18122:2016, and PN EN ISO 18123:2016, respectively. The higher heating values were determined using a KL-10 bomb calorimeter, according to PN-ISO 1928:2002. Mass and energy yields of hydrochars were determined according to [45].

2.3.1. Combustion Performance and Behaviour

The TGA analysis, in an air atmosphere (40 mL min^{-1}), and at a heating rate of 10, 20, and 30 K·min^{-1} using the Netzsch STA 449 F3 Jupiter apparatus, was performed. The samples (10 mg) were heated in alumina crucibles (capacity 70 µL) from an ambient temperature up to 700 °C. Based on TG and DTG results, performed at a heating rate of 10 K min^{-1}, the combustion characteristic temperatures including T_i—the ignition temperature, T_{max}—maximum peak temperature, and T_b—burnout temperature, were determined. The ignition (D_i), burnout (D_f), combustion stability (H_f) and comprehensive combustion (S) indices were also calculated to evaluate the combustion performance [34].

2.3.2. Kinetic Analysis

Three isoconversion methods: Friedman, Kissinger-Akahira-Sunose and Flynn_Wall_Ozawa, were employed to estimate the most significant factor of kinetics: the activation energy. These methods were adopted due to their reliability and currency by other researchers [46]. The activation energy was determined by measuring the temperatures corresponding to fixed values of α from experiments at different heating rates from the slope of a plot of lnβ vs. $1/T$, where α was a conversion rate for the sample

$$\alpha = \frac{m_{i0} - m_a}{m_{i0} - m_f} \tag{1}$$

and m_{i0}, m_a, and m_f were the initial mass of the sample, the actual mass, and mass after combustion in g. The ratio of solid-state reaction rate was described by the following equation, where t and T were the time and temperature of the process, respectively, and k the rate constant:

$$\frac{d\alpha}{dt} = k(T)f(\alpha) \tag{2}$$

A number of approximation formulas and different methods were applied to calculate the activation energy. The Friedman method was based on the following equation:

$$\ln\left[\beta\frac{d\alpha}{dT}\right] = \ln[A_\alpha f(\alpha)] - \frac{E_{a,\alpha}}{RT_\alpha} \tag{3}$$

The Kissinger-Akahira-Sunose method was expressed by

$$\ln\left(\frac{\beta}{T_\alpha^2}\right) = \ln\left(\frac{A_\alpha R}{F_{a\alpha} g(\alpha)}\right) - \frac{E_{a\alpha}}{RT_\alpha} \tag{4}$$

The Flynn-Wall-Ozawa method was based on:

$$\ln\beta = \ln\left[\frac{0.0048E_a}{Rg(\alpha)}\right] - 1.0516\frac{E_a}{RT} \tag{5}$$

2.3.3. Structural Analysis

Scanning electron microscopy supported by energy dispersive spectroscopy was investigated by an FEI Inspect S50 microscope. Additionally, Fourier Transformation Infrared Spectroscopy was performed by Brucker spectroscope, which investigated the wavelengths of studied samples in the range 400–4000 cm^{-1}. Both methods were used to determine changes in the structure and surface of the digestates and their hydrothermally carbonized samples [9].

2.4. Methods for Analysis Methods of the Liquid Phase

The liquid phase, derived from the filtration process after the hydrothermal carbonization test, was non-translucent and of a dark colour; thus, the distillation process, performed under low pressure, was applied for purification [6]. For both the liquid phase and its distillate, the following analyses were conducted: chemical oxygen demand (COD) analysis, according to PN-ISO 6060:2006, and pH, conductivity, and biological oxygen demands (BOD) were measured by the Multifunction Laboratory Meter CX-505 ELMETRON in order to make a comparative analysis. Additionally, an acid-base titration of the distillate was carried out to determine the concentration of acetic acid. All measurements were repeated at least three times. The remaining dark solid part from the distillation process was dried at 105 °C, weighed and then analysed, using the Elemental Analyser Truespec Leco 628 according to PKN-ISO/TS 12902:2007, for its carbon content to determine the carbon balance.

3. Results and Discussion

The results of ultimate and proximate analyses supported by higher and lower heating values (HHV and LHV) are summarized in Table 1.

Table 1. Characteristics of feedstocks and hydrothermally carbonized digestates, db.

	D1	HTCD1_1	HTCD1_2	HTCD1_3	HTCD1_4	D2	HTCD2_1	HTCD2_2
Temperature, °C		**200**	**200**	**220**	**220**		**200**	**200**
Digestate to water ratio		**1:8**	**1:11**	**1:8**	**1:11**		**1:8**	**1:8**
Residence time		**4**	**4**	**4**	**4**		**2**	**4**
Ultimate analysis								
C, %	29.6	26.2	26.50	26.70	26.00	32.8	30.4	30.8
H, %	4.3	3.14	3.34	3.14	3.05	4.69	3.83	3.84
N, %	4.35	2.32	2.14	2.16	1.95	4.74	2.55	2.52
S, %	1.58	1.42	1.32	1.36	1.51	1.61	1.43	1.55
O, %	20.11	11.12	12.34	9.44	11.21	22.85	12.67	11.93
Proximate analysis								
FC, %	8.16	9.01	10.45	10.96	11.88	8.66	9.42	9.24
VM, %	50.45	34.03	33.08	31.18	30.58	53.83	39.59	39.46
Ash, %	40.06	55.80	54.36	57.20	56.28	33.31	49.12	49.36
M, %	1.33	1.16	2.11	0.66	1.26	4.20	1.89	1.94
Fuel ratio FC/VM	0.16	0.26	0.32	0.35	0.39	0.16	0.24	0.23
Heating values								
HHV, MJ/kg	14.34	11.40	11.40	13.41	11.41	14.66	13.76	15.12
LHV, MJ/kg	13.57	10.82	11.45	12.73	10.80	13.75	13.09	14.46

The chemical changes in the properties of all hydrothermally carbonized digestates were compared to raw materials and depicted in a van Krevelen diagram visualized in Figure 1. Due to the decarboxylation, dehydration and demethanation reactions, which occurred in the aqueous environment under temperature and pressure, the molar ratios of O/C and H/C were much lower in comparison with the raw materials. Consequently, they moved into a more coal-like zone towards the lower end of the scheme. Organoleptic analysis confirmed that all hydrothermally carbonized materials were more carbonaceous, brittle and easier to grind than raw dried digestates. Regarding both raw materials, the digestates derived from sewage sludge of different origins, and the ultimate and proximate analysis indicated that they were similar from a chemical and physical point of view. Carbon and volatile matter contents were slightly higher in D2 when compared to D1, whereas ash content, conversely, was lower. The changes in the chemical properties of the samples presented in Figure 1 clearly indicated that the temperature and digestate to water ratio had an impact on the properties of hydrothermally treated material; the HTCD1_3 samples derived at 220 °C and a digestate to water ratio of 1:8, moved significantly to the left side of the scheme than the other wet torrefied samples from D2. A similar tendency was observed regarding the increased residence time of the process: HTCD2_2 at 4 h also moved to the more carbonaceous zone. These results are consistent with those presented by

He et al. [47]. Even though the higher temperature was applied to produce HTCD1_3, the location of this sample was closer to HTCD2_2, which was produced at a lower temperature but with the same residence time and digestate to water ratio. This is due to the slightly higher carbon content in D2 and its hydrothermally carbonized samples.

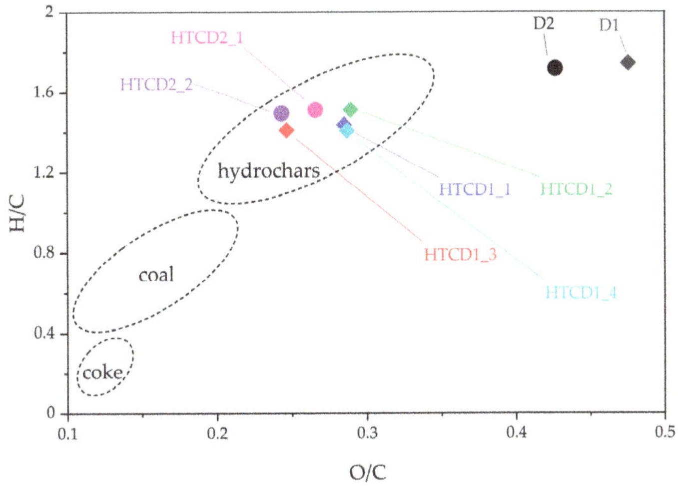

Figure 1. Van Krevelen diagram.

The fuel ratio (FC/VM) grades the hydrochars as an alternative carbonaceous fuel [41]. Since for HTCD1 the values of FC increased and VM decreased with dilution of feedstock for the HTC process and carbonization temperature, the fuel ratio increased from 0.16 to 0.39. In the case of HTCD2, the fuel ratio also increased by about 50% due to an increase of FC and a decrease of VM caused by the HTC process. However, the increase of carbonization time from 2 to 4 h did not affect the fuel ratio, which is contrary to the results of He et al. [47].

The ultimate analysis also indicated changes in the nitrogen and sulphur contents caused by the HTC process. The significant removal of nitrogen content from digestates was observed due to the degradation of proteins containing N which occurs above 150 °C, and decomposition of organic-N to NH+4-N [7]. The higher dilution (1:11) and higher temperature (220 °C) applied in the HTC of D1 resulted in 50 and 55% of N release into gases and liquid phases. In the case of D2, the increase of reaction time was not significant and the removal of N was 46% for both HTCD2_1 and HTCD2_2. Regarding sulphur, the HTC process caused a slight removal, but increased dilution and temperature did not indicate a clear tendency. The increase of carbonization time caused an increase of sulphur in the hydrochar, which is consistent with Aragón-Briceño et al. [39]. In conclusion, the NOx and SOx emissions from the combustion of hydrochars will be lower than in the case of digestates derived from sewage sludge, proving that hydrochar is the more favourable fuel. The physical properties are presented in Figure 2. The mass yields for HTCD1 differ between 52 and 64%, which is related to the applied conditions. The highest value of 64% was obtained for HTCD1_1 performed at 200 °C and a digestate to water ratio of 1:8, whereas the lowest was for HTCD1_2 performed at 200 °C and a digestate to water ratio of 1:11. Energy yields were almost the same for samples studied in different digestate to water ratios, and the temperature impact on these properties was not visible. The fixed carbon increased slightly with an increase in temperature, which corresponded to a decrease in volatile matter content in the samples. The more diluted material and higher temperature of the process, the less volatile matter and higher fixed carbon in the samples were found. In the case of HTCD2, the highest mass yield was found in the sample conducted for 4 h of residence time, and fixed carbon, found for 2 h of residence time, which was lower than for

4 h, where it was slightly higher than in the case of D2. When comparing hydrothermally carbonized samples derived under the same conditions, namely 200 °C of temperature, digestate to water ratio of 1:8 and 4 h of residence time, the mass yield of HTCD1_1 was slightly higher and differed by only 3% from HTCD2_2. However, energy yields were lower, around 23% when compared with HTCD2_2, indicating that this material reacted differently despite being under the same conditions of the process. Higher heating values were depicted for all samples, confirming the above statement. In the case of D1, the hydrothermal treatment caused a decrease in HHV, whereas with D2, there was a slight increase. The results were similar to those presented by [35].

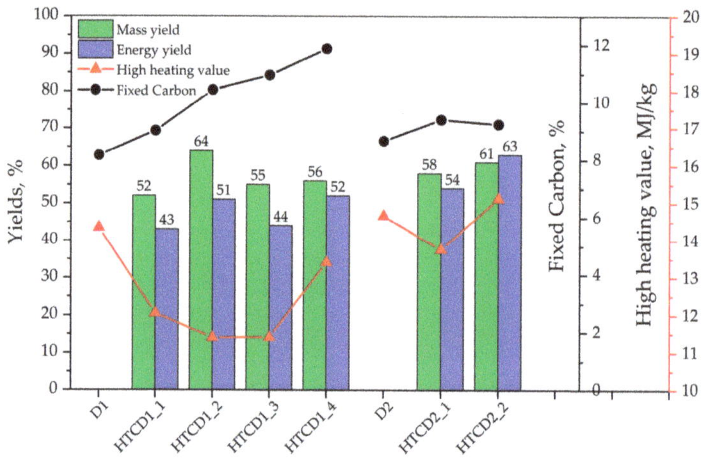

Figure 2. Chemical and physical properties.

Thermal analysis was conducted in order to study the combustion profiles of the samples. The results were depicted in the forms of TG/DTG/DSC curves (Figures 3a and 4a) for D1 and D2. The combustion process was observed on TG profiles. It was divided into three distinct stages: release of the moisture content, release and combustion of volatile matters, and, finally, combustion of the char. A lack of changes in the mass of solid residue indicated the end of the process. The combustion of the two raw digestates was similar. However, the TG curves were the same in shape and character; D2 initiated the combustion slightly later and combusted longer at a higher temperature (504 °C) for 3 min when compared to D1 (488 °C). The combustion process was much more visible with DTG curves, which occurred with two peaks: one c.a. 270 °C, and, more noticeably, the second highest weight loss was observed around 480 °C. The char combustion was confirmed by the DSC peak, which corresponded with the second DTG peak.

The comparison between TG curves of hydrothermally carbonized products derived from D1 and D2 are presented in Figures 3b and 4b. The TG profiles of carbonized samples differed and moved slightly towards higher temperatures. In both cases, a sharp decrease was observed between 250–300 °C and a much higher quantity of solid residue was collected after the treatment. Concerning HTCD1 samples, the TG profiles of HTCD1_1 and HTCD1_2 differed when compared to HTCD1_3 and HTCD1_4. Both HTCD1_1 and HTCD1_2, showed greater decreases and higher rates of volatile matter release represented by DTG1 curves. The location of DTG1 for HTCD1_2 suggested that its volatile matter was more violently combusted at a lower temperature, 272 °C, than in the case of HTCD1_1, namely 280 °C. The other two samples, HTCD1_3 and HTCD1_4, derived at a higher temperature, 220 °C, initiated and finalized the combustion at the same moment, and also combusted with two peaks of DTG, but with less intensity as in the case of the previously mentioned HTCD1_1 and HTCD1_2.

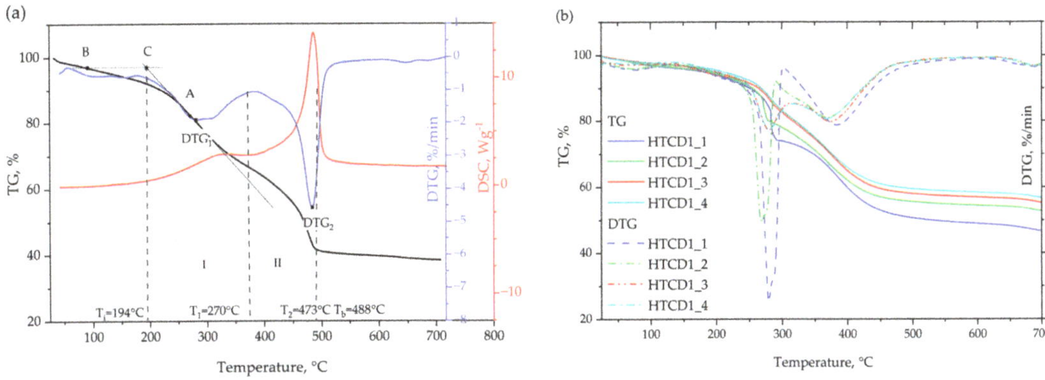

Figure 3. (**a**) TG/DTG/DSC of D1 and (**b**) TG curves of HTCD1.

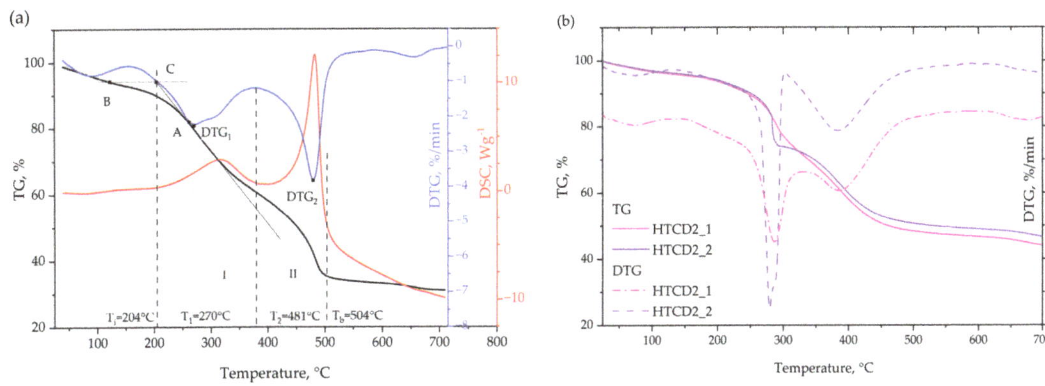

Figure 4. (**a**) TG/DTG/DSC of D2 and (**b**) TG curves of HTCD2.

In addition, based on the TGA results, the most important combustion parameters including characteristic temperatures and combustibility indexes, were determined and summarized in Table 2. The following temperatures were found: T_i—ignition temperature, which defined the beginning of the combustion, T_1—was the maximum peak temperature, and T_b—burnout temperature indicated the temperature at which the fuel was completely combusted [47]. In both cases, the hydrothermal carbonization caused the shift of T_i temperatures in HTCD1 and HTCD2 samples towards higher temperatures, due to an intensive volatile release confirmed by D_i. The Tb occurred earlier in the case of pretreated samples, indicating that they were combusted in a shorter time, and according to a higher value of S index, with a higher intensity and probably more easily due to a slightly better combustion performance. The H_f represents the rate and intensity of the combustion process. Additionally, the results proved that the hydrothermally carbonized process unified the properties of the pretreated samples conducted under the same conditions giving very similar values of combustion characteristic temperatures, indexes and profiles for HTCD1_1 and HTCD2_2. For instance, the ignition indexes (D_i) were 1.41 and 1.34 (%/min^3), respectively. The comprehensive combustion (S) index differed only by 1.7 (%/(min·K^2) and combustion stability (H_f) index by 5 (%^{2}/(min^2·K^3)). A higher temperature of the process (220 °C) caused a decrease in S, providing results approximately half lower (c.a. S = 6.5 (%/(min·K^2)) and an increase of about 40–50 in the H_f value. In conclusion, according to Song et al. [48] the results indicated that hydrochars from digestates performed better ignition and combustion characteristics in comparison to both digestates due to a larger D_i index and a higher value of the S index.

Table 2. The combustion characteristics parameters of raw and hydrothermally carbonized digestates.

Material	D1	HTCD1_1	HTCD1_2	HTCD1_3	HTCD1_4	D2	HTCD2_1	HTCD2_2
Temperature, °C		**200**	**200**	**220**	**220**		**200**	**200**
Digestate to water ratio		**1:8**	**1:11**	**1:8**	**1:11**		**1:8**	**1:8**
Residence time		**4**	**4**	**4**	**4**		**2**	**4**
T_i, °C	194	250	255	225	225	204	250	257
T_b, °C	488	438	423	427	423	504	442	438
T_1, °C	270	280	272	280	282	270	290	280
DTG_1, %/min	−4.00	−7.67	−5	−2.27	−2.25	−2.72	−4.58	−7.46
T_2, °C	473	384	363	374	370	481	384	382
DTG_2, %/min	−5.50	−2.54	−2.36	−2.02	−1.94	−5.29	−2.54	−2.16
D_i, $\%/min^3 \cdot 10^{-2}$	0.63	1.41	0.94	0.48	0.47	0.66	0.82	1.34
D_b, $\%/min^4 \cdot 10^{-5}$	9.5	30.9	22.4	9.3	10.3	12.3	18.0	30.9
S, $\%^2/(min^2 \cdot °C^3) \cdot 10^{-8}$	11.7	21.1	12.1	6.6	6.4	12.4	12.9	19.4
H_f, $\%/(min \cdot °C^2)$	875	972	973	1026	1014	813	992	967

The activation energy was estimated by three isoconversional model-free methods, Friedman, Kissinger-Akahira-Sunose, and Flynn-Wall-Ozawa [49] for D1 and D2 and their pretreated samples derived under the same conditions, 200 °C, 4 h of residence time and 1:8 of digestate to water ratio, namely HTCD1_1 and HTCD2_2 (Table 3). An example for E_a determination is presented in Figure 5, where the Kissinger-Akahira-Sunose model was used. Different heating rates (β = 10, 20, 30 K/min) were applied and the slope of $\ln(\beta/T2)$ for different conversion rates (α) was depicted. For the evaluation, only the fittings with a high correlation coefficient were considered. Raw samples of D1 and D2 showed nearly parallel fittings for most of the conversion rates, indicating that activation energy changes slightly with ongoing combustion. HTCD samples display a greater variation, suggesting a multistep kinetic mechanism [50,51]. All methods provided very similar results, especially in the case of D2 and its hydrothermally carbonized product, which differed by only 1 to 12%. The activation energy for raw D1 was lower, c.a 30%, when compared to pretreated material, and this was probably the result of the degradation of organic compounds in sewage sludge.

Table 3. Activation energy estimated by Friedman, Kissinger-Akahira-Sunose and Flynn-Wall-Ozawa.

Material	E_a, kJ/kmol		
	Friedman	**Kissinger-Akahira-Sunose**	**Flynn-Wall-Ozawa**
D1	132	179	152
HTCD1	169	180	182
D2	165	166	153
HTCD2	168	169	152

FTIR spectra of D1 and D2 digestates and their hydrochars HTCD1 and HTCD2 are depicted in Figure 6a,b. There were no significant changes found in the digestates after hydrothermal treatment, indicating that the hydrothermal carbonization process slightly affected the chemical bond of digestates, which corresponds with Peng's investigations [52]. Both digestates were similar in shape, but the intensity of peaks differed a little. For instance, with D2 the peak at 3400 cm^{-1}, which is attributed to the vibration of -OH stretching found in cellulose, decreased slightly with a rise in temperature and dilution, suggesting the ability for dehydration, whereas, in the case of D1, it was weaker for a shorter period of the hydrothermal carbonization process. The wavelength range of 2800–3000 cm^{-1} is assigned to aliphatic hydrocarbons, most likely aliphatic carbon C-H and the symmetrical stretching of methylene groups, represented by 2923 and 2853 cm^{-1}, respectively [53], which were slightly affected by temperature and time of the process. Two peaks found in the region of

1350–1800 cm^{-1} at 1658 and 1442 cm^{-1} were probably attributed to the stretching vibration of C=N amides [52,54] and the presence of olefins components [55]. In both studied cases they were slightly weaker after hydrothermal conditioning. The strong and broad peak at 1007 (D1) or 1011 cm^{-1} (D2) is connected to -C-OH, attributed to carboxylic acids or alcohols. The other hypothesis is that these peaks are assigned to Si-O stretching and Si-O-Si bonds, confirming the presence of Si in digestates from sewage sludge. According to Wang et al. [56], polysaccharides can be detected in the region of 1000–1100 cm^{-1} after decomposition caused by the hydrothermal carbonization process. There are also some subtle peaks below the region 600–400 cm^{-1} slightly affected by the hydrothermal process, which are probably caused by Si-O or Al-O bending vibration [47].

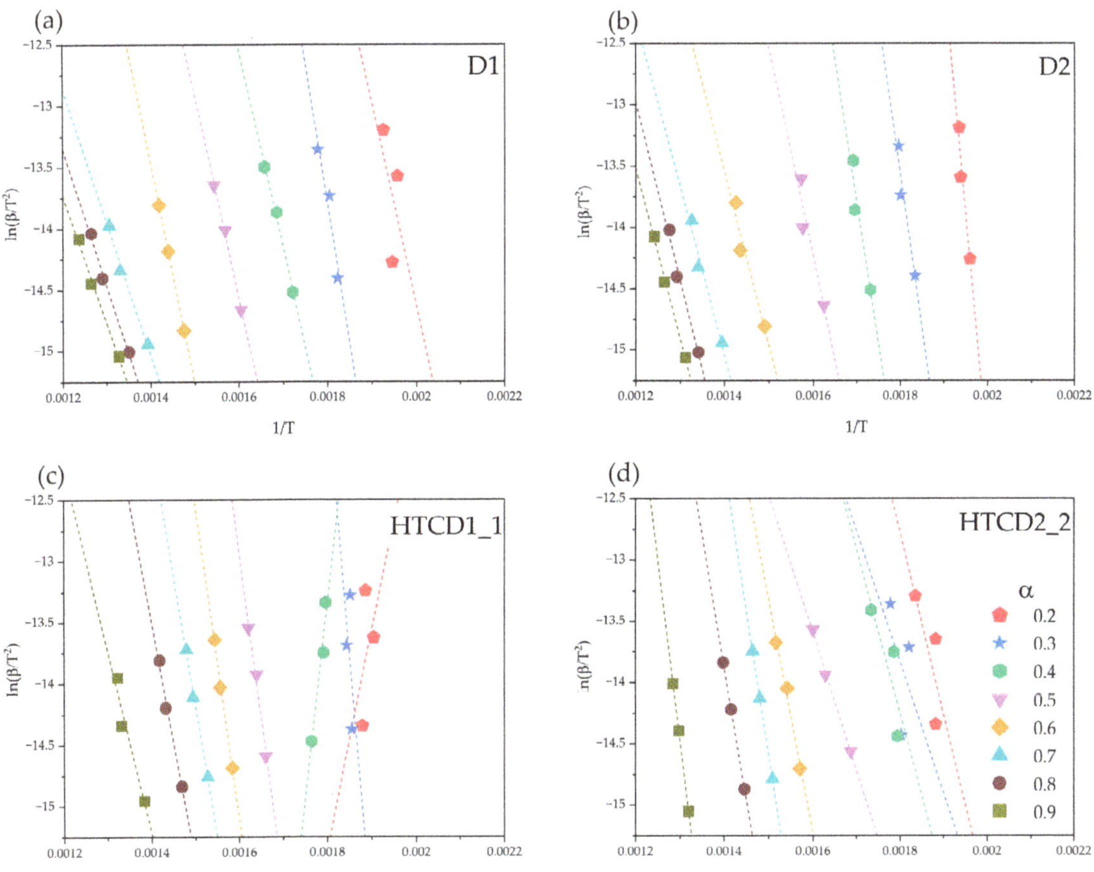

Figure 5. Linear fitting to the Kissinger-Akahira-Sunose kinetic model for various conversion during combustion for: (**a**) raw D1, (**b**) raw D2, (**c**) HTCD1_1, and (**d**) HTCD2_2.

SEM-EDS results depicted in Figures 7 and 8 confirmed the degradation observed by FTIR analysis caused by the hydrothermal carbonization process of pretreated digestates. In both cases defragmentation was also found, and smaller particles were observed after the process.

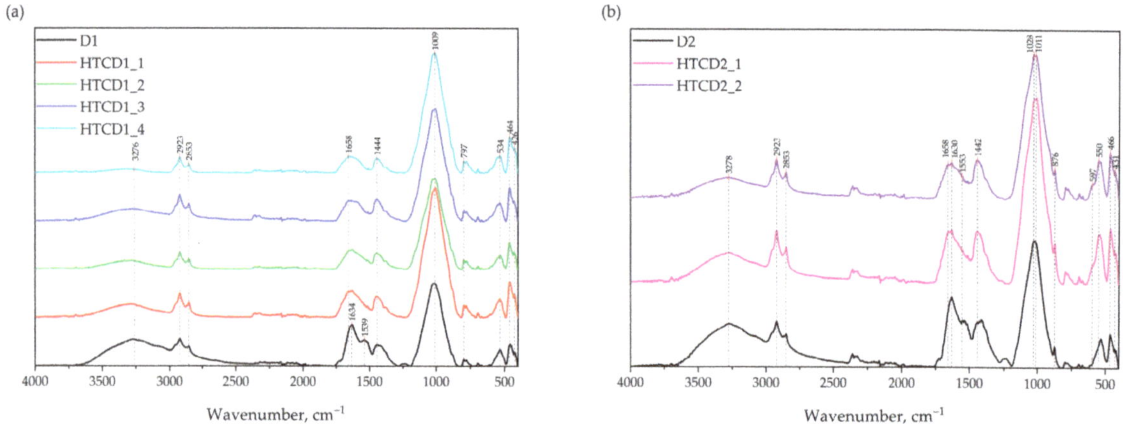

Figure 6. FTIR results for (**a**) D1 and HTCD1 and (**b**) D2 and HTCD2.

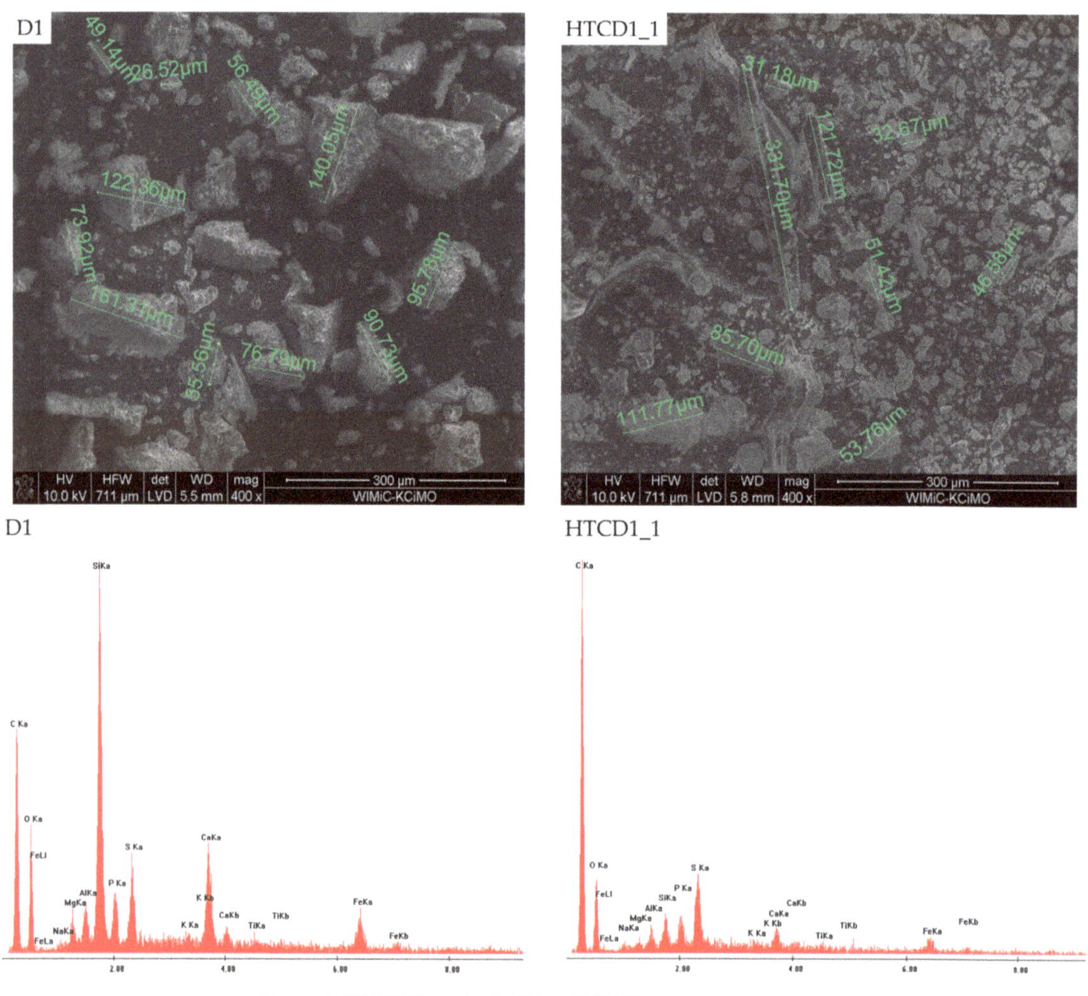

Figure 7. SEM-EDS results for D1 and HTCD1_1.

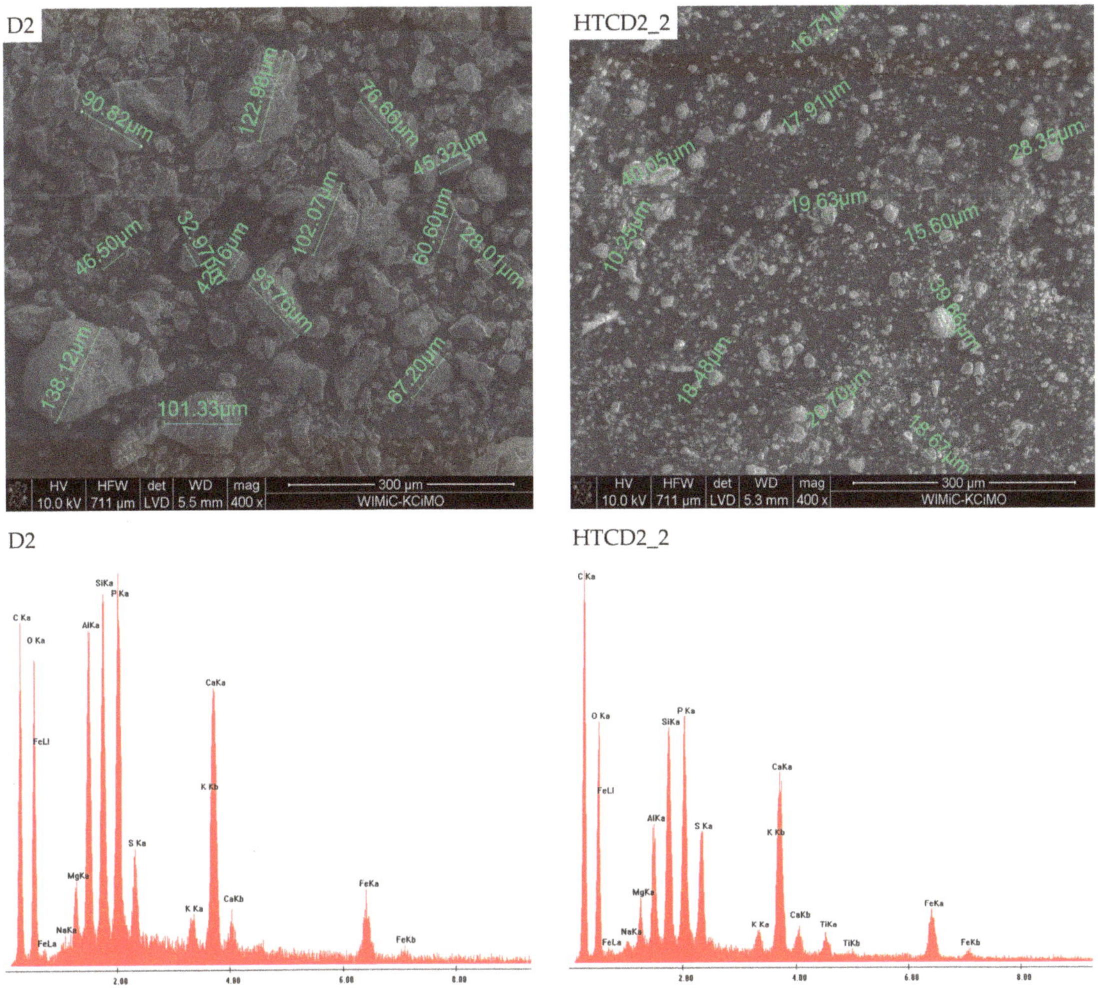

Figure 8. SEM-EDS results for D2 and HTCD2_2.

Liquid contains a high quantity of dissolved total organic carbon content and nutrients and necessitates adequate disposal. Accordingly, the preliminary study of vacuum distillation of the post-processing liquid phase was tested as a suitable method for purification. This method proved to be a good solution, not only for the elimination of odour and colour from non-translucent and toxic liquid, but also as an effective method in COD, BOD and conductivity reduction [6]. Table 4 summarized the results presenting that COD was decreased by ten times, BOD by two times and conductivity by four times, even though pH increased by 30%.

Table 4. Liquid filtrate and its distillate characteristics.

Material	HTCD1_1	HTCD1_2	HTCD1_3	HTCD1_4	HTCD2_1	HTCD2_2
pH	7.33	7.00	7.74	7.46	7.09	7.13
pH_d	9.5	9.18	8.87	8.95	9.09	8.94
Conductivity, mS	13.95	10.36	16.84	12.89	11.27	12.68
$Conductivity_d$, mS	3.09	2.28	8.88	7.09	3.85	5.20
COD, gO_2 L^{-1}	30.3	22.62	29.67	23.85	28.72	26.76
COD_d, gO_2 L^{-1}	3.16	1.49	4.57	3.33	2.77	2.37
BOD, gO_2 L^{-1}	1.76	1.04	1.32	1.52	1.08	1.12
BOD_d, gO_2 L^{-1}	0.8	0.72	0.72	0.84	0.60	0.48

d—distillate.

4. Conclusions

The digestates from two wastewater treatment plants were hydrothermally carbonized at 200 and 220 °C, 2 and 4 h of time, and 1:8 and 1:11 of D/W ratios. The changes in the physical and chemical properties of hydrochars were confirmed by proximate and ultimate analysis, thermal analysis and FTIR and SEM-EDS methods. Hydrothermal carbonization caused an increase of up to 50% of ash content and a decrease c.a. 6–12% of carbon content, which corresponded to a decrease c.a. 20% of the higher heating value except for a slight increase to 15 kJ/kg at 200 °C and 4 h in hydrochars. The degradation of morphology and structure caused by rigid conditions were confirmed by FTIR and SEM-EDS analysis. Energy and mass yields varied 52 to 64 and 43 to 63%, respectively. Thermal analysis presented combustion profiles of hydrochars which moved towards higher temperatures (225–257 °C) and finished earlier at lower temperatures (423–438 °C) in comparison to digestates. An important guidance for industry regarding the energy application of digestate from sewage sludge is provided. Indeed, one of the most significant aspects of the results is that the hydrochar properties are mainly dependent on the process conditions (200 °C, 4 h, and 1:8 D/W) rather than on the feedstock characteristics. This is very promising from an industrial point of view, because a single carbonization plant can process feedstock arriving from different waste water treatment plants, providing an homogenous and constant product output. In particular, the following properties were found to be unified for the two digestates: fixed carbon (c.a. 9.12%), combustion characteristic temperatures (T_i c.a. 254 °C, T_b = 438 °C, T_1 = 280 °C), indices (D_i c.a. 1.37 $min^2 \cdot °C^3 \cdot 10^{-2}$, D_b = 30.9 $min^2 \cdot °C^3 \cdot 10^{-2}$, S c.a. 20.25% $2/(min^2 \cdot °C^3) \cdot 10^{-8}$, H_f c.a. 970 $min^2 \cdot °C^2$) and E_a = 169 $kJ \cdot kmol^{-1}$ calculated by the Friedman method. Moreover, to manage the problem of polluted process water, the vacuum distillation process is proposed as a successful disposal of the liquid phase from the hydrothermal carbonization process, thereby giving a spectacular reduction in the main toxicity indicators (two times lower BOD values, up to fifteen times lower COD values, and c.a. three times lower conductivity).

Author Contributions: Conceptualization, M.W.; data curation, M.W., M.G., M.Ś.; formal analysis, M.W., M.G., M.Ś., K.C. and L.L.; resources, M.W.; methodology, M.W., M.G., M.Ś.; investigation, M.W., M.G., M.Ś. and K.C.; funding acquisition, M.W., project administration, M.W., validation, M.W., M.G. and M.Ś.; visualization, M.W., M.G. and M.Ś.; supervision, M.W.; writing—original draft, M.W.; writing review & editing, M.W. and L.L. All authors have read and agreed to the published version of the manuscript.

Funding: This research was funded by the programme „Excellence initiative—research university" for the AGH University of Science and Technology, Poland [grant no 501.696.7995].

Institutional Review Board Statement: Not applicable.

Informed Consent Statement: Not applicable.

Data Availability Statement: Not applicable.

Acknowledgments: The authors would like to express their thanks to the proprietor of the experimental apparatus EKOPROD Ltd. in Bytom. The authors also wish to thank Józef Marszałek for his exceptional experimental help.

Conflicts of Interest: The authors declare that they have no conflict of interest. The funders had no role in the design of the study; in the collection, analyses, or interpretation of data; in the writing of the manuscript; or in the decision to publish the results.

References

1. Lee, M.; Hidaka, T.; Hagiwara, W.; Tsuno, H. Comparative performance and microbial diversity of hyperthermophilic and thermophilic co-digestion of kitchen garbage and excess sludge. *Bioresour. Technol.* **2009**, *100*, 578–585. [CrossRef] [PubMed]
2. Lastella, G.; Testa, C.; Cornacchia, G.; Notornicola, M.; Voltasio, F.; Sharma, V.K. Anaerobic digestion of semi-solid organic waste: Biogas production and its purification. *Energy Convers. Manag.* **2002**, *43*, 63–75. [CrossRef]
3. Themelis, N.J.; Ulloa, P.A. Methane generation in landfills. *Renew. Energy* **2007**, *32*, 1243–1257. [CrossRef]
4. Dahlin, J.; Herbes, C.; Nelles, M. Biogas digestate marketing: Qualitative insights into the supply side. *Resour. Conserv. Recycl.* **2015**, *104*, 152–161. [CrossRef]
5. Werle, S. Sewage sludge-to-energy management in Eastern Europe: A Polish Perspective. *Ecol. Chem. Eng. S* **2015**, *22*, 459–469. [CrossRef]
6. Wilk, M.; Magdziarz, A.; Jayaraman, K.; Szymańska-Chargot, M.; Gökalp, I. Hydrothermal carbonization characteristics of sewage sludge and lignocellulosic biomass. A Comparative Study. *Biomass Bioenergy* **2019**, *120*, 166–175. [CrossRef]
7. Djandja, O.S.; Duan, P.G.; Yin, L.X.; Wang, Z.C.; Duo, J. A Novel machine learning-based approach for prediction of nitrogen content in hydrochar from hydrothermal carbonization of sewage sludge. *Energy* **2021**, *232*, 121010. [CrossRef]
8. Wilk, M.; Śliz, M.; Gajek, M. The Effects of hydrothermal carbonization operating parameters on high-value hydrochar derived from beet pulp. *Renew. Energy* **2021**, *177*, 216–228. [CrossRef]
9. Basso, D.; Patuzzi, F.; Castello, D.; Baratieri, M.; Rada, E.C.; Weiss-Hortala, E.; Fiori, L. Agro-industrial waste to solid biofuel through hydrothermal carbonization. *Waste Manag.* **2016**, *47*, 114–121. [CrossRef]
10. Pham, T.P.T.; Kaushik, R.; Parshetti, G.K.; Mahmood, R.; Balasubramanian, R. Food waste-to-energy conversion technologies: Current status and future directions. *Waste Manag.* **2015**, *38*, 399–408. [CrossRef]
11. Volpe, M.; Fiori, L. From olive waste to solid biofuel through hydrothermal carbonisation: The role of temperature and solid load on secondary char formation and hydrochar energy properties. *J. Anal. Appl. Pyrolysis* **2017**, *124*, 63–72. [CrossRef]
12. Saqib, N.U.; Sharma, H.B.; Baroutian, S.; Dubey, B.; Sarmah, A.K. Valorisation of food waste via hydrothermal carbonisation and techno-economic feasibility assessment. *Sci. Total Environ.* **2019**, *690*, 261–276. [CrossRef] [PubMed]
13. González-Arias, J.; Gómez, X.; González-Castaño, M.; Sánchez, M.E.; Rosas, J.G.; Cara-Jiménez, J. Insights into the product quality and energy requirements for solid biofuel production: A comparison of hydrothermal carbonization, pyrolysis and torrefaction of olive tree pruning. *Energy* **2022**, *238*, 122022. [CrossRef]
14. Wilk, M.; Magdziarz, A.; Kalemba-Rec, I.; Szymańska-Chargot, M. Upgrading of green waste into carbon-rich solid biofuel by hydrothermal carbonization: The effect of process parameters on hydrochar derived from acacia. *Energy* **2020**, *202*, 117717. [CrossRef]
15. Yuan, C.; Wang, S.; Qian, L.; Barati, B.; Gong, X.; Abomohra, A.E.; Wang, X.; Esakkimuthu, S.; Hu, Y.; Liu, L. Effect of cosolvent and addition of catalyst (HZSM-5) on hydrothermal liquefaction of macroalgae. *Int. J. Energy Res.* **2019**, *43*, 4843. [CrossRef]
16. Krylova, A.Y.; Zaitchenko, V.M. Hydrothermal carbonization of biomass: A review. *Solid Fuel Chem.* **2018**, *52*, 91–103. [CrossRef]
17. Zhou, S.; Liang, H.; Han, L.; Huang, G.; Yang, Z. The influence of manure feedstock, slow pyrolysis, and hydrothermal temperature on manure thermochemical and combustion properties. *Waste Manag.* **2019**, *88*, 85–95. [CrossRef]
18. Lee, J.; Lee, K.; Sohn, D.; Kim, Y.M.; Park, K.Y. Hydrothermal carbonization of lipid extracted algae for hydrochar production and feasibility of using hydrochar as a solid fuel. *Energy* **2018**, *153*, 913–920. [CrossRef]
19. Pawlak-Kruczek, H.; Niedzwiecki, L.; Sieradzka, M.; Mlonka-Mędrala, A.; Baranowski, M.; Serafin-Tkaczuk, M.; Magdziarz, A. Hydrothermal carbonization of agricultural and municipal solid waste digestates—Structure and energetic properties of the solid products. *Fuel* **2020**, *275*, 117837. [CrossRef]
20. Fang, J.; Zhan, L.; Ok, Y.S.; Gao, B. Minireview of potential applications of hydrochar derived from hydrothermal carbonization of biomass. *J. Ind. Eng. Chem.* **2018**, *57*, 15–21. [CrossRef]
21. Azzaz, A.A.; Khiari, B.; Jellali, S.; Ghimbeu, C.M.; Jeguirim, M. Hydrochars production, characterization and application for wastewater treatment: A review. *Renew. Sustain. Energy Rev.* **2020**, *127*, 109882. [CrossRef]
22. Wang, T.; Zhai, Y.; Zhu, Y.; Li, C.; Zeng, G. A review of the hydrothermal carbonization of biomass waste for hydrochar formation: Process conditions, fundamentals, and physicochemical properties. *Renew. Sustain. Energy Rev.* **2018**, *90*, 223–247. [CrossRef]
23. Tasca, A.L.; Puccini, M.; Gori, R.; Corsi, I.; Galletti, A.M.R.; Vitolo, S. Hydrothermal carbonization of sewage sludge: A critical analysis of process severity, hydrochar properties and environmental implications. *Waste Manag.* **2019**, *93*, 1–13. [CrossRef] [PubMed]

24. Zhai, Y.; Liu, X.; Zhu, Y.; Peng, C.; Wang, T.; Zhu, L.; Li, C.; Zeng, G. Hydrothermal carbonization of sewage sludge: The effect of feed-water pH on fate and risk of heavy metals in hydrochars. *Bioresour. Technol.* **2016**, *218*, 183–188. [CrossRef]
25. Werle, S.; Dudziak, M. Evaluation of toxicity of sewage sludge and gasification waste-products. *Przem. Chem.* **2013**, *92*, 1350–1353.
26. Shi, W.; Liu, C.; Ding, D.; Lei, Z.; Yang, Y.; Feng, C.; Zhang, Z. Immobilization of heavy metals in sewage sludge by using subcritical water technology. *Bioresour. Technol.* **2013**, *137*, 18–24. [CrossRef]
27. Berslin, D.; Reshmi, A.; Sivaprakash, B.; Rajamohan, N.; Kumar, P.S. Remediation of emerging metal pollutants using environment friendly biochar- Review on applications and mechanism. *Chemosphere* **2022**, *290*, 133384. [CrossRef]
28. Masoumi, S.; Borugadda, V.B.; Nanda, S.; Dalai, A.K. Hydrochar: A Review on Its Production Technologies and Applications. *Catalysts* **2021**, *11*, 939. [CrossRef]
29. Nandhini, R.; Berslin, D.; Sivaprakash, B.; Rajamohan, N.; Vo, D.V.N. Thermochemical conversion of municipal solid waste into energy and hydrogen: A review. *Environ. Chem. Lett.* **2022**, *20*, 1645–1669. [CrossRef]
30. Monisha, R.S.; Mani, R.L.; Sivaprakash, B.; Rajamohan, N.; Vo, D.V.N. Green remediation of pharmaceutical wastes using biochar: A review. *Environ. Chem. Lett.* **2022**, *20*, 681–704. [CrossRef]
31. Shi, W.; Liu, C.; Shu, Y.; Feng, C.; Lei, Z.; Zhang, Z. Synergistic Effect of rice husk addition on hydrothermal treatment of sewage sludge: Fate and environmental risk of heavy metals. *Bioresour. Technol.* **2013**, *149*, 496–502. [CrossRef] [PubMed]
32. Sharma, H.B.; Sarmah, A.K.; Dubey, B. Hydrothermal carbonization of renewable waste biomass for solid biofuel production: A discussion on process mechanism, the influence of process parameters, environmental performance and fuel properties of hydrochar. *Renew. Sustain. Energy Rev.* **2020**, *123*, 109761. [CrossRef]
33. Titirici, M.M.; Funke, A.; Kruse, A. Hydrothermal Carbonization of Biomass. In *Recent Advances in Thermo-Chemical Conversion of Biomass*; Elsevier: Amsterdam, The Netherlands, 2015; pp. 325–352.
34. Wilk, M.; Śliz, M.; Lubieniecki, B. Hydrothermal co-carbonization of sewage sludge and fuel additives: Combustion performance of hydrochar. *Renew. Energy* **2021**, *178*, 1046–1056. [CrossRef]
35. Gao, N.; Li, Z.; Quan, C.; Miskolczi, N.; Egedy, A. A new method combining hydrothermal carbonization and mechanical compression in-situ for sewage sludge dewatering: Bench-scale verification. *J. Anal. Appl. Pyrolysis* **2019**, *139*, 187–195. [CrossRef]
36. Breulmann, M.; van Afferden, M.; Müller, R.A.; Schulz, E.; Fühner, C. Process conditions of pyrolysis and hydrothermal carbonization affect the potential of sewage sludge for soil carbon sequestration and amelioration. *J. Anal. Appl. Pyrolysis* **2017**, *124*, 256–265. [CrossRef]
37. De la Rubia, M.A.; Villamil, J.A.; Rodriguez, J.J.; Mohedano, A.F. Effect of inoculum source and initial concentration on the anaerobic digestion of the liquid fraction from hydrothermal carbonisation of sewage sludge. *Renew. Energy* **2018**, *127*, 697–704. [CrossRef]
38. Wang, S.; Persson, H.; Yang, W.; Jönsson, P.G. Pyrolysis Study of hydrothermal carbonization-treated digested sewage sludge using a py-gc/ms and a bench-scale pyrolyzer. *Fuel* **2020**, *262*, 116335. [CrossRef]
39. Aragón-Briceño, C.; Ross, A.B.; Camargo-Valero, M.A. Evaluation and comparison of product yields and bio-methane potential in sewage digestate following hydrothermal treatment. *Appl. Energy* **2017**, *208*, 1357–1369. [CrossRef]
40. Parmar, K.R.; Ross, A.B. Integration of hydrothermal carbonisation with anaerobic digestion; opportunities for valorisation of digestate. *Energies* **2019**, *12*, 1586. [CrossRef]
41. Cao, Z.; Jung, D.; Olszewski, M.P.; Arauzo, P.J.; Kruse, A. Hydrothermal carbonization of biogas digestate: Effect of digestate origi5 and process conditions. *Waste Manag.* **2019**, *100*, 138–150. [CrossRef]
42. Werle, S. A reburning process using sewage sludge-derived syngas. *Chem. Pap.* **2012**, *66*, 99–107. [CrossRef]
43. Cao, Z.; Hülsemann, B.; Wüst, D.; Illi, L.; Oechsner, H.; Kruse, A. Valorization of maize silage digestate from two-stage anaerobic digestion by hydrothermal carbonization. *Energy Convers. Manag.* **2020**, *222*, 113218. [CrossRef]
44. Wilk, M. A Novel Method of Sewage Sludge Pre-Treatment-HTC. In Proceedings of the 1st International Conference on the Sustainable Energy and Environment Development (SEED 2016), Kraków, Poland, 17–19 May 2016; Volume 10.
45. Śliz, M.; Tuci, F.; Czerwińska, K.; Fabrizi, S.; Lombardi, L.; Wilk, M. Hydrothermal carbonization of the wet fraction from mixed municipal solid waste: Hydrochar characteristics and energy balance. *Waste Manag.* **2022**, *151*, 39–48. [CrossRef] [PubMed]
46. Ozawa, T. Estimation of activation energy by isoconversion methods. *Thermochim. Acta* **1992**, *203*, 159–165. [CrossRef]
47. He, C.; Giannis, A.; Wang, J.Y. Conversion of sewage sludge to clean solid fuel using hydrothermal carbonization: Hydrochar fuel characteristics and combustion behavior. *Appl. Energy* **2013**, *111*, 257–266. [CrossRef]
48. Song, C.-Z.; Wen, J.-H.; Li, Y.-Y.; Dan, H.; Shi, X.-Y.; Xin, S. Thermogravimetric assessment of combustion characteristics of blends of lignite coals with coal gangue. In Proceedings of the 3rd Annual International Conference on Mechanics and Mechanical Engineering (MME 2016), Chengdu, China, 16–18 December 2016; Volume 105, pp. 490–495.
49. Magdziarz, A.; Wilk, M.; Straka, R. Combustion Process of torrefied wood biomass: A kinetic study. *J. Therm. Anal. Calorim.* **2017**, *127*, 1339–1349. [CrossRef]
50. Barzegar, R.; Yozgatligil, A.; Olgun, H.; Atimtay, A.T. TGA and kinetic study of different torrefaction conditions of wood biomass under air and oxy-fuel combustion atmospheres. *J. Energy Inst.* **2020**, *93*, 889–898. [CrossRef]
51. Ren, S.; Lei, H.; Wang, L.; Bu, Q.; Chen, S.; Wu, J. Thermal Behaviour and kinetic study for woody biomass torrefaction and torrefied biomass pyrolysis by TGA. *Biosyst. Eng.* **2013**, *116*, 420–426. [CrossRef]

52. Peng, C.; Zhai, Y.; Zhu, Y.; Xu, B.; Wang, T.; Li, C.; Zeng, G. Production of char from sewage sludge employing hydrothermal carbonization: Char properties, combustion behavior and thermal characteristics. *Fuel* **2016**, *176*, 110–118. [CrossRef]
53. Zheng, X.; Jiang, Z.; Ying, Z.; Song, J.; Chen, W.; Wang, B. Role of feedstock properties and hydrothermal carbonization conditions on fuel properties of sewage sludge-derived hydrochar using multiple linear regression technique. *Fuel* **2020**, *271*, 117609.
54. Lin, Y.; Wang, D.; Wang, T. Ethanol Production from pulp & paper sludge and monosodium glutamate waste liquor by simultaneous saccharification and fermentation in batch condition. *Chem. Eng. J.* **2012**, *191*, 31–37.
55. Zhuang, X.; Zhan, H.; Song, Y.; He, C.; Huang, Y.; Yin, X.; Wu, C. Insights into the evolution of chemical structures in lignocellulose and non-lignocellulose biowastes during hydrothermal carbonization (HTC). *Fuel* **2019**, *236*, 960–974. [CrossRef]
56. Wang, L.; Chang, Y.; Li, A. Hydrothermal carbonization for energy-efficient processing of sewage sludge: A review. *Renew. Sustain. Energy Rev.* **2019**, *108*, 423–440. [CrossRef]

energies

Article

Hydrothermal Carbonization of the Wet Fraction from Mixed Municipal Solid Waste: A Fuel and Structural Analysis of Hydrochars

Maciej Śliz [1,*], Klaudia Czerwińska [1], Aneta Magdziarz [1], Lidia Lombardi [2] and Małgorzata Wilk [1]

[1] Department of Heat Engineering and Environment Protection, AGH University of Science and Technology, Mickiewicza 30 Av., 30-059 Krakow, Poland
[2] Niccolò Cusano University, Via Don Carlo Gnocchi 3, 00166 Rome, Italy
* Correspondence: msliz@agh.edu.pl

Abstract: One of the by-products of a mechanical-biological waste treatment plant is the under-sieve fraction, which requires separation prior to further processing of municipal mixed waste. This stream usually follows the fate of landfilling. Instead, it could be introduced to the hydrothermal carbonization (HTC) process to improve its fuel properties and become solid recovered fuel. The organic fraction and high moisture content (approximately 26%) of under-sieve fraction are favorable properties for the HTC process. In this study, hydrochars, the solid product of HTC, were produced at 200 and 220 °C with residence times of 1, 4, and 8 h. The main aim of this investigation was to establish the influence of different process parameters on hydrochars' fuel properties. Thermogravimetric analysis (TGA), Fourier-transform infrared spectroscopy (FTIR), and scanning electron microscopy (SEM) were employed in the analyses. The results confirmed the positive effects of hydrothermal carbonization on the under-sieve fraction of municipal mixed waste properties. The ignition temperature increased from 247 °C to 288 °C and burnout temperature decreased to 443 °C from 489 °C after hydrothermal carbonization, causing a shorter combustion process. The determined key combustion parameters were: $S = 12.4 \times 10^{-8}\%{\cdot}\mathrm{min}^{-2}{\cdot}{}^{\circ}\mathrm{C}^{-3}$, $H_f = 1174.9$ °C and $Di = 0.0075\%{\cdot}\mathrm{min}^{-3}$, which in comparison to USF decreased by 44%, increased by 33%, and decreased by 29%, respectively, and became closer to those of coal. Furthermore, the identified structural changes indicate that hydrochars could be successfully used in energy production. The most promising results were found for hydrochar produced at 220 °C for 1 h, leading to a better combustion performance and providing a more stable and a less violent flame.

Keywords: hydrothermal carbonization; municipal solid waste; TGA; FTIR; SEM-EDS

Citation: Śliz, M.; Czerwińska, K.; Magdziarz, A.; Lombardi, L.; Wilk, M. Hydrothermal Carbonization of the Wet Fraction from Mixed Municipal Solid Waste: A Fuel and Structural Analysis of Hydrochars. *Energies* **2022**, *15*, 6708. https://doi.org/10.3390/en15186708

Academic Editor: Attilio Converti

Received: 17 August 2022
Accepted: 8 September 2022
Published: 14 September 2022

1. Introduction

Recently, hydrothermal carbonization has more often been brought to attention as a process that can change and improve the properties of various feedstock. Due to the reactions achieved in an aqueous environment, it is a suitable treatment for wet materials with unfavorable, high organic content, e.g., vegetation, food and paper waste, sewage sludge, or mixed solid waste [1–9]. Generally, the latter, municipal mixed waste, is treated in a mechanical-biological treatment plant where the recycling of metal from waste is processed and then the remaining residue is aerobically stabilized and landfilled. However, according to European legislation, the quantity of waste that is landfilled must decrease to a limit of 10% by 2035, hence, other solutions are being sought [10]. Accordingly, hydrothermal carbonization is suggested as a suitable method to improve dewatering, grinding, handling, transportation, and storage of the wet feedstock. Moreover, the elevated temperature and pressure guarantee an altered structure of the material as well as limiting the biodegradation process. Furthermore, the reactions that occur during hydrothermal treatment, e.g., hydrolysis, dehydration, decarboxylation, polymerization and aromatization, provide lower oxygen and higher carbon content in the final product [11]. Those properties, supported by an increased calorific value, create an environmentally-friendly solid recovered fuel, hydrochar, which could be successfully applied to energy systems, e.g., in an incineration plant. Additionally, the hydrothermal treatment of post-processing organic waste is in concordance with the waste-to-energy and minimizing of waste at source European environmental policy [12].

Consequently, the hydrothermal carbonization process is often studied to establish the possible application of hydrochars and to find the best process parameters for the process, which could minimize costs and maximize possible benefits [13,14]. For instance, Santos et al. [15] studied hydrochars derived from the hydrothermal carbonization process of low-grade organic fraction of municipal solid waste (OFMSW) at 180 °C for 2 h as a novel, sustainable, low-CO_2 building material for thermal insulation panels. In addition, Ischia [16] studied this material under similarly mild conditions (180 °C) but applied thermal and chemical extraction (1:4 methanol: dichloromethane) on hydrochars. The removal of reactive volatile matter in hydrochars was investigated during thermal treatment and chemical extraction, suggesting a potential pathway for the recovery of fatty acids and condensation of fuel molecules in the solid hydrochar. Furthermore, the relevant properties of hydrochar could be used as a soil amendment or an environmental adsorbent as well as simultaneously extracting valuable biodiesel and biofuel precursors [16]. Aragon-Briceno et al. [17] studied the optimum hydrothermal carbonization process conditions for recovering water from the MSW digestate at three temperatures 180, 200, and 230 °C and three residence times 30, 60, and 120 min and proved that the most suitable conditions were 200 °C with a residence time of 120 min, resulting in 47.2% of water recovery. Based on the net energy balance (5.6 MJ) and the applied Aspen model, the integration of HTC as part of the dewatering system of an MSW treatment plant was proposed and provided very promising results. Espro et al. [18] investigated hydrochars derived from orange peels in sensing applications. Hydrochars produced at 300 °C were used to fabricate high performance conductometric sensors for the monitoring of NO_2.

Furthermore, Roman et al. [19] studied different residence times (1–5 h), on a quantity of compost derived from MSW (1–4 g in 30 mL) and at different temperatures (150–230 °C) regarding the mass and energy yield of hydrochars and, most importantly, on postprocessing water disposal, while searching for the catalytic behavior of both products. On the contrary, Lin et al. [20] studied the behavior of four typical components of municipal solid waste (MSW), such as waste textile, wood, paper, and food under hydrothermal and co-hydrothermal carbonization. The blends of MSW and waste textile derived at 240 °C and 90 min with a 1:12 solid/liquid ratio exhibited synergistic effects during the co-HTC of typical MSW components. The co- and hydrochars behavior and characteristics' parameters of combustion were studied by profound analysis of TG and DTG data. Moreover, it was found that hydrochars more stably combusted and effectively fitted the nth-order kinetic model of the combustion reaction process. Venna et al. [21] also confirmed the potential of hydrochar derived from municipal solid waste, namely yard and food waste, as a solid biofuel. In comparison to distilled water, a landfill leachate was applied as a water carrier in the HTC process, but fuel characteristics of the hydrochars did not significantly change. TGA was implemented to describe the combustion behavior and parameters of the derived hydrochars. Moreover, the morphology and surface chemistry of the hydrochars was applied to detect the morphological changes caused by the HTC process. SEM images depicted microspheres, higher porosity, and surface roughness after hydrothermal degradation resulted in better fuel characteristics. The changes in functional groups in hydrochar in comparison to raw samples were determined by FTIR examination. These techniques were also used by Wang et al. [22] to study food waste and their hydrochars produced within a temperature range of 180–260 and 60 min of residence time in terms of solid biofuels characteristics. Akarsu et al. [23], Saqib et al. [24], and Nguyen et al. [25] also studied food waste and their hydrochars to confirm their ability in energy production by emphasizing their physicochemical, structural, and combustion performance. Sharma et al. [26] studied in detail the mechanism and process parameters of the hydrothermal carbonization of municipal solid waste as a potential biofuel taking into account the environmental impact. It was suggested that the optimized parameters of the process can ensure the improved properties of hydrochars identified by advanced instrumental methods for a net renewable energy production potential value.

The need for a profound discussion concerning the hydrothermal carbonization of municipal solid waste in order to fully understand its behavior was underlined by many researchers [5,27–31]. Accordingly, in this study, the hydrothermal carbonization of municipal solid waste is investigated in detail, in particular under-sieve mixed waste, mainly focused on the fuel properties of hydrochar derived under various conditions of the process.

2. Materials and Methods

The stream of municipal mixed waste, investigated in this research, was collected from a rotating sieve drum at an Italian mechanical-biological treatment plant. The particles of material were smaller than 60 mm and referred to as under-sieve fraction (USF)—Figure 1. The USF was grey, fluffy, and resembled dust. A composition analysis of USF identified some paper, plastics, glass, wood, textiles, food waste, coffee pods, inert materials, and metals [32]. However, approximately 75% of the USF proved to be indistinguishable due to its small size—Figure 2. To simplify handling of the samples the material was dried and milled.

Figure 1. Under-sieve fraction of municipal mixed waste prior to and after hydrothermal carbonization.

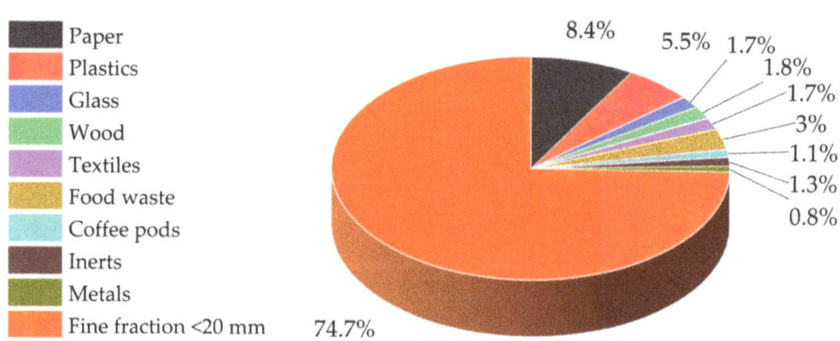

Figure 2. Composition of under-sieve fraction of municipal mixed waste [32].

For hydrothermal carbonization, a 1000 mL, stainless steel, Zipperclave® Stirred Reactor (Parker Autoclave Engineers, Erie, PA, USA) was used. The schematic of the experimental set-up is presented in Figure 3. The USF was mixed with deionized water prior to the experiment, achieving 0.15 dry solids to water ratio. The experiments were conducted at 200 and 220 °C with reaction times of 1, 4, and 8 h measured from the moment the reactor was heated to the desired temperature. During the process, the mixture was constantly stirred and afterwards immediately cooled down by feeding cold water through the coil inside the reactor. The separation of solid and liquid products was conducted via vacuum filtration. The produced hydrochars were dried at 105 °C in an oven overnight and stored in sealed containers prior to further analyses. The hydrochars were brown in color and with an increase in the process temperature or residence time the color became darker. After hydrothermal carbonization the material can be easily ground after drying. Additional information about the USF and hydrochars can be found in a previous publication by Śliz et al. [32].

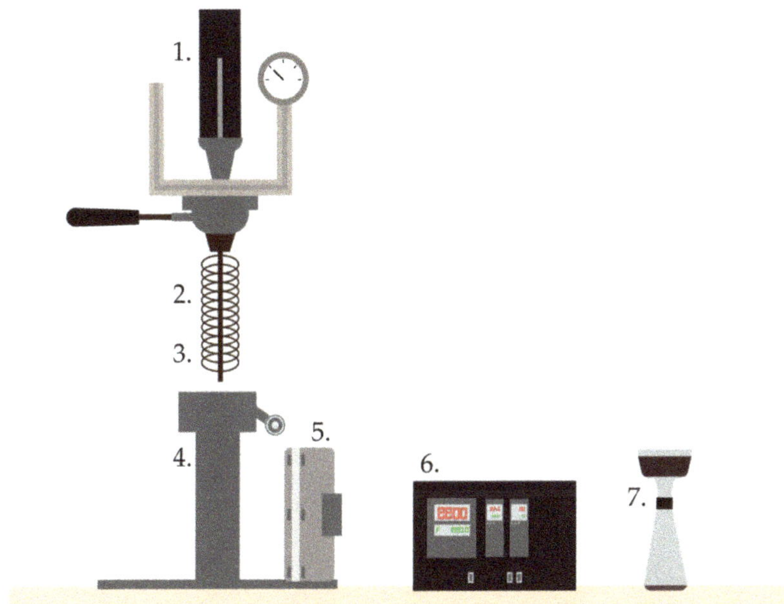

1—MagneDrive, 2—cooling coil, 3—stirrer, 4—reactor, 5—heating mantle, 6—control panel, 7—filtration apparatus: Büchner funnel and flask.

Figure 3. Experimental setup used for hydrothermal carbonization and filtration [32].

The proximate and ultimate analyses were performed by a muffle furnace according to European Standards (EN 15414–3:2011, EN 15403:2011, and EN 15402:2011) and a Truspec LECO 628 CHNS Analyzer (PKN-ISO/TS 12902:2007), respectively. Additionally, the high heating value determination was carried out by means of a LECO AC500 Isoperibol Calorimeter (DIN 51,900 and ISO 1928).

The combustion process was investigated by means of thermal analysis in an air atmosphere. Samples deposited in the crucible were heated from room temperature up to 700 °C at a heating rate of 10 °C/min. This method enables TG, DTG, and DSC curves for USF and hydrochars to be presented.

Moreover, based on those results, the characteristic parameters of the combustion process were determined. Among them an ignition temperature, T_i, and a burnout temperature, T_b, should be distinguished. According to the literature, the ignition temperature (point C) can be found at the intersection of the horizontal line drawn from point B, when devolatilization started, and a tangent line drawn from point A, where the vertical line from the first DTG peak crosses the TG line. The burnout temperature was found at the mass stabilization point where the mass loss rate decreased to below 1%/min [33,34]. Combustibility indices, such as the ignition index (D_i), were also included:

$$D_i = \frac{DTG_1}{t_1 \cdot t_i}$$

where DTG_1 is the value of the highest mass loss rate (first peak in the DTG curve), t_1 and t_i are the times of occurrence of this first peak and when the ignition temperature was reached, respectively. The sooner the ignition started and the highest mass loss rate was achieved, the higher the value of the D_i index reached. This corresponds with an easy release of volatile matter and simple ignition. Next, is the burnout index (D_b):

$$D_b = \frac{DTG_1}{\Delta t_{0.5} \cdot t_1 \cdot t_b}$$

where $\Delta t_{0.5}$ is the time at which the mass loss rate reached half of its maximum value and t_b is the time the combustion process takes to reach burnout temperature. The lower value of D_b required less time to complete the combustion process. The comprehensive combustion index (*CCI* or *S*) follows:

$$S = \frac{DTG_1 \cdot DTG_{mean}}{T_i^2 \cdot T_b}$$

where DTG_{mean} is the mean value of the mass loss rate. It reflects the ignition, combustion, and burnout properties of the fuel. Higher values of the *S* index indicate easy combustion at an early stage [35]. The combustion stability index (*CSI* or H_f) is calculated according to the following equation:

$$H_f = T_1 \cdot \ln\left(\frac{\Delta t_{0.5}}{DTG_{mean}}\right)$$

where T_1 is the temperature corresponding to DTG_1. Index H_f expresses the rate and intensity of the combustion process in which lower values are preferred. Combustibility indices were calculated based on equations presented by Mureddu et al. [36,37].

To investigate the morphology of the samples, a scanning electron microscope, (SEM) FEI Inspect S50, was employed and supported by the energy dispersive X-ray spectroscopy (EDS). Fourier-transform infrared spectroscopy was conducted with the Nicolet 6700 FTIR (Thermo Scientific, Waltham, MA, USA), including the attenuated total reflectance module to assess changes in bonds of municipal mixed solid waste.

3. Results and Discussions

To summarize the results of proximate analysis, a Tanner diagram is presented in Figure 4. It has been established that for the combustion of municipal solid waste certain requirements have to be met. Namely, the moisture content cannot exceed 50%, ash content cannot be higher than 60%, and combustible fraction has to be at least 25% [38]. The materials which meet those restrictions are depicted in the bottom right corner of the triangle diagram (grey area). Herein, the hydrochars were dried after the HTC process and easily fulfilled those requirements. However, an increase in the moisture content would cause them to move up and slightly to the left side of the diagram (simultaneously, the ash and combustible fraction content would decrease), which should retain them in the highlighted area of the autothermic combustion. Similar conclusions were presented by Komilis et al. for municipal solid waste [39]. The results of the proximate and ultimate analyses can be found in Table 1. The detailed description, presented in previous publication concerns the under-sieve fraction of municipal mixed solid waste [32]. Nevertheless, to summarize the key changes in the properties of hydrochars, it could be concluded that with a longer residence time and a higher reaction temperature, it would cause a higher value of high heating value and a lower mass yield (ratio of hydrochar mass to feedstock mass). An increase in the ash content with a longer reaction time was also a noticeable trend. The correlation between the process parameters and the analysis results is also in accordance with general trends found by various researchers investigating the hydrothermal carbonization process [7,40].

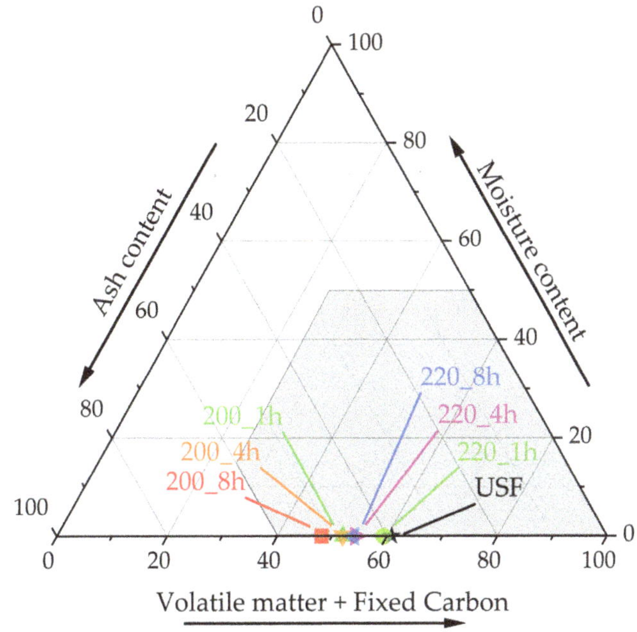

Figure 4. Composition of the under-sieve fraction of municipal mixed waste.

Table 1. Results of ultimate and proximate analysis [32].

	C,%	H,%	N,%	S,%	O,%	A,%	VM,%	FC,%	HHV, MJ/kg
USF	36.0	5.2	1.3	0.2	18.1	39.2	51.7	9.2	14.6
200_1 h	32.6	4.2	0.8	0.5	13.9	48.0	45.9	6.1	13.5
200_4 h	33.7	4.2	0.9	0.5	12.6	48.1	45.0	6.9	14.2
200_8 h	32.3	4.0	1.0	0.5	10.2	52.0	42.7	5.3	14.4
220_1 h	38.3	4.7	0.9	0.1	15.3	40.7	48.4	11.0	16.4
220_4 h	38.2	4.5	1.1	0.2	10.3	45.7	42.0	12.3	16.7
220_8 h	40.5	4.8	1.3	0.3	7.0	46.1	42.9	11.0	18.5

3.1. *Thermogravimetric Analysis*

The results of the thermogravimetric analysis are presented in Figures 5 and 6. The combustion profiles are depicted in the form of TG and DTG curves. The shape of those curves clearly represents the influence of the reaction time on the combustion performance of hydrochars. A typical combustion process can be divided into three stages [12,41]. The first stage, which starts almost immediately after increasing the temperature to above room temperature and lasts to circa 150 °C, is connected with moisture release. Herein, this stage was omitted as the samples were dried after the separation of the phases following the experiments. The beginning of the next stage is marked by the ignition temperature. In Figure 5, depicted in green, the volatile matter is released and combusted. For USF, the T_i equals 247 °C and at this stage the highest mass drop occurs. The first DTG peaks have a value of 5.973%/min and were found at 289 °C. In the next step, the mass loss rate decreases, and at a higher temperature the last stage concerning the combustion of char is distinguished—marked as stage II and depicted in blue. The second DTG peak is much lower (2.332%/min at 463 °C for USF). The burnout temperature was 489 °C and subsequently the mass loss stabilized. Hydrochars in general, have a higher mass of residue after the combustion process and a lower volatile matter content in comparison to an untreated sample, as a mass loss begins at higher temperatures. After hydrothermal carbonization the first DTG peak becomes lower and switches to higher temperatures in the range of 325 to 332 °C for all of the hydrochars [42,43]. Based on these results, it could be stated that the HTC process evidently causes the decrease in the combustion time. The ignition temperature rises due to the decrease in the volatile matter content. The samples 200_1 h and 220_1 h presented the highest ignition temperatures, 288 °C and 282 °C, respectively. However, after a longer reaction time of 8 h, the ignition temperatures decreased reaching 273 °C for hydrochar derived at 200 °C and 256 °C for hydrochar treated at 220 °C. Simultaneously, burnout occurs at lower temperatures, 443 °C for samples carbonized at 200 °C and c.a. 463–477 °C for those pretreated at 220 °C. Moreover, the mass rate loss in the first stage of combustion was maintained only for samples treated for 1 h and is close to 6%/min. Samples treated for 4 and 8 h of residence time dipped to 3.8 and 2.7%/min at 200 °C, and to 3.6 and 3.2 at 220 °C.

Furthermore, the temperatures in the second stage of combustion decreased. For USF it was 409 °C, while for samples retained at 200 °C it was closer to 375 °C, and for those treated at 220 °C it was 10 °C higher at 385 °C.

The analysis of TG and DTG profiles not only presents the combustion performance but are also the basis for assessing the key combustion factors summarized in Table 2. Among them combustion indices should be distinguished. A lower ignition index (D_i), which dropped from 0.0106%·min^{-3} for USF to the lowest value of 0.0036%·min^{-3} for sample 220_8 h, corresponds to a lower volatile matter content and a higher ignition temperature. Index S connects the ease of ignition, burning velocity and burnout temperature. The highest value for USF, 22.0 × 10^{-8}%·min^{-2}·°C^{-3}, indicates that it ignites faster and more intensively than in the case of hydrochars—a lower ignition temperature and a high mass loss rate [44]. Conversely, the lowest value, 4.6 × 10^{-8}%·min^{-2}·°C^{-3}, was reached for hydrochar derived at 220_8 h. Similarly, the lowest value of the H_f index for USF suggests better combustion properties [35,36]. A few trends can be noticed when comparing results calculated for hydrochars obtained at different parameters. Prolonging the reaction time from 1 to 8 h causes a decrease in fuel properties estimated by combustibility indices. By comparing the processing temperature, the following observation could be made: when a higher temperature was applied, stronger effects were observed. In addition, it should be noted that the HTC process increased the ignition temperature but a longer residence time caused a decrease in its value, whereas the burnout temperatures were lower for hydrochars than for feedstock. At the same time, the first DTG peak occurred at a higher temperature for hydrochars, approximately 330 °C, than for feedstock, 289 °C. A similar observation on the influence of HTC on the combustion process was presented by Arellano et al. [45]. In summary, hydrothermal carbonization does not significantly improve the fuel properties of the under-sieve fraction of municipal solid waste, but it probably slightly improves the stability of the combustion process due to the release of a lower content of volatile matter.

Table 2. Combustion parameters estimated by means of thermogravimetric analysis.

	T_i °C	t_i min	T_b °C	t_b min	T_1 °C	t_1 min	DTG_1 %/min	$t_{0.5}$ min	DTG_{mean} %/min	D_i %·min^{-3}	D_b %·min^{-4}	S %·min^{-2}·°C^{-3}	H_f °C
USF	247	21.77	489	45.47	289	25.85	5.973	23.08	1.097	0.0106	22.0 × 10^{-5}	22.0 × 10^{-8}	880.3
200_1 h	288	26.32	443	41.80	327	30.10	5.934	27.80	1.007	0.0083	18.0 × 10^{-5}	16.7 × 10^{-8}	1069.4
200_4 h	276	25.17	443	41.85	329	30.33	3.822	26.85	0.782	0.0053	10.9 × 10^{-5}	8.7 × 10^{-8}	1146.2
200_8 h	273	24.85	442	41.72	330	30.45	2.697	26.32	0.945	0.0048	10.6 × 10^{-5}	9.7 × 10^{-8}	1057.0
220_1 h	282	25.12	464	42.98	326	29.38	6.099	26.78	0.765	0.0075	17.0 × 10^{-5}	12.4 × 10^{-8}	1174.9
220_4 h	265	23.45	463	42.87	325	29.28	3.624	26.58	0.684	0.0050	11.2 × 10^{-5}	7.8 × 10^{-8}	1207.3
220_8 h	256	22.60	477	44.32	332	29.87	3.207	22.80	0.565	0.0036	8.1 × 10^{-5}	4.6 × 10^{-8}	1267.5

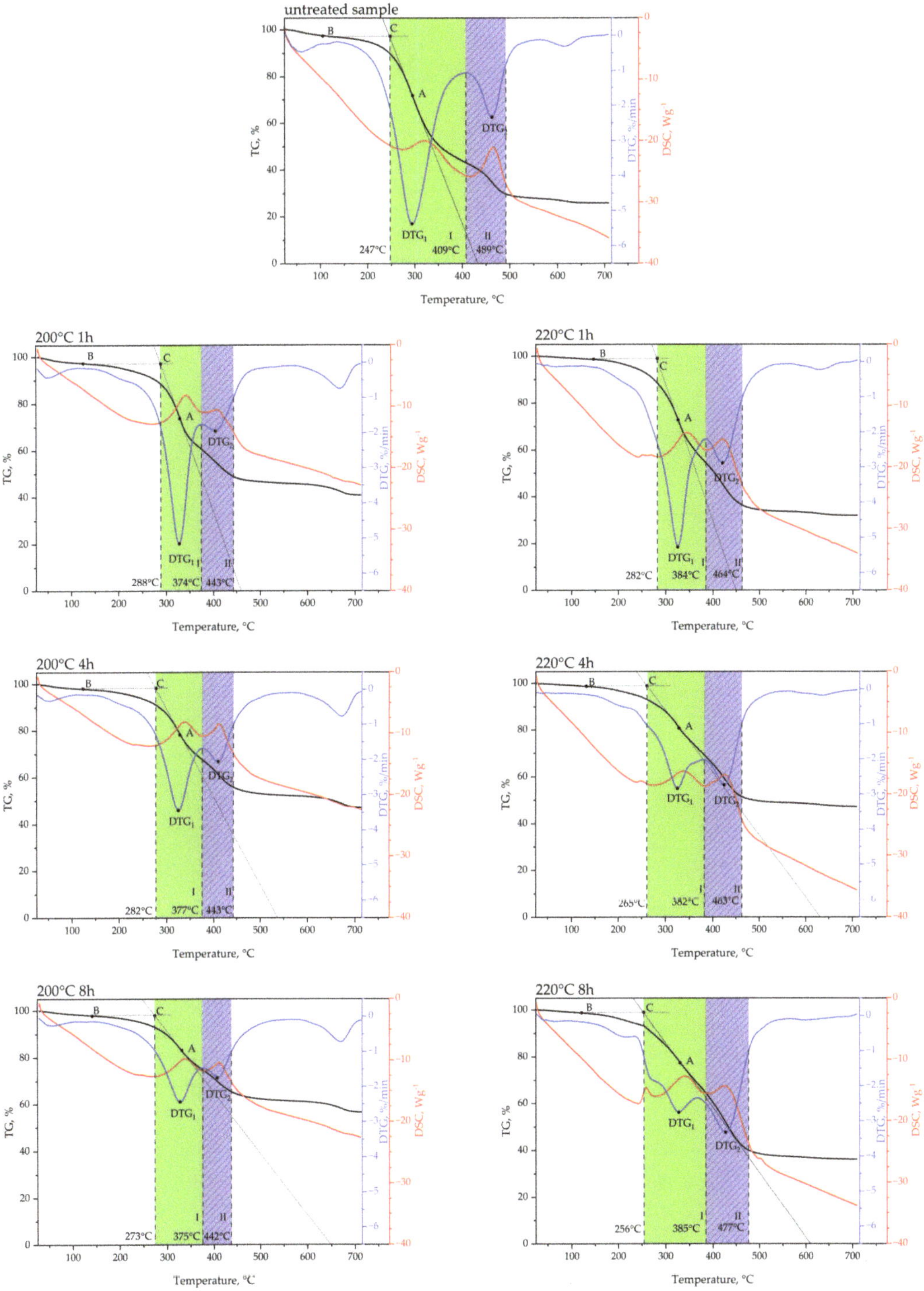

Figure 5. Summary of TG and DTG curves for each experiment conducted at 200 °C and 220 °C for 1, 4, and 8 h.

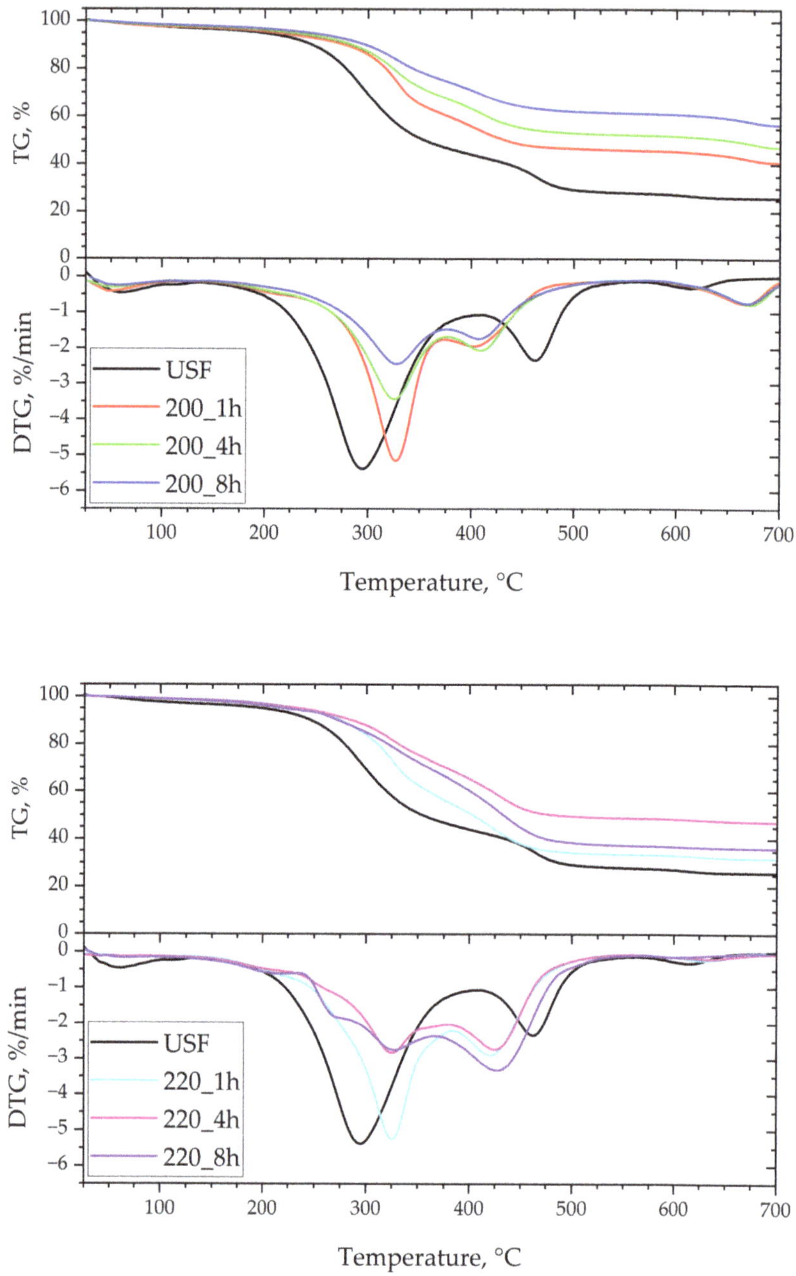

Figure 6. Comparison of TG and DTG curves for samples hydrothermally treated at 200 and 220 °C.

3.2. Fourier-Transform Infrared Spectroscopy

Hydrochars present a similarly shaped curve on the FTIR diagram (Figure 7) as untreated USF. The most important peaks with corresponding bonds are illustrated in Table 3. The increase of the reaction time caused a decrease in the peaks, which became less distinguishable. Nevertheless, a wide mound at approximately 3100–3600 cm^{-1} that spiked at approximately 3330–3340 cm^{-1} is identified as a characteristic stretching O–H vibration in hydroxyl or carboxyl groups. Its decrease with a longer reaction time and higher process temperature indicates that those factors facilitate dehydration [46,47]. Double peaks, in the range of 2860–2930 cm^{-1} represent the stretching vibration of the C–H bond in aromatic and aliphatic compounds. Their presence indicates unsaturation [21,48]. The disappearing peak at 1740 cm^{-1} is associated with the stretching of C=O in alkali esters in hemicellulose. The decreasing signal at approximately 874 cm^{-1} supports degradation of hemicellulose in HTC conditions. Moreover, the peaks at 1516 cm^{-1} and 1454 cm^{-1} are clearly visible for the sample after hydrothermal carbonization and prove the presence of lignin, which has a higher decomposition temperature than hemicellulose. The degrading peak at 1102 cm^{-1} corresponds to C-O vibrations in cellulose. Additionally, the change in the peaks at approximately 1155 cm^{-1} suggests its mild degradation. The highest peak, which for all samples has a maximum of approximately 1030 cm^{-1} and merges with the signal at 1051 cm^{-1}, is from the stretching vibration of the C–O bond in aliphatic ether and alcohol, respectively [49]. The results confirm that USF has some organic origin. The decrease in intensity of the signal for the hydrochars can be associated with the fact that some of the bonds were broken during the hydrothermal process.

3.3. *Scanning Electron Microscopy with Energy Dispersive X-ray Spectroscopy*

In the images received via scanning electron microscopy (Figure 8) a fibrous structure of untreated under-sieve fraction of municipal mixed waste can be seen. The hydrothermal carbonization process significantly alters the structure and morphology of the material. In Figure 8b there are still many elongated particles visible, however they are much shorter than those presented for untreated USF in Figure 8a. In the following Figure 8c,d, samples after a longer residence time are depicted. The open, circular pores in Figure 8c prove that some of the volatile matter is released during the HTC process, which may cause higher surface degradation but also a higher specific surface area. It can be seen that with an increase in the residence time of the process, the quantity of fine particles also increases. This can influence not only the combustion behavior of hydrochar but also its dewaterability. EDS analysis (Table 4) distinguished mainly carbon and oxygen, but calcium, silicon, and sulphur can also be observed, among other elements. This provides valuable information concerning potential challenges, such as slagging and fouling or corrosion, which may occur in a scaled-up process [50,51].

Table 3. Indicator bands, their vibrations, and corresponding functional groups.

Wavenumber (cm^{-1})	Vibration	Functional Group or Component
3600–3100	O-H stretch	hydroxyl or carboxyl group
2930–2860	C-H stretch	aliphatic methylene group
1740	C=O	alkali esters in hemicellulose
~1515	C=O	lignin
~1455	C-H	lignin
1102	C-O	cellulose
1050	C-O stretch	alcohol
1030	C-O stretch	aliphatic ether

Table 4. Results of EDS analysis (ZAF method), wt%.

	C K	N K	O K	F K	Cu L	Na K	Mg K	Al K	Si K	P K	S K	Cl K	K K	Ca K	Ti K	Fe K
USF	66.31	1.24	28.87	0.15	0.19	0.33	0.17	0.33	0.74	0.00	0.17	0.20	0.09	0.86	0.08	0.27
220_1 h	66.95	1.45	28.29	0.08	0.08	0.28	0.21	0.38	1.21	0.06	0.16	0.05	0.05	0.52	0.14	0.10
220_4 h	75.21	0.99	20.94	0.08	0.13	0.19	0.16	0.32	1.09	0.02	0.17	0.03	0.00	0.39	0.08	0.20
220_8 h	75.37	1.20	19.83	0.08	0.20	0.16	0.16	0.40	1.21	0.06	0.27	0.08	0.10	0.47	0.16	0.25

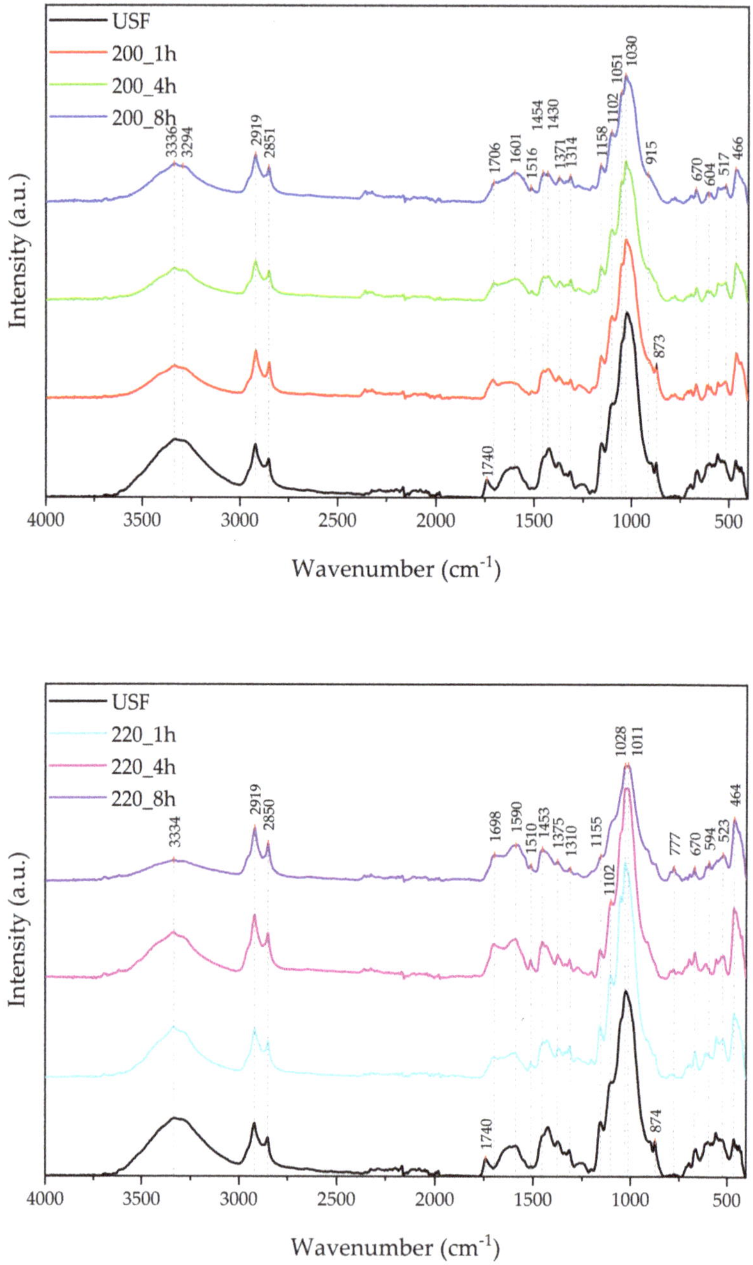

Figure 7. FTIR spectra of the USF and hydrochars.

Figure 8. SEM images of (**a**) USF and hydrochars produced at 220 °C for (**b**) 1 h; (**c**) 4 h; (**d**) 8 h.

4. Conclusions

Hydrothermal carbonization of the under-sieve fraction of municipal mixed solid waste was performed. Conducted analyses proved that the hydrothermal process altered the structure and other physical and chemical properties of USF. The most promising results were found for hydrochar produced at 220 °C for 1 h leading to a better combustion performance and providing a more stable and a less violent flame. This was confirmed by combustibility indices $S = 12.4 \times 10^{-8}\%\cdot\text{min}^{-2}\cdot{}^{\circ}\text{C}^{-3}$, $H_f = 1174.9$ °C, and $D_i = 0.0075\%\cdot\text{min}^{-3}$, which in comparison to USF decreased by 44%, increased by 33%, and decreased by 29%, respectively. The correlation between the process parameters and the analysis results was proved. The changed morphology after the hydrothermal process results in easier grinding, handling, and dewatering of hydrochars. Further, detailed analysis may aid understanding in how process parameters influence the specific properties of hydrochars. Nevertheless, an improvement in the performance of mixed solid waste as a fuel by transforming it into hydrochar was confirmed.

Author Contributions: Conceptualization, L.L. and M.W.; methodology, L.L., M.W. and A.M.; validation, M.Ś., K.C. and A.M.; investigation, M.Ś., K.C. and A.M.; resources, L.L. and M.W.; data curation, M.Ś., K.C. and A.M.; writing—original draft preparation, M.Ś. and M.W.; writing—review and editing, L.L. and M.W.; visualization, M.Ś.; supervision, L.L. and M.W. All authors have read and agreed to the published version of the manuscript.

Funding: The research was financed by the Ministry of Science and Higher Education [AGH UST grant no. 16.16.110.663].

Institutional Review Board Statement: Not applicable.

Informed Consent Statement: Not applicable.

Data Availability Statement: The data presented in this study are available on request from the corresponding author.

Acknowledgments: The authors would like to thank the proprietor of the experimental apparatus EKOPROD Ltd. in Bytom.

Conflicts of Interest: The authors declare no conflict of interest. The funders had no role in the design of the study; in the collection, analyses, or interpretation of data; in the writing of the manuscript; or in the decision to publish the results.

References

1. Wilk, M.; Śliz, M.; Gajek, M. The Effects of Hydrothermal Carbonization Operating Parameters on High-Value Hydrochar Derived from Beet Pulp. *Renew. Energy* **2021**, *177*, 216–228. [CrossRef]
2. Saqib, N.U.; Sharma, H.B.; Baroutian, S.; Dubey, B.; Sarmah, A.K. Valorisation of Food Waste via Hydrothermal Carbonisation and Techno-Economic Feasibility Assessment. *Sci. Total Environ.* **2019**, *690*, 261–276. [CrossRef]
3. Gupta, D.; Mahajani, S.M.; Garg, A. Effect of Hydrothermal Carbonization as Pretreatment on Energy Recovery from Food and Paper Wastes. *Bioresour. Technol.* **2019**, *285*, 121329. [CrossRef]
4. Wilk, M.; Śliz, M.; Lubieniecki, B. Hydrothermal Co-Carbonization of Sewage Sludge and Fuel Additives: Combustion Performance of Hydrochar. *Renew. Energy* **2021**, *178*, 1046–1056. [CrossRef]
5. Lin, Y.; Ma, X.; Peng, X.; Yu, Z. Hydrothermal Carbonization of Typical Components of Municipal Solid Waste for Deriving Hydrochars and Their Combustion Behavior. *Bioresour. Technol.* **2017**, *243*, 539–547. [CrossRef]
6. Triyono, B.; Prawisudha, P.; Aziz, M.; Mardiyati; Pasek, A.D.; Yoshikawa, K. Utilization of Mixed Organic-Plastic Municipal Solid Waste as Renewable Solid Fuel Employing Wet Torrefaction. *Waste Manag.* **2019**, *95*, 1–9. [CrossRef]
7. Czerwińska, K.; Śliz, M.; Wilk, M. Hydrothermal Carbonization Process: Fundamentals, Main Parameter Characteristics and Possible Applications Including an Effective Method of SARS-CoV-2 Mitigation in Sewage Sludge. A Review. *Renew. Sustain. Energy Rev.* **2022**, *154*, 111873. [CrossRef]
8. Merzari, F.; Goldfarb, J.; Andreottola, G.; Mimmo, T.; Volpe, M.; Fiori, L. Hydrothermal Carbonization as a Strategy for Sewage Sludge Management: Influence of Process Withdrawal Point on Hydrochar Properties. *Energies* **2020**, *13*, 2890. [CrossRef]
9. Wilk, M. A Novel Method of Sewage Sludge Pre-Treatment—HTC. *E3S Web Conf.* **2016**, *10*, 00103. [CrossRef]
10. Jakubus, M.; Stejskal, B. Municipal Solid Waste Management Systems in Poland and the Czech Republic. A Comparative Study. *Environ. Prot. Eng.* **2020**, *46*, 61–78. [CrossRef]
11. Lin, Y.; Ma, X.; Peng, X.; Yu, Z.; Fang, S.; Lin, Y.; Fan, Y. Combustion, Pyrolysis and Char CO_2-Gasification Characteristics of Hydrothermal Carbonization Solid Fuel from Municipal Solid Wastes. *Fuel* **2016**, *181*, 905–915. [CrossRef]
12. Magdziarz, A.; Mlonka-Mędrala, A.; Sieradzka, M.; Aragon-Briceño, C.; Pożarlik, A.; Bramer, E.A.; Brem, G.; Niedzwiecki, Ł.; Pawlak-Kruczek, H. Multiphase Analysis of Hydrochars Obtained by Anaerobic Digestion of Municipal Solid Waste Organic Fraction. *Renew. Energy* **2021**, *175*, 108–118. [CrossRef]
13. Heidari, M.; Dutta, A.; Acharya, B.; Mahmud, S. A Review of the Current Knowledge and Challenges of Hydrothermal Carbonization for Biomass Conversion. *J. Energy Inst.* **2019**, *92*, 1779–1799. [CrossRef]
14. Lühmann, T.; Wirth, B. Sewage Sludge Valorization via Hydrothermal Carbonization: Optimizing Dewaterability and Phosphorus Release. *Energies* **2020**, *13*, 4417. [CrossRef]
15. Santos, M.M.; Diez, M.A.; Suárez, M.; Centeno, T.A. Innovative Particleboard Material from the Organic Fraction of Municipal Solid Waste. *J. Build. Eng.* **2021**, *44*, 103375. [CrossRef]
16. Ischia, G.; Fiori, L.; Gao, L.; Goldfarb, J.L. Valorizing Municipal Solid Waste via Integrating Hydrothermal Carbonization and Downstream Extraction for Biofuel Production. *J. Clean. Prod.* **2021**, *289*, 125781. [CrossRef]
17. Aragon-Briceño, C.; Pożarlik, A.; Bramer, E.; Brem, G.; Wang, S.; Wen, Y.; Yang, W.; Pawlak-Kruczek, H.; Niedźwiecki, Ł.; Urbanowska, A.; et al. Integration of Hydrothermal Carbonization Treatment for Water and Energy Recovery from Organic Fraction of Municipal Solid Waste Digestate. *Renew. Energy* **2022**, *184*, 577–591. [CrossRef]
18. Espro, C.; Satira, A.; Mauriello, F.; Anajafi, Z.; Moulaee, K.; Iannazzo, D.; Neri, G. Orange Peels-Derived Hydrochar for Chemical Sensing Applications. *Sens. Actuators B Chem.* **2021**, *341*, 130016. [CrossRef]
19. Roman, F.F.; Diaz De Tuesta, J.L.; Praça, P.; Silva, A.M.T.; Faria, J.L.; Gomes, H.T. Hydrochars from Compost Derived from Municipal Solid Waste: Production Process Optimization and Catalytic Applications. *J. Environ. Chem. Eng.* **2021**, *9*, 104888. [CrossRef]
20. Lin, Y.; Ge, Y.; Xiao, H.; He, Q.; Wang, W.; Chen, B. Investigation of Hydrothermal Co-Carbonization of Waste Textile with Waste Wood, Waste Paper and Waste Food from Typical Municipal Solid Wastes. *Energy* **2020**, *210*, 118606. [CrossRef]
21. Venna, S.; Sharma, H.B.; Reddy, P.H.P.; Chowdhury, S.; Dubey, B.K. Landfill Leachate as an Alternative Moisture Source for Hydrothermal Carbonization of Municipal Solid Wastes to Solid Biofuels. *Bioresour. Technol.* **2021**, *320*, 124410. [CrossRef] [PubMed]

22. Wang, T.; Zhai, Y.; Zhu, Y.; Gan, X.; Zheng, L.; Peng, C.; Wang, B.; Li, C.; Zeng, G. Evaluation of the Clean Characteristics and Combustion Behavior of Hydrochar Derived from Food Waste towards Solid Biofuel Production. *Bioresour. Technol.* **2018**, *266*, 275–283. [CrossRef] [PubMed]
23. Akarsu, K.; Duman, G.; Yilmazer, A.; Keskin, T.; Azbar, N.; Yanik, J. Sustainable Valorization of Food Wastes into Solid Fuel by Hydrothermal Carbonization. *Bioresour. Technol.* **2019**, *292*, 121959. [CrossRef] [PubMed]
24. Saqib, N.U.; Baroutian, S.; Sarmah, A.K. Physicochemical, Structural and Combustion Characterization of Food Waste Hydrochar Obtained by Hydrothermal Carbonization. *Bioresour. Technol.* **2018**, *266*, 357–363. [CrossRef]
25. Nguyen, D.; Zhao, W.; Mäkelä, M.; Alwahabi, Z.T.; Kwong, C.W. Effect of Hydrothermal Carbonisation Temperature on the Ignition Properties of Grape Marc Hydrochar Fuels. *Fuel* **2022**, *313*, 122668. [CrossRef]
26. Sharma, H.B.; Sarmah, A.K.; Dubey, B. Hydrothermal Carbonization of Renewable Waste Biomass for Solid Biofuel Production: A Discussion on Process Mechanism, the Influence of Process Parameters, Environmental Performance and Fuel Properties of Hydrochar. *Renew. Sustain. Energy Rev.* **2020**, *123*, 109761. [CrossRef]
27. Yang, Y.; Wang, J.; Chong, K.; Bridgwater, A.V. A Techno-Economic Analysis of Energy Recovery from Organic Fraction of Municipal Solid Waste (MSW) by an Integrated Intermediate Pyrolysis and Combined Heat and Power (CHP) Plant. *Energy Convers. Manag.* **2018**, *174*, 406–416. [CrossRef]
28. Zhou, H.; Long, Y.; Meng, A.; Li, Q.; Zhang, Y. Classification of Municipal Solid Waste Components for Thermal Conversion in Waste-to-Energy Research. *Fuel* **2015**, *145*, 151–157. [CrossRef]
29. Wang, T.; Zhai, Y.; Zhu, Y.; Li, C.; Zeng, G. A Review of the Hydrothermal Carbonization of Biomass Waste for Hydrochar Formation: Process Conditions, Fundamentals, and Physicochemical Properties. *Renew. Sustain. Energy Rev.* **2018**, *90*, 223–247. [CrossRef]
30. Basso, D.; Patuzzi, F.; Castello, D.; Baratieri, M.; Rada, E.C.; Weiss-Hortala, E.; Fiori, L. Agro-Industrial Waste to Solid Biofuel through Hydrothermal Carbonization. *Waste Manag.* **2016**, *47*, 114–121. [CrossRef]
31. Berge, N.D.; Ro, K.S.; Mao, J.; Flora, J.R.V.; Chappell, M.A.; Bae, S. Hydrothermal Carbonization of Municipal Waste Streams. *Environ. Sci. Technol.* **2011**, *45*, 5696–5703. [CrossRef] [PubMed]
32. Śliz, M.; Tuci, F.; Czerwińska, K.; Fabrizi, S.; Lombardi, L.; Wilk, M. Hydrothermal Carbonization of the Wet Fraction from Mixed Municipal Solid Waste: Hydrochar Characteristics and Energy Balance. *Waste Manag.* **2022**, *151*, 39–48. [CrossRef] [PubMed]
33. Parshetti, G.K.; Kent Hoekman, S.; Balasubramanian, R. Chemical, Structural and Combustion Characteristics of Carbonaceous Products Obtained by Hydrothermal Carbonization of Palm Empty Fruit Bunches. *Bioresour. Technol.* **2013**, *135*, 683–689. [CrossRef] [PubMed]
34. Lu, J.J.; Chen, W.H. Investigation on the Ignition and Burnout Temperatures of Bamboo and Sugarcane Bagasse by Thermogravimetric Analysis. *Appl. Energy* **2015**, *160*, 49–57. [CrossRef]
35. Sieradzka, M.; Gao, N.; Quan, C.; Mlonka-Mędrala, A.; Magdziarz, A. Biomass Thermochemical Conversion via Pyrolysis with Integrated CO_2 Capture. *Energies* **2020**, *13*, 1050. [CrossRef]
36. Mureddu, M.; Dessì, F.; Orsini, A.; Ferrara, F.; Pettinau, A. Air- and Oxygen-Blown Characterization of Coal and Biomass by Thermogravimetric Analysis. *Fuel* **2018**, *212*, 626–637. [CrossRef]
37. Wang, L.; Chang, Y.; Zhang, X.; Yang, F.; Li, Y.; Yang, X.; Dong, S. Hydrothermal Co-Carbonization of Sewage Sludge and High Concentration Phenolic Wastewater for Production of Solid Biofuel with Increased Calorific Value. *J. Clean. Prod.* **2020**, *255*, 120317. [CrossRef]
38. Dolgen, D.; Sarptas, H.; Alpaslan, N.; Kucukgul, O. Energy Potential of Municipal Solid Wastes. *Energy Sources* **2005**, *27*, 1483–1492. [CrossRef]
39. Komilis, D.; Kissas, K.; Symeonidis, A. Effect of Organic Matter and Moisture on the Calorific Value of Solid Wastes: An Update of the Tanner Diagram. *Waste Manag.* **2014**, *34*, 249–255. [CrossRef]
40. Nizamuddin, S.; Baloch, H.A.; Griffin, G.J.; Mubarak, N.M.; Bhutto, A.W.; Abro, R.; Mazari, S.A.; Ali, B.S. An Overview of Effect of Process Parameters on Hydrothermal Carbonization of Biomass. *Renew. Sustain. Energy Rev.* **2017**, *73*, 1289–1299. [CrossRef]
41. Wang, J.; Liu, H.; Deng, H.; Jin, M.; Xiao, H.; Yao, H. Deep Dewatering of Sewage Sludge and Simultaneous Preparation of Derived Fuel via Carbonaceous Skeleton-Aided Thermal Hydrolysis. *Chem. Eng. J.* **2020**, *402*, 126255. [CrossRef]
42. Zheng, C.; Ma, X.; Yao, Z.; Chen, X. The Properties and Combustion Behaviors of Hydrochars Derived from Co-Hydrothermal Carbonization of Sewage Sludge and Food Waste. *Bioresour. Technol.* **2019**, *285*, 121347. [CrossRef] [PubMed]
43. Pawlak-Kruczek, H.; Niedzwiecki, L.; Sieradzka, M.; Mlonka-Mędrala, A.; Baranowski, M.; Serafin-Tkaczuk, M.; Magdziarz, A. Hydrothermal Carbonization of Agricultural and Municipal Solid Waste Digestates—Structure and Energetic Properties of the Solid Products. *Fuel* **2020**, *275*, 117837. [CrossRef]
44. Lin, Y.; Ma, X.; Ning, X.; Yu, Z. TGA-FTIR Analysis of Co-Combustion Characteristics of Paper Sludge and Oil-Palm Solid Wastes. *Energy Convers. Manag.* **2015**, *89*, 727–734. [CrossRef]
45. Arellano, O.; Flores, M.; Guerra, J.; Hidalgo, A.; Rojas, D.; Strubinger, A. Hydrothermal Carbonization of Corncob and Characterization of the Obtained Hydrochar. *Chem. Eng. Trans.* **2016**, *50*, 235–240. [CrossRef]
46. He, C.; Zhao, J.; Yang, Y.; Wang, J.Y. Multiscale Characteristics Dynamics of Hydrochar from Hydrothermal Conversion of Sewage Sludge under Sub- and near-Critical Water. *Bioresour. Technol.* **2016**, *211*, 486–493. [CrossRef]
47. Lang, Q.; Guo, Y.; Zheng, Q.; Liu, Z.; Gai, C. Co-Hydrothermal Carbonization of Lignocellulosic Biomass and Swine Manure: Hydrochar Properties and Heavy Metal Transformation Behavior. *Bioresour. Technol.* **2018**, *266*, 242–248. [CrossRef]

48. Nizamuddin, S.; Jaya Kumar, N.S.; Sahu, J.N.; Ganesan, P.; Mubarak, N.M.; Mazari, S.A. Synthesis and Characterization of Hydrochars Produced by Hydrothermal Carbonization of Oil Palm Shell. *Can. J. Chem. Eng.* **2015**, *93*, 1916–1921. [CrossRef]
49. PetroviĽ, J.; PerišiĽ, N.; MaksimoviĽ, J.D.; MaksimoviĽ, V.; KragoviĽ, M.; StojanoviĽ, M.; LauševiĽ, M.; MihajloviĽ, M. Hydrothermal Conversion of Grape Pomace: Detailed Characterization of Obtained Hydrochar and Liquid Phase. *J. Anal. Appl. Pyrolysis* **2016**, *118*, 267–277. [CrossRef]
50. Ghosh, D.; Mitra, S.K. High Temperature Corrosion Problem of Boiler Components in Presence of Sulfur and Alkali Based Fuels. *High Temp. Mater. Processes* **2011**, *30*, 81–85. [CrossRef]
51. Kou, X.; Jin, J.; Wang, Y.; Li, Y.; Hou, F. Understanding the Effect of Calcium Containing Compounds on Ash Deposition during Boiler Operation: Experiment Study and Dynamics Calculation. *Environ. Sci. Pollut. Res.* **2022**, *1*, 1–14. [CrossRef] [PubMed]

energies MDPI

Article

Possibilities of Using Zeolites Synthesized from Fly Ash in Adsorption Chillers

Agata Mlonka-Mędrala [1,*], Tarikul Hasan [1,2], Wojciech Kalawa [1], Marcin Sowa [1], Karol Sztekler [1], Moises Luzia Pinto [2] and Łukasz Mika [1]

1 Faculty of Energy and Fuels, AGH University of Science and Technology, Mickiewicza 30, 30-059 Krakow, Poland
2 Department of Chemical Engineering, Instituto Superior Tecnico, Av. Rovisco Pais 1, 1049-001 Lisbon, Portugal
* Correspondence: amlonka@agh.edu.pl

Abstract: Adsorption chillers produce cold energy, using heat instead of electricity, thus reducing electrical energy consumption. A major industrial waste, fly ash, can be converted to zeolite and used in adsorption chillers as an adsorbent. In this research, three different types of zeolites were synthesised from fly ash via a hydrothermal reaction in an alkaline solution (NaOH). The obtained samples (Na-A zeolites) were modified with K_2CO_3 to increase the water adsorption capacity of these samples. Phase and morphology analyses shows that desired zeolites formed properly but other crystalline phases also exist along with nonporous amorphous phases. The determined specific surface areas for Na-A zeolite (12 h) and Na-A zeolite (24 h) are 45 m^2/g and 185 m^2/g respectively, while the specific surface area for synthesized 13X zeolite is almost negligible. Water-isotherm for each of these samples was measured. Considering the application of adsorption chillers, average adsorption capacity was very low, 1.73% and 1.27%, respectively, for the two most probable operating conditions for synthesized 13X zeolite, whereas no water was available for the evaporation from Na-A zeolite (12 h) and Na-A zeolite (24 h). This analysis implies that among the synthesized materials only 13X zeolite has a potential as an adsorber in sorption chillers.

Keywords: adsorption chiller; fly ash; Zeolite; XRD; specific surface area; water adsorption

Citation: Mlonka-Mędrala, A.; Hasan, T.; Kalawa, W.; Sowa, M.; Sztekler, K.; Pinto, M.L.; Mika, Ł. Possibilities of Using Zeolites Synthesized from Fly Ash in Adsorption Chillers. *Energies* **2022**, *15*, 7444. https://doi.org/10.3390/en15197444

Academic Editor: Gabriele Di Giacomo

Received: 19 August 2022
Accepted: 4 October 2022
Published: 10 October 2022

1. Introduction

Modern civilization is driven mostly by energy. The development and application of the latest technologies largely depend on non-renewable energy. The main sources of this energy are fossil fuels, such as natural gas, oil and coal. There are two main problems associated with the consumption of fossil fuels. The first one is the scarcity of these sources, as they are running out quickly. Secondly, the consumption of fossil fuels is the main reason for the greatest environmental problems associated with releases of greenhouse and hazardous combustion by various products, including CO_2, NO_x, SO_2, heavy metals and dust. Moreover, the demand for energy is increasing day by day. Therefore, the development of sustainable energy systems is both needed and required.

Due to serious climate changes and an urgent need to increase thermal comfort in indoor spaces, the demand for cooling is increasing rapidly. Common air-conditioning and refrigeration systems are usually driven by electricity. Between 2003 and 2004, the electricity used for air conditioning in USA constituted 15.4% of total electricity consumption. In Shanghai city, China, the electricity used for air conditioning constituted 45 to 56% of total electricity consumption in the summer of 2010 [1]. In 2019, 8.5% of total generated electricity was used for space cooling. Energy demand for this purpose has increased more than three times from 1990 [2]. It demonstrates how fast the demand for space cooling is growing and will grow in the future. It has been projected that the number of air-conditioning units will increase up to 5600 million in 2050 from almost 1600 million in 2016, with significant increases in countries such as China and India [3]. So, electricity consumption will also increase for the purpose of space cooling. This electricity mainly comes from fossil fuels. In the year 2018, coal, natural gas and oil constituted 39%, 26% and 3% of total electricity generation [4]. Additionally, electricity production has low efficiency, around 40–50%, and the temperature of the waste heat ranges between 70–200 °C [1]. It ultimately affects the environment. Furthermore, chlorofluorocarbons (CFC), which was used for refrigeration, caused significant damage to ozonosphere, which stops the dangerous ultraviolet rays that originate from the sun. Hydrochlorofluorocarbons (HCFC) are only a temporary alternative because they also create a greenhouse effect and cause damage to the ozonosphere [1].

Renewable energy resources, such as solar energy and geothermal energy, together with waste energy resources with low to very low temperature ranges are available. An adsorption chiller can be employed to utilize these small temperature range heat resources. There are existing sorption refrigeration and heat pump technologies that can be run with such low-grade heat and also use environmentally friendly refrigerants [1]. Adsorption refrigeration technologies can use different adsorbents which can be operated with low-grade heat, for instance, the zeolite–water pair works in the range of 70–250 °C, silica gel–water pair works in the range of 55–120 °C, the activated carbon–methanol pair the works up to a temperature of 120 °C and the activated carbon–ammonia pair usually works up to temperatures of 150 °C (can be used even for 200 °C or more) [1,5]. The commonly used refrigerants are water, ammonia, methanol, etc. These are called green refrigerants because they have no ozonosphere depletion potential (ODP) or greenhouse warming potential (GWP). Additionally, the adsorption refrigeration system has other advantages over absorption systems, e.g., this equipment does not include a solution pump and rectification equipment. Moreover, solution crystallization is absent for such a system. However, there are additional drawbacks as well. Adsorption refrigeration is not as efficient as absorption refrigeration and the volume and mass of this system is also large [1].

The silica gel–water adsorption chiller has been marketed successfully, but there is still much space for improvement in the adsorption refrigeration system. The focus should be given to the factors that can improve the performance of adsorption refrigeration technology, such as adsorption working pairs and their working mechanisms, the structure and design of adsorption refrigeration devices, the improvement of heat and the mass transfer of the adsorption/regeneration bed, etc. [1].

The presented study is dedicated to the analysis of fly ash-derived zeolites as potential candidates for an adsorption chiller. Zeolites are microporous materials that contain aluminium, silicon and oxygen in their main three-dimensional, crystalline structure and also carry cations and water [6]. The general chemical formula of zeolite is as follows: $M_{y/n}[(AlO_2)_y(SiO_2)_m]_zH_2O$ [1]. The water contained in pores and cavities in the structural composition of zeolites can be removed upon the application of heat [7]. High thermal and mechanical stability make zeolites proper candidates for adsorption chiller adsorbents, usually in a working pair with water. This working pair is characterised by a heat of adsorption equal to 3200–4200 kJ/kg [1].

In the tetrahedral structure, zeolite includes silicon cations (Si^{4+}) and aluminium cations (Al^{3+}). These ions are bonded with four oxygen anions (O^{2-}). Thus, oxygen ions connect the tetrahedrons. The ratio of Si:O is 1:2 [7].

The Na-A zeolite is one of the aluminosilicate molecular sieves with a Si/Al ratio of 1, also known as Linde Type A with the composition $(Na^{+}{}_{12}(H_2O)_{27}\,|_{8}[Al_{12}Si_{12}O_{48}]_8$ [8]. The framework includes β-cages (SOD) and α-cages (supercages). Four-membered rings (D4R) link the β-cages with each other and form an inner cavity [9]. The diameter of the super cage is 11.4 Å and the aperture size is 4.1 Å. The Na+ ion can be replaced with K^+, reducing the pore opening, and replaced by Ca^{2+}, increasing the pore opening [8]. It can be used as a desiccant, catalyst and also in cation exchange [8,10].

The Na-X zeolite has a Si/Al ratio that varies from 1 to 1.5 with a faujasite-type (FAU) structure that has the chemical formula $(Ca, Mg, Na_2)_{29}\,(H_2O)_{240}\,|\,[Al_{58}Si_{134}O_{384}]$ [11]. The framework includes β-cages (SOD) and α-cages (supercages). Double six-membered rings

(D6R) link the β-cages with each other and form an inner cavity [9]. The diameter of the super cage is 13 Å and the aperture size is 8 Å [9,12]. There are several uses of the FAU type zeolites, such as ion exchangers, adsorbents or catalysts, in chemical or oil refining; natural gas purification from sulphur compounds; drying oils and gases; separation of hydrocarbon mixtures; sorption of radio nuclides; etc. [13].

The evaluation of synthetic zeolite production from different materials, such as clay minerals, low carbon materials and fly ashes, allowed the production of new value-added products and their use in industry, in particular the agriculture, biochemical and chemical sectors. The utilisation of fly ash as a feedstock for zeolite synthesis subscribes to waste minimization strategies and circular economy concepts, but it requires the elaboration of special procedures. The most often used and well-known synthesis methods are hydrothermal synthesis, molten salt method, alkali activation and microwave synthesis [14,15].

The performance of the adsorption chiller largely depends on the performance of the adsorption and desorption of the adsorbent, namely the amount of adsorbed refrigerant, the time required to reach equilibrium, the temperature of adsorption and desorption, the heat of adsorption and desorption, the heat and mass transfer of the adsorber bed, etc. [1]. The purposes of the present work included the synthesis of Na-A and 13X zeolite from fly ash and the analysis of the potential application of these adsorbents compared to the performance of silica gel with regard to the adsorption chiller.

2. Material and Methods

2.1. Zeolites Preparation Methods

The fly ash used for zeolite synthesis was collected from a power plant located in Poland. The composition of the fly ash was as follows: SiO_2 = 45.50%, Al_2O_3 = 23.10%, Fe_2O_3 = 7.38%, CaO = 6.30%, MgO = 4.22%, Active CaO + MgO = 0.55%, S = 0.54%, SO_4^{2-} = 1.62%, TiO_2 = 0.72%, P_2O_5 = 0.29%, Mn_3O_4 = 0.17, Na_2O = 1.55%, K_2O = 2.96 and Combustion Loss = 6.42%. Three types of zeolites were prepared and modified with K_2CO_3 to further develop the adsorption capacity of these materials.

The procedure of the Na-A zeolite (12 h) preparation was conducted using dry fly ash, which was mixed with NaOH into a mass ratio of 5:6. The mixture was milled and exposed to a temperature of 550 °C in a muffle furnace for 1 h. The mixture after heating was cooled down to room temperature and milled for 1 h. It was then mixed with distilled water at a 4:1 ratio. The obtained slurry was stirred at room temperature for 12 h. After this stage, the mixture (solid and liquid phases) sample crystallization was obtained at 100 °C for 12 h. The obtained zeolite was washed with distilled water to remove NaOH and reach c.a. pH = 7 and filtered. The obtained material was dried at 100 °C for 12 h.

The procedure of the Na-A zeolite (24 h) preparation was similar to the methodology for the Na-A zeolite sample (12 h), but the crystallization time at 100 °C was extended up to 24 h.

Both Na-A zeolites were modified with K_2CO_3. In this step, the obtained zeolite was mixed with K_2CO_3 at a ratio of 2:1 (by weight) and dissolved in distilled water at a ratio of 1 g to 5 mL. The obtained slurry was stirred at room temperature for 24 h. Modified zeolite was filtered and dried at 60 °C for 12 h. The last step was the calcination, which was performed in a muffle furnace at 300 °C for 4 h.

During 13X Zeolite preparation, fly ash was first calcined in a muffle furnace at 800 °C to remove the remaining carbon and volatile matter. To enhance the activity, thermal stability and acidity of the zeolite, the removal of Al and Fe was performed by treatment with hydrochloric acid. The calcined sample was treated with 1 M HCl. Dried pre-treated fly ash was then mixed with NaOH at a ratio of 1.5:1 (by weight). This mixture was exposed to 550 °C for 1 h and later cooled down to room temperature. The obtained mixture was milled and distilled water was added at a ratio of 10 g fly ash/100 mL water. The obtained mixture was stirred for 12 h. Then, it was allowed to settle at 90 °C for 6 h. After that, distilled water was used to wash the sample and filtered to remove the remaining sodium hydroxide and was later dried at 100 °C for 12 h.

2.2. Surface Area Analysis

The nitrogen gas adsorption method was used at −196 °C using Gemini V 2.00; model 2380. All the samples were degassed overnight at 250 °C under vacuum before measurement. The surface area of the samples was determined using the BET model.

2.3. Phase Analysis

The analysis of the crystalline phases was carried out by X-ray diffraction using an XRD DX-27mini manufactured by HAOYUAN. Analyses were performed at a 2θ angle measuring range from 5° to 110° with a step of 0.02° 2θ. The tube voltage and the tube current were 35 kV and 12 mA, respectively. The percentage of crystalline phases was determined by a semi-quantitative method and was implemented to Malvern Panalytical's HighScore Plus software using PDF4+ database (2012 version). The relative intensity of the reflections of the phases was compared with references included in the ICDD database and the phase composition was determined for studied samples.

2.4. Morphology Analysis and Semi-Quantitative Analysis of Synthesised Zeolites Using Scanning Electron Microscopy with Energy Dispersive Spectroscopy (SEM–EDS)

The surface structure of the samples and the chemical composition of selected areas on their surfaces were analyzed using the scanning electron microscopy with energy dispersive spectroscopy (SEM–EDS method). The images and chemical composition analysis (EDS) were performed using a Prisma E scanning electron microscope manufactured by Thermo Scientfic, Waltham, MA, USA.

2.5. Water Adsorption Isotherm Determination

The amount of adsorbed water, so-called water uptake was measured using a gravimetric method in a DVS vacuum-surface measurement system presented in Figure 1. It is able to perform multi-component experiments using vapor and/or gas sorbate molecules with the in situ samples, degassing up to 400 °C and high vacuum [16]. This device is used to analyse the performance of porous materials, such as zeolites, porous polymers, composites, aluminophosphates (AlPOs) and silica aluminophosphates (SAPOs), silica gels, activated carbons and metal organic frameworks (MOFs) [16].

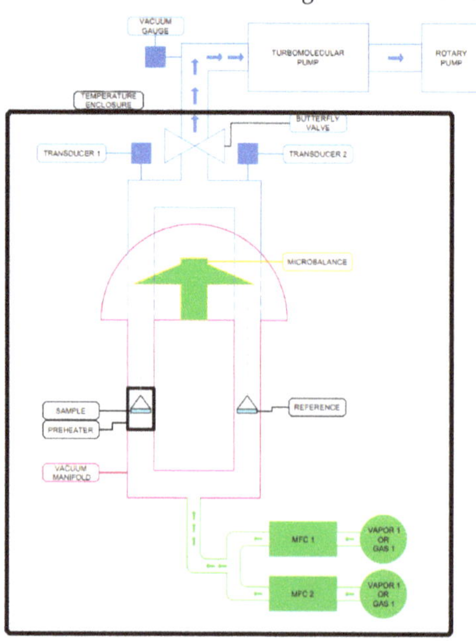

Figure 1. DVS Vacuum Surface Measurement System.

About 25 to 30 mg of the sample was analysed during the sorption capacity tests. In this study, the adsorption and desorption processes were analysed in the water vapor P/P_0 range from 0 up to 90%. Five process temperatures were selected in this study: 25, 35, 45, 55 and 65 °C; relative pressure step (P/P_0) was set to 10%; and time duration for each step was 20 min. The activation temperature of 170 °C was set for 90 min to dry and degas the analysed sample prior to the experiment; the incubator temperature (same as system temperature) and vapor flow rate were 25/65 °C and 50 cm, respectively.

3. Results and Discussion

The results of the performed experiments are presented and discussed in detail in this section. Adsorbents were compared according to their thermal and sorption properties. An additional comparison with previously published studies was performed.

3.1. Sample Characterization—Surface Area Analysis

Low-temperature gas adsorption results obtained from the studied samples allow for the determination the porosity of the samples (Figure 2). The results for the silica gel sample confirmed the mesoporous nature of the material with the highest adsorption capacity up to 0.4 P/P_0, a specific surface area for this sample was 613 m^2/g, and a total porous volume of 0.314 cm^3/g. For the synthesized Na-A zeolite (12 h), the observed isotherm demonstrated lower adsorption capacity than expected for pure NaA, and the obtained specific surface area was only 45 m^2/g. The specific surface area for commercial A-type zeolite, as noted in the literature, is around 850 m^2/g [11], suggesting that the synthesized sample has a lower than expected share of zeolites in its structure. Thus, this result indicates that the conversion of fly ash to zeolite was very low. From XRD analysis, it has also been found that the desired zeolite was formed along with additional non-porous crystalline phases. These phases can block void spaces. Additionally, there are present amorphous phases which are apparently nonporous. This resulted in a lower specific area. Additionally, it has been found that the total porous volume is 0.109 cm^3/g for the Na-A zeolite (12 h). The synthesis for 24 h resulted in a sample with better adsorption properties. The obtained specific surface area of the Na-A zeolite (24 h) is 185 m^2/g, which is 78% lower compared to commercial A zeolite, suggesting that the synthesized sample has a smaller share of zeolite but is also significantly better that the sample obtained within a 12-h synthesis. Similar to the previous case, phase analysis shows that the desired zeolite was formed along with some nonporous crystalline phases, resulting in a lower specific surface area. The total porous volume of this sample was 0.171 cm^3/g.

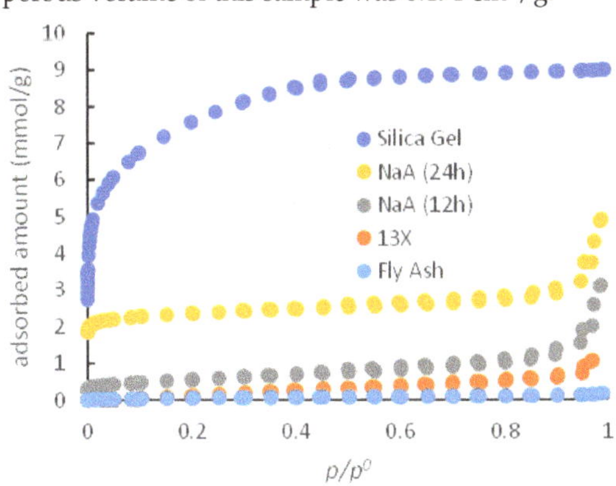

Figure 2. Adsorption Isotherm for Different Samples (at −196 °C).

For both synthesized 13X zeolite and fly ash, which was the raw material for zeolite synthesis, the obtained specific surface area was almost negligible. This suggests that these are almost nonporous materials. However, phase analysis results showed that desired zeolite was formed. Based on phase analysis and surface area analysis it can be seen that only a very small portion was converted to zeolite, and the sample contains a significant amount of non-porous amorphous mass.

3.2. Sample Characterization—Morphology Analysis and Semi-Quantitative Analysis of Synthesised Zeolites Using Scanning Electron Microscopy with Energy Dispersive Spectroscopy (SEM–EDS)

The structural and morphological analysis of obtained zeolites was performed and the images are presented in Figure 3. The obtained materials are composed of irregular particles of different sizes and shapes. The microscopic photographs enable the analysis of the crystalline structure of obtained zeolites. In the case of Na-A zeolite (12 h) and 13X zeolite synthesized from fly ash, a well-developed, regular structure, characteristic for zeolites was observed, whereas in the case of the Na-A zeolite (24 h) sample, the structure was more irregular and crystals were present only locally. All samples present a mesoporous structure with a possible presence of micropores.

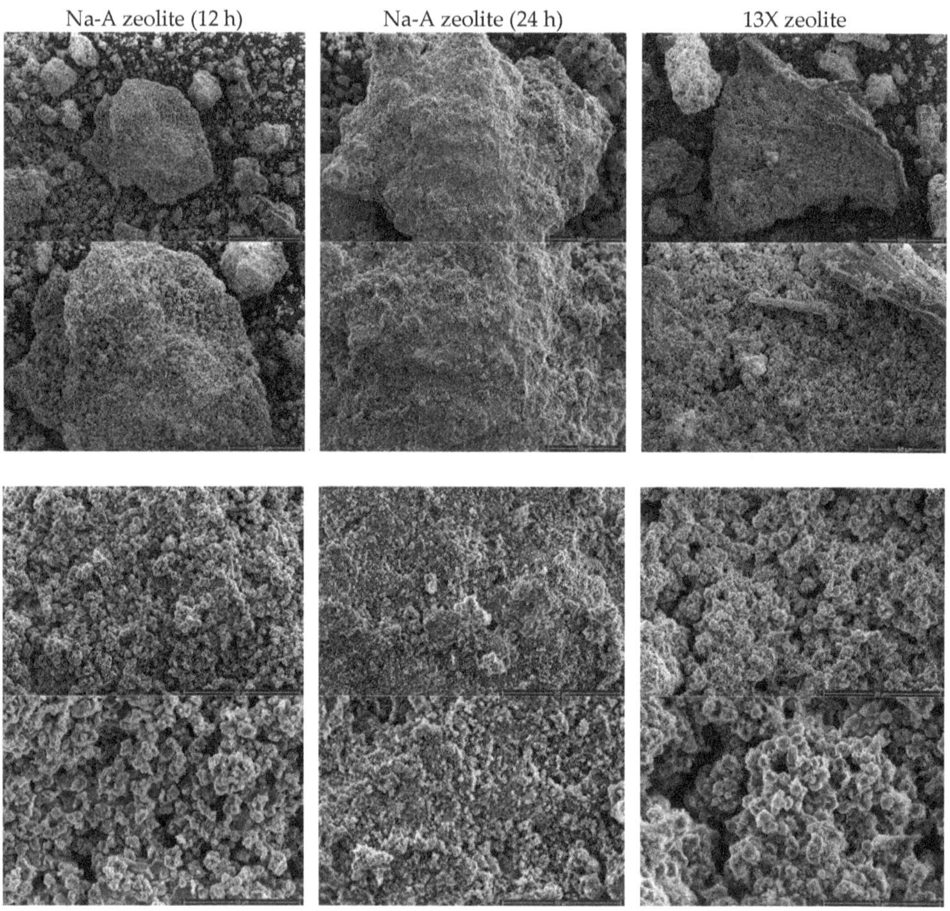

Figure 3. Morphology of synthesized zeolites.

The energy dispersive spectroscopy (EDS) technique is generally used for the qualitative analysis of materials, but in this study semi-quantitative results for major components was provided to define the Si/Al ratio for the synthesised materials. The results are presented in Table 1. The type of zeolites depend on the content of SiO_2 (the molar ratio of Si/Al). The ratio of Si/Al determines the properties of zeolite. Low-silica zeolites are characterized by a higher acid proofness, noticeable stability at higher temperatures and hydrophilicity. On the other hand, high-silica zeolites are more hydrophobic and have high ions exchange features.

Table 1. Semi-quantitative analysis of selected components of produced zeolites.

	Na-A Zeolite (12 h), mol%	Na-A Zeolite (24 h), mol%	13X Zeolite, mol%
Na	30.6	22.1	22.7
Si	28.4	19.5	37.8
Al	24.3	16.1	21.7
Si/Al	1.17	1.21	1.74

The highest molar Si/Al ratio was obtained for 13X zeolites, treated with HCl before crystallization to minimize Fe and Al content. A similar Si/Al ratio was achieved for Na-A zeolites whose crystallization time was different and does not influence the relationship between silicon and aluminium content.

3.3. Sample Characterization—Phase Analysis

The X-ray diffractograms of two commercial zeolites, 4A and 5A, are presented and compared with the obtained Na-A zeolites in Figure 4. For commercial 4A zeolite, numerous peaks were noted at 7.2, 10.18, 12.48, 16.12, 21.7, 24.02, 27.14, 29.98, 32.6 and 34.22, in the case of the second commercial zeolite (5A), similar peaks were noted. For synthesized Na-A zeolite (12 h), reflections were noted at 13.0, 24.22, 28.06, 31.46, 34.56, 42.7, 51.9, 58.08, 60.12, 61.96, 63.88, 69.38 and 78.2. The similarity between commercial and synthesized Na-A zeolite in the case of phase analysis was achieved only to a minor extent. Other peaks were noted as well, which indicate the presence of other crystalline phases. Other detected phases are potassium aluminium silicon oxide and sodium calcium aluminium silicate hydrate. Zeolite constitutes 66% of the total crystalline phase and the rest of the crystalline phases constitute 34%.

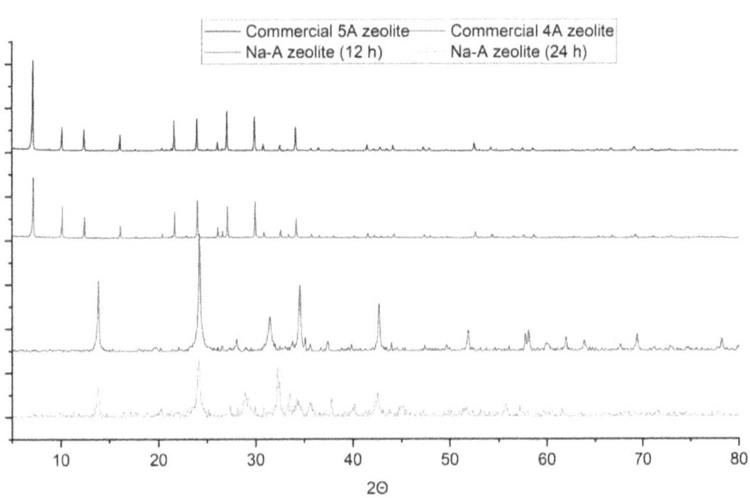

Figure 4. Phase Analisis Result Comparison, Commercial 4A Zeolite, Commercial 5A Zeolite, Synthesized Na-A Zeolite and Synthesized Na-A Zeolite (24 h).

For the synthesized Na-A zeolite (24 h), several reflections were noted at 13.88, 24.16, 28.94, 32.24, 32.5, 37.82, 40.2, 42.54 and 55.78 and a higher amorphous phase content was also noted in this case. Other peaks were observed as well, which indicate the presence of other crystalline phases. Other detected phases are calcium hydroxide and potassium hydrogen carbonate. Zeolite constitutes 47% of the total crystalline phase and the rest of the crystalline phases constitute 53%. Both Na-A synthesized zeolites were characterised by a very similar composition.

The X-ray diffractograms of 13X commercial zeolite is presented and compared with obtained 13X zeolite in Figure 5. For commercial zeolite, peaks are present at 6.13, 10.02, 11.75, 15.45, 18.46, 20.12, 23.34, 26.70, 30.94, 31, 32 and 33.65. For synthesised 13X zeolite, reflections were noted at 6.16, 10, 11.78, 13.92, 15.46, 18.08, 20.08, 23.34, 24.3, 26.7, 29, 31, 32 and 33.86, which almost superimposes on the reflection peaks of commercial zeolite. Some additional peaks are present as well, which indicate the presence of other crystalline phases as well. Other detected phases are sodium hydrogen carbonate, $CaCO_3$ and sodium aluminium silicon carbonate oxide. Zeolite constitutes 60% of the total crystalline phase and the rest of the crystalline phases constitutes 40%.

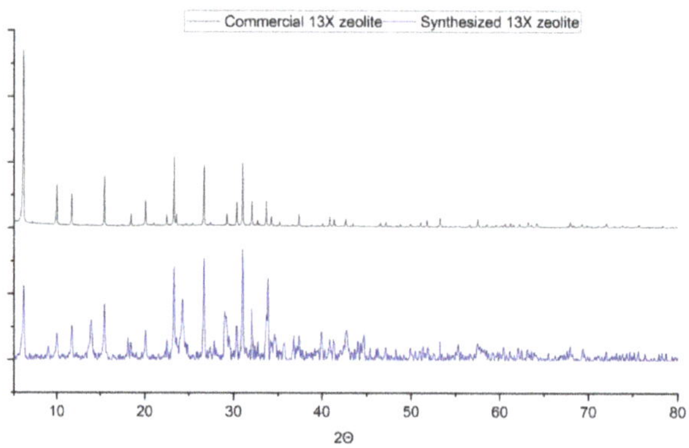

Figure 5. Phase Analisis Result Comparison: Commercial 13X Zeolite and Synthesized 13X Zeolite.

3.4. Adsorption Isotherm and Potential Use in Adsorption Chiller

The water intake was tested in the temperature range of 25–65 °C, the sorption isotherms are shown in Figure 6 for silica gel as a reference material and for synthesized zeolites (Figures 7–9).

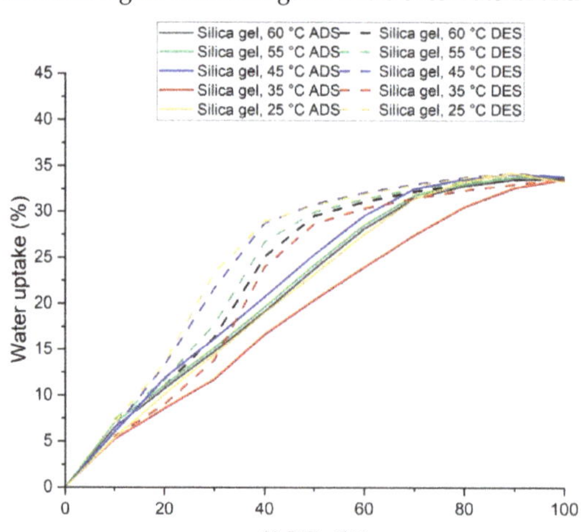

Figure 6. Adsorption and Desorption Isotherms for Silica Gel (at five process temperatures 25, 35, 45, 55 and 60 °C.

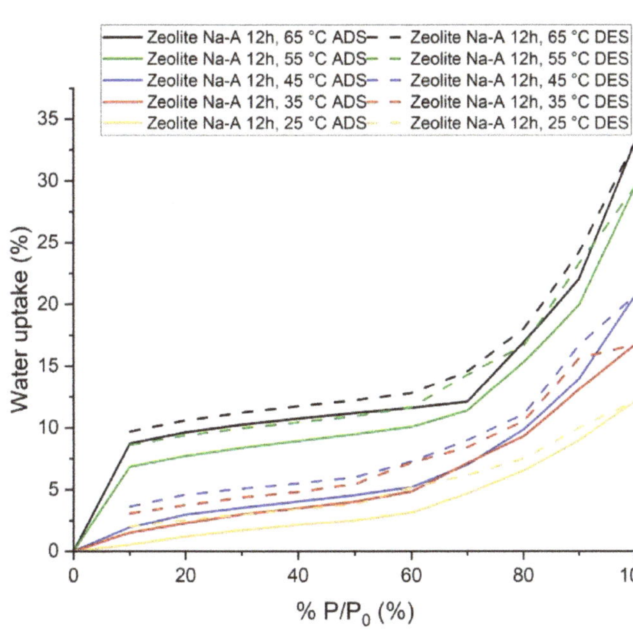

Figure 7. Adsorption and Desorption Isotherms for Synthesized Na-A Zeolite (12 h) at five process temperatures: 25, 35, 45, 55 and 65 °C.

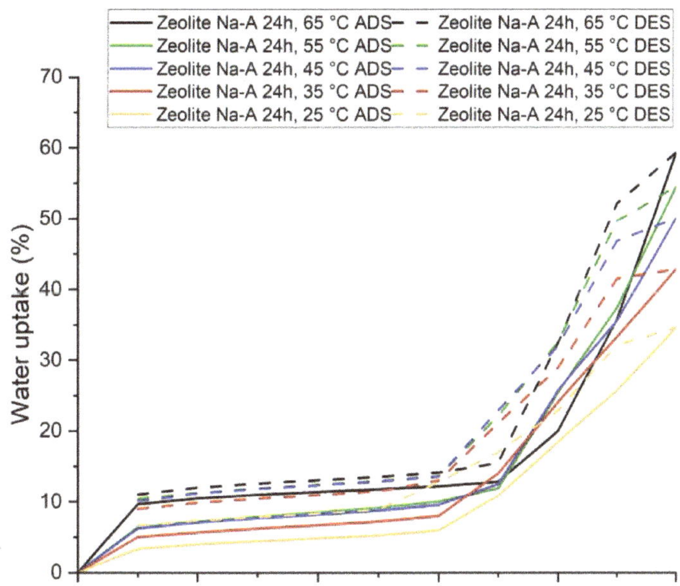

Figure 8. Adsorption and Desorption Isotherms for Synthesized Na-A Zeolite (24 h) at five process temperatures: 25, 35, 45, 55 and 65 °C.

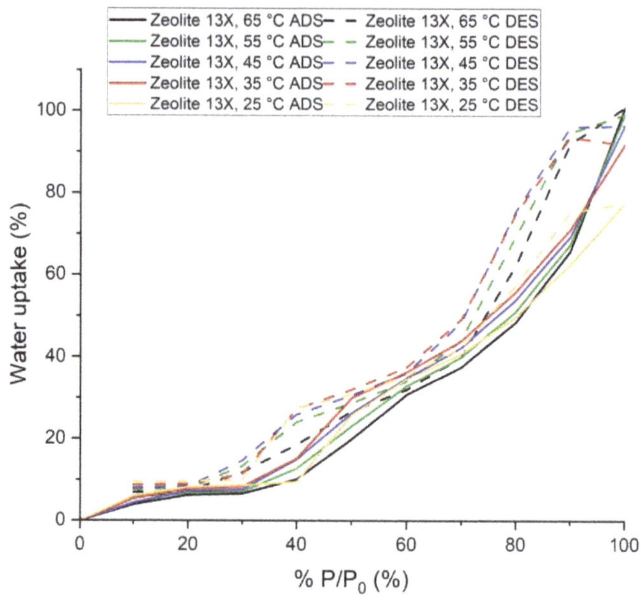

Figure 9. Adsorption and Desorption Isotherms for Synthesized 13X at five process temperatures: 25, 35, 45, 55 and 65 °C.

3.4.1. Silica Gel

Silica gel adsorbed 33.25% water vapor for P/P_0 = 100% at 25 °C and 33.69% for P/P_0 = 100% at 65 °C. This amount was adsorbed within 25 min. Within this time, silica gel almost reached its highest adsorption capacity. However, adsorption kinetics suggest that it could have adsorbed and desorbed (for lower P/P_0) more if the time for each step had been increased. This would result in smaller hysteresis.

From the shape of the isotherm in Figure 5, it can be concluded that it is a type IV isotherm [17,18]. It is a multilayer adsorption process. It also has a significant hysteresis issue. The observed type H2(b) hysteresis usually results from pore neck blocking [18]. The hysteresis loop becomes smaller with increased temperatures.

3.4.2. Synthesized Na-A Zeolite (12 h)

Synthesized Na-A zeolite (12 h) adsorbed 12.09% water vapor at P/P_0 = 100% and 25 °C and 33.07% at P/P_0 = 100% and 65 °C. This amount of water was adsorbed within 60 min. The adsorption kinetics showed that for higher pressures, if the time of adsorption had been increased, this sample would have been able to adsorb more water vapor. However, for lower P/P_0 (which is actually the working condition for adsorption chillers [19–31]), equilibrium was almost reached in less than 20 min.

The shape of the isotherm is type IV. The flat region indicates monolayer adsorption, followed by a rise which indicates multilayer adsorption. It is characteristic of mesoporous materials [17,18], but this does not seem to be the case here. The adsorption in the lower pressure region occurred due to the presence of zeolite. The increase in adsorption in higher pressure regions occurred due to capillary condensation in intergranular voids. Additionally, hydration probably played a part. It also has significant hysteresis.

The usual shape for commercial Na-A zeolite is Type I [32] and the adsorption capacity is also very high at lower P/P_0 compared to synthesized Na-A zeolite (12 h). For commercial Na-A zeolite and synthesized Na-A zeolite (12 h), the adsorption capacity at 25 °C and P/P_0 = 20% is 28% and 1.66%, respectively. XRD and surface area analysis suggested that other non-porous crystalline phases and amorphous phases are present, which explains the reason for having different types of isotherms and lower adsorption capacity.

Potential Use in Adsorption Chillers: The performance of the adsorption refrigeration system is determined by COP and P. These two parameters largely depend on the adsorbent–adsorbate working pair performance, such as how quickly the adsorbent can adsorb and desorb the refrigerant and what amount of adsorbate will be taken by the absorbent under given working conditions (pressure and temperature) [1].

As explained before, for this sample, the adsorption took 60 min. However, based on adsorption kinetics for Na-A zeolite (12 h), it actually took less than 20 min to reach equilibrium (almost) at lower P/P_0, whereas for silica gel it was around 25 min. This implies that the sample will have a better adsorption refrigeration cycle time compared to silica gel.

To determine the potential of the Na-A zeolite (12 h) as an adsorbent in the sorption chiller in working pair with water, conditions and performance of adsorption chillers were taken from previously published papers [19–31]. For those chillers, during adsorption, the pressure of the adsorbent bed is always lower, usually above 15% of the saturation pressure, and can sometimes be as high as 45% [22–32]. The temperature usually ranges between 25–40 °C [19–31]. During the desorption of the refrigerant the temperature varies significantly (usually more than 65 °C and can be as high as 120 °C for solar energy as heat resource), whereas the pressure is usually less than 5% of the saturation pressure of the respective temperature and can sometimes be more than 10% as well [19–31]. Based on these research works, two different working conditions have been considered.

First, for adsorption, 25 °C temperature and 30% of saturation pressure (for water at 25 °C) were considered. In these conditions, silica gel (14.37%) adsorbed much more water vapor than Na-A zeolite (12 h) (1.66%). For desorption, if the temperature 65 °C and the pressure of 10% of the saturation pressure were considered, the water uptake at this condition was 9.68% and 6.99% for synthesized zeolite and silica gel, respectively. Therefore, under these conditions, the average adsorption capacity in the case of our sample and silica gel are 0% and 7.38%, respectively. So, produced zeolite cannot produce any cold energy and replace silica gel as an adsorbent for the adsorption chiller.

Second, for adsorption, 25 °C temperature and 20% of saturation pressure (for water at 25 °C) were considered. In these conditions, silica gel (10.02%) adsorbed much more water vapor than synthesized zeolite (1.16%). For desorption, at a temperature of 65 °C and a pressure of 10% of the saturation pressure, the adsorbed amount at this condition is 9.68% and 6.99%. Therefore, under these conditions, the amount of water refrigerant available for evaporation at the evaporator in the case of the Na-A zeolite (12 h) and silica gel would be 0% and 3.03%, respectively. Thus, our sample cannot produce any cold energy and replace silica gel as an adsorbent for the sorption chiller.

3.4.3. Synthesized Na-A Zeolite (24 h)

Synthesized Na-A zeolite (24 h) adsorbed 34.57% water vapor for P/P_0 = 100% at 25 °C and 59.33% for P/P_0 = 100% at 65 °C (Figure 7). This amount of adsorption was achieved in 60 min. The adsorption kinetics suggests that for higher pressure if we increased the duration of adsorption, this sample would have been able to adsorb a little more water vapor. However, for lower P/P_0 (which is actually the working condition for adsorption chiller [19–31]), equilibrium was almost reached in less than 20 min.

The shape of the isotherm is a type IV isotherm. The flat region indicates monolayer adsorption, followed by a rise which indicates multilayer adsorption [17,18]. The adsorption in the lower pressure region occurred due to the presence of zeolite. The increase in adsorption in the region of higher pressure occurred due to capillary condensation in intergranular voids. It seems $KHCO_3$ also contributed, while hydration probably also played a part. It also has a significant hysteresis.

The usual shape for commercial Na-A zeolite is Type I and the adsorption capacity is also very high at lower P/P_0 compared to synthesized Na-A zeolite (24 h). For commercial Na-A zeolite and synthesized Na-A zeolite (24 h), the adsorption capacity at 25 °C and

P/P_0 = 20% is 28% and 3.90%, respectively. The reasons again here are the presence of non-porous crystalline phases and amorphous phases.

Potential Use in Adsorption Chillers: For this sample, the whole adsorption cycle took 60 min. However, the adsorption kinetics of the sample showed that it actually took less than 20 min to reach equilibrium at lower P/P_0, whereas for silica gel it was around 25 min. This implies that the sample has better adsorption refrigeration cycle times compared to silica gel.

Similar to the previous cases, two working scenarios were taken into consideration.

First, for adsorption, 25 °C temperature and 30% of saturation pressure (for water at 25 °C) were considered. In these conditions, silica gel (14.37%) adsorbed much more water vapor than the Na-A zeolite (24 h) sample (4.35%). For desorption, considering the temperature of 65 °C and pressure of 10% of the saturation pressure, the adsorption amounts at these conditions were 11.03% and 6.99% for synthesized zeolite and silica gel, respectively. Therefore, under these conditions, the average adsorption capacity in the case of our sample and silica gel are 0% and 7.38%, respectively. Thus, synthesized zeolite cannot produce any cold energy and replace silica gel as an adsorbent for the adsorption chiller.

Second, for adsorption, the temperature of 25 °C and 20% of saturation pressure (for water at 25 °C) were considered. In these conditions, silica gel (10.02%) adsorbed much more water vapor than the zeolite sample (3.90%) at this temperature and pressure. For desorption, if we consider the temperature is 65 °C and the pressure is 10% of the saturation pressure, the adsorption amount under these conditions was 11.03% and 6.99%. Therefore, under these conditions, the possible amount of water refrigerant available for vaporization at the evaporator in the case of our sample and silica gel should be 0% and 3.03%, respectively. Additionally, in this case, the synthesized sample cannot produce any cold energy and replace silica gel as an adsorbent for the adsorption chiller.

3.4.4. Synthesized 13X Zeolite

Synthesized 13-X zeolite (12 h) adsorbed 77.64% water vapor for P/P_0 = 100% at 25 °C and 100.9% for P/P_0 = 100% at 65 °C. This amount of adsorption was achieved in 60 min. The adsorption kinetics suggests that for higher pressure by increasing the adsorption time, this sample would have been able to adsorb a little more water vapor. However, for lower P/P_0 (which is actually the working condition for adsorption chiller [19–31]), equilibrium was almost reached in less than 25 min.

The shape of the isotherm for this sample is a type VI isotherm. Similarly to the previous case, the flat region indicates monolayer adsorption, followed by a rise in two different steps. The adsorption in the lower pressure region occurred due to the presence of synthesized zeolite. The increase in adsorption in the region of higher pressure occurred due to capillary condensation in intergranular voids. It seems that $NaHCO_3$ also contributed, while hydration probably played a part.

The usual shape for commercial Na-X zeolite is Type I [32] and the adsorption capacity is also very high at lower P/P_0 compared to the synthesized zeolite. For commercial zeolite and synthesized zeolite, the adsorption capacity at 25 °C and P/P_0 = 20% is 34% and 8.27%, respectively. XRD and surface analysis suggests that there are other non-porous crystalline phases and amorphous phases present, which explains the reason for having a different type of isotherm and lower adsorption capacity. Additionally, we can say that only a small amount of raw material is converted to zeolite.

Potential Use in Adsorption Chillers: As presented before, for this sample, the adsorption took 60 min. However, according to the adsorption kinetics, it can be concluded that it actually took less than 25 min to reach equilibrium at lower P/P_0. In fact, for higher temperatures, it was less than 20 min. For silica gel, it was around 25 min. This implies that for this sample adsorption refrigeration cycle, the time will be similar or less when compared to silica gel.

At 25 °C and 30% of saturation pressure (for water at 25 °C) silica gel (14.37%) adsorbed much more water vapor than the synthesized 13X zeolite (8.73%). For desorption, taking into consideration a temperature of 65 °C and a pressure of 10% of the saturation pressure, the adsorbed amount of water vapor under these conditions was 7% and 6.99% for zeolite and silica gel, respectively. Therefore, under these conditions, the average adsorption capacity for 13X zeolite and silica gel would be 1.73% and 7.38%, respectively. The zeolite–water vapor working pair produces cold energy, but it cannot replace silica gel as an adsorbent for the adsorption chiller. In these working conditions, silica gel has the potential to produce almost 4.3 times more cold energy than our samples.

Taking into consideration the decrease of pressure to 20% of saturation pressure (for water at 25 °C), silica gel (10.02%) adsorbed much more water vapor than analysed zeolite (8.27%). For desorption, if a temperature of 65 °C and a pressure of 10% of the saturation pressure were considered, the adsorption amount would be 7% and 6.99%. So, under these conditions, the amount of water refrigerant available for vaporization at the evaporator for 13X zeolite and silica gel would be 1.27% and 3.03%, respectively. Similarly to the previous case, zeolite will be able to produce cold energy, but it cannot replace silica gel as an adsorbent for the adsorption chiller. For such working conditions, silica gel has the potential to produce almost 2.4 times more cold energy than synthesized zeolite.

4. Conclusions

The results of the phase analysis showed that the targeted compound, such as Na-A zeolite and 13X zeolite, were produced successfully from the fly ash using the applied hydrothermal treatment. However, the presence of other non-porous crystalline phases and amorphous phases was evident as well. In the respective samples, Na-A zeolite (12 h), Na-A zeolite (24 h) and 13X zeolite constitute 66%, 47% and 60% of the total crystalline phase. Through the impregnation of K_2CO_3 other phases were developed. Additionally, the conversion percentage was low and for 13X zeolite it was very low. However, the applied treatments increased the water adsorption capacity of the raw material.

The morphology of the samples analysed using a scanning electron microscope showed that a well-developed crystalline structure was observed in the case of Na-A (12 h) and 13X zeolites. A high-silicone 13X zeolite was obtained through the treatment of a sample with HCl prior crystallization.

Considering the standard operating condition of an adsorption chiller, it has been found that synthesized Na-A zeolite samples cannot produce any cold energy, whereas the synthesized 13X zeolite has the potential to create cold energy. However, it cannot replace silica gel (which is used commercially) as an adsorbent in an adsorption chiller. The performance of zeolite is 2–4 times lower when compared with silica gel.

Different approaches can be used to obtain better zeolite synthesis from fly ash in order to impregnate zeolite with K_2CO_3 without forming any other type of compound and to improve the adsorption and desorption performances of these synthesized materials so that they are suitable for adsorption chillers. Additionally, they could be used for other purposes, such as water desiccant and CO_2 capture. However, in order to test feasibility, these possibilities will require further study.

Author Contributions: Conceptualization, T.H., K.S. and M.L.P.; methodology, W.K. and M.S.; validation, A.M.-M. and K.S.; formal analysis, A.M.-M.; investigation, T.H. and A.M.-M.; resources, K.S., M.L.P. and Ł.M.; data curation, W.K. and M.S.; writing—original draft preparation, T.H.; writing—review and editing, A.M.-M., K.S. and M.L.P.; visualization, A.M.-M.; supervision, K.S., M.L.P. and Ł.M.; project administration, K.S.; funding acquisition, A.M.-M., K.S., M.L.P. and Ł.M. All authors have read and agreed to the published version of the manuscript.

Funding: This research was funded by Ministry of Science and Higher Education, Poland, grant AGH no 16.16.210.476, and partly supported by the program "Excellence initiative—research university" for the AGH University of Science and Technology. A.M.-M. is supported by the Foundation for Polish Science (FNP).

Institutional Review Board Statement: Not applicable.

Informed Consent Statement: Not applicable.

Conflicts of Interest: The authors declare no conflict of interest.

References

1. Wang, R.; Wang, L.; Wu, J. *Adsorption Refrigeration Technology: Theory and Application*, 1st ed.; Wiley: Wiley, NJ, USA, 2014. [CrossRef]
2. IEA. Cooling. Available online: https://www.iea.org/reports/cooling (accessed on 23 January 2021).
3. IEA. The Future of Cooling. Available online: https://www.iea.org/reports/the-future-of-cooling (accessed on 23 January 2021).
4. IEA. Explore Energy Data by Category, Indicator, Country or Region. Available online: https://www.iea.org/data-and-statistics?country=WORLD&fuel=Electricity%20and%20heat&indicator=ElecGenByFuel (accessed on 23 January 2021).
5. Wolak, E.; Kraszewski, S. An overview of adsorptive processes in refrigeration systems. In Proceedings of the 1st International Conference on the Sustainable Energy and Environment Development (SEED), Krakow, Poland, 17–19 May 2016; Volume 10, p. 00104. [CrossRef]
6. Lenntech. Zeolite Structure. Available online: https://www.lenntech.pl/zeolites-structure-types.htm#:~{}:text=Zeolites%20are%20three%2Ddimensional%2C%20microporous,other%20through%20shared%20oxygen% (accessed on 23 January 2021).
7. Moshoeshoe, M.; Tabbiruka, M.S.N.; Obuseng, V. A Review of the Chemistry, Structure, Properties and Applications of Zeolites. *Am. J. Mater. Sci.* **2017**, *7*, 196–221. [CrossRef]
8. Julbe, A.; Drobek, M. Zeolite A Type. In *Encyclopedia of Membranes*; Drioli, E., Giorno, L., Eds.; Springer: Berlin/Heidelberg, Germany, 2016. [CrossRef]
9. Ríos, C.A.; Williams, C.D.; Castellanos, O.M. Crystallization of low silica Na-A and Na-X zeolites from transformation of kaolin and obsidian by alkaline fusion. *Ing. Compet.* **2012**, *14*, 9–22.
10. Fruijtier-Pölloth, C. The safety of synthetic zeolites used in detergents. *Arch. Toxicol.* **2009**, *83*, 23–35. [CrossRef] [PubMed]
11. Rahmati, M.; Modarress, H. The effects of structural parameters of zeolite on the adsorption of hydrogen: A molecular simulation study. *Mol. Simul.* **2012**, *38*, 1038–1047. [CrossRef]
12. Julbe, A.; Drobek, M. Zeolite X: Type. In *Encyclopedia of Membranes*; Drioli, E., Giorno, L., Eds.; Springer: Berlin/Heidelberg, Germany, 2014. [CrossRef]
13. Georgiev, D.; Bogdanov, B.; Markovska, I.; Hristov, Y. A Study on the Synthesis and Structure of Zeolite NaX. *J. Chem. Technol. Metall.* **2013**, *48*, 168–173.
14. Ma, B.; Lothenbach, B. Synthesis, characterization, and thermodynamic study of selected Na-based zeolites. *Cem. Concr. Res.* **2020**, *135*, 106111. [CrossRef]
15. Indira, V.; Abhitha, K. A review on recent developments in Zeolite A synthesis for improved carbon dioxide capture: Implications for the water-energy nexus. *Energy Nexus* **2022**, *7*, 100095. [CrossRef]
16. DVS Vacuum-Surface Measurement System (Device Brochure). Available online: https://www.surfacemeasurementsystems.com/wpcontent/uploads/2017/08/DVS_Vacuum.pdf (accessed on 14 April 2021).
17. Yahia, M.B.; Torkia, Y.B.; Knani, S.; Hachicha, M.A.; Khalfaoui, M.; Lamine, A.B. Models for Type VI Adsorption Isotherms from a Statistical Mechanical Formulation. *Adsorpt. Sci. Technol.* **2013**, *31*, 341–357. [CrossRef]
18. Sotomayor, F.J.; Cychosz, K.A.; Thommes, M. Characterization of Micro/Mesoporous Materials by Physisorption: Concepts and Case Studies. *Acc. Mater. Surf. Res.* **2018**, *3*, 34–50.
19. Ambarita, H.; Kawai, H. Experimental study on solar-powered adsorption refrigeration cycle with activated alumina and activated carbon as adsorbent. *Case Stud. Therm. Eng.* **2016**, *7*, 36–46. [CrossRef]
20. Wang, X.; He, Z.; Chua, H.T. Performance simulation of multi-bed silica gel-water adsorption chillers. *Int. J. Refrig.* **2015**, *52*, 32–41. [CrossRef]
21. Sayfikar, M.; Behbahani-nia, A. Study of the performance of a solar adsorption cooling system. *Energy Equip. Syst.* **2013**, *1*, 75–90. [CrossRef]
22. Brites, G.; Costa, J.; Costa, V. Sustainable Refrigeration Based on The Solar Adsorption Cycle. Available online: https://inovenergy.inovcluster.pt/media/29412/SUSTAINABLE_REFRIGERATION_BASED_ON_THE_SOLAR_ADSORPTION_CYCLE.pdf (accessed on 14 April 2021).
23. Uyun, A.S.; Akisawa, A.; Miyazaki, T.; Ueda, Y.; Kashiwagi, T. Numerical analysis of an advanced three-bed mass recovery adsorption refrigeration cycle. *Appl. Therm. Eng.* **2009**, *29*, 2876–2884. [CrossRef]
24. Saha, B.B.; Koyama, S.; Lee, J.B.; Kuwahara, K.; Alam, K.C.A.; Hamamoto, Y.; Akiswa, A.; Kashiwagi, T. Performance evaluation of a low-temperature waste heat driven multi-bed adsorption chiller. *Int. J. Multiph. Flow* **2003**, *29*, 1249–1263. [CrossRef]
25. Sharafian, A.; Mehr, S.M.N.; Huttema, W.; Bahrami, M. Effects of different adsorber bed designs on in-situ water uptake rate measurements of AQSOA FAM-Z02 for vehicle air conditioning applications. *Appl. Therm. Eng.* **2016**, *98*, 568–574. [CrossRef]
26. Khanam, M.; Jribi, S.; Miyazaki, T.; Saha, B.B.; Koyama, S. Energy Analysis and Performance Evaluation of the Adsorption Refrigeration System. *Energies* **2018**, *11*, 1499. [CrossRef]
27. Sitorus, T.B.; Napitupulu, F.H.; Ambarita, H. A Study on Adsorption Refrigerator Driven by Solar Collector Using Indonesian Activated Carbon. *J. Eng. Technol. Sci.* **2017**, *49*, 657–670. [CrossRef]
28. Sharafian, A.; Dan, P.C.; Huttema, W.; Bahrami, M. Performance analysis of a novel expansion valve and control valves designed for a waste heat-driven two-adsorber bed adsorption cooling system. *Appl. Therm. Eng.* **2016**, *100*, 1119–1129. [CrossRef]

29. Myat, A.; Choon, N.K.; Thu, K.; Kim, Y.D. Experimental investigation on the optimal performance of Zeolite–water adsorption chiller. *Appl. Energy* **2013**, *102*, 582–590. [CrossRef]
30. Schawe, D. Theoretical and Experimental Investigations of an Adsorption Heat Pump with Heat Transfer between Two Adsorbers. Ph.D. Dissertation, Process and Biotechnology, Faculty of Energy, University of Stuttgart, Stuttgart, Germany, 2001. [CrossRef]
31. Trindade, M.V. Modelling and Optimization of an Adsorption Cooling System for Automotive Applications. Ph.D. Dissertation, Doctoral Program of Energy Technology, Universidad Politécnica De Valencia, Valencia, Spain, 2015. Available online: https://riunet.upv.es/handle/10251/54120 (accessed on 9 September 2022).
32. Sharma, P.; Song, J.S.; Han, M.; Cho, C.H. GIS-NaP1 zeolite microspheres as potential water adsorption material: Influence of initial silica concentration on adsorptive and physical/topological properties. *Sci. Rep.* **2016**, *6*, 22734. [CrossRef] [PubMed]

energies

MDPI

Article

Treatment of Liquid Fraction of Digestate by Integrated Process Struvite Precipitation—Forward Osmosis

Agnieszka Urbanowska [1], Izabela Polowczyk [2] and Małgorzata Kabsch-Korbutowicz [1,*]

[1] Department of Environment Protection Engineering, Faculty of Environmental Engineering, Wroclaw University of Science and Technology, Wybrzeze Wyspianskiego 27, 50-370 Wroclaw, Poland
[2] Technology of Polymer and Carbon Materials, Department of Process Engineering, Faculty of Chemistry, Wroclaw University of Science and Technology, Wybrzeze Wyspianskiego 27, 50-370 Wroclaw, Poland
* Correspondence: malgorzata.kabsch-korbutowicz@pwr.edu.pl

Abstract: The research undertaken in this paper was aimed at determining the effect of struvite precipitation, one of the potential products that can be obtained during digestate management, on the performance of the non-pressurized membrane process—forward osmosis (FO). The effect of using an integrated struvite precipitation—forward osmosis process to treat the digestate liquid on the changes in the properties of organic substances present in the treated solution (particle size distribution, ζ-potential) was analysed as well. The study was conducted for the liquid fraction of municipal waste biogas plant digestate. The obtained results demonstrate the suitability of this process for recovering water from liquid digestate. It was found that forward osmosis conducted for a digestate pre-treated by chemical struvite precipitation leads to higher water flux values and increased salt concentration in the receiving solution (from 0.5 to 3 mol/dm^3 NaCl). There is practically no concomitant infiltration of organic substances into the receiving solution. Therefore, the use of 3 mol/dm^3 NaCl as a draw solution results in the recovery of the highest volume of water per unit of time. An analysis of the particle size distribution shows that the removal of the macromolecular fraction of organic compounds from the digestate mainly takes place simultaneously with the chemical precipitation of struvite. It was found that an increase in the concentration of the draw solution, which allows for greater water recovery, resulted in the aggregation of the concentrated organic molecules.

Keywords: digestate; municipal waste biogas plant; struvite precipitation; forward osmosis; size distribution; zeta potential

Citation: Urbanowska, A.; Polowczyk, I.; Kabsch-Korbutowicz, M. Treatment of Liquid Fraction of Digestate by Integrated Process Struvite Precipitation—Forward Osmosis. *Energies* **2023**, *16*, 47. https://doi.org/10.3390/en16010047

Academic Editor: José Carlos Magalhães Pires

Received: 26 November 2022
Revised: 13 December 2022
Accepted: 15 December 2022
Published: 21 December 2022

1. Introduction

The operation of a biogas plant, regardless of type, is associated with the formation of both biogas and a large amount of digestate as one of the products of organic substance processing. Management of the resulting digestate is troublesome, which is why the technology of waste processing in biogas plants needs to be improved, along with both existing and developing methods of digestate management [1]. Because digestate is a fertilizer and a source of water, it is extremely important to choose the right method of purification to ensure that all valuable components can be recovered from the substance previously treated as a troublesome waste. This is very important from the point of view of implementing the circular economy.

Forward osmosis (FO), i.e., the diffusion of a solvent through a semi-permeable membrane separating two solutions of different concentrations, may be an alternative method of purifying the liquid fraction of the digestate to the processes currently in use. At atmospheric pressure, water passes through a synthetic membrane from the solution being treated to a concentrated receiving solution (the draw solution). The driving force of the process is produced naturally, and is the result of the difference in the osmotic pressure of the solutions on both sides of the membrane. The flow of water is spontaneous [2].

This process does not require external energy other than the energy associated with the circulation of solutions on both sides of the membrane, and continues until the osmotic pressures of the solutions on both sides of the membrane equalize. The efficiency of FO is highly dependent upon, among other things, the properties of the receiving solution. This solution should be characterized by high osmotic pressure at the lowest possible concentration, low viscosity, ease of recovery, and low purchase/production cost, and must not exhibit toxicity. The most common receiving solutions are [3]:

- gases dissolved in water (SO_2 or a mixture of NH_3 and CO_2)
- sugars (glucose, fructose, sucrose)
- inorganic salts (NaCl, $MgCl_2$, $CaCl_2$, $Al_2(SO_4)_3$)
- organic salts (Na^+ and Mg^{2+} salts of formic, acetic or propionic acid)
- hydrophilic magnetic nanoparticles.

FO uses non-porous asymmetric membranes manufactured from hydrophilic polymers. The most common are membranes made of cellulose triacetate or composite membranes containing a polyamide active layer [3]. Because the transport of water through the membrane does not require hydrostatic pressure, the membranes used in FO can be thinner than those used in reverse osmosis. To reduce flow resistance, the pores in the support layer should be low in tortuosity [4]. FO is increasingly used in many fields of environmental engineering, including those related to water and wastewater treatment [5–15]. While there are many reports on the subject in the literature, at present there is relatively little information on the direct use of this process to treat the liquid fraction of the digestate. As an example, only the research presented by Wu et al. [10] shows that it is possible to use FO for water recovery and struvite precipitation from agricultural digestate. In the experiments described therein, more than 50% of water recovery was achieved, along with a 99% removal of phosphate and 93% removal of ammonium nitrogen. In addition, the studies by Camilleri-Rumbau et al. [16] indicate the high potential of this process for purifying the liquid fraction of agricultural digestate. Regardless of the process conditions used, ammonium nitrogen removal was at the 95.5% level. These authors suggest that FO can be extremely useful for purifying the digestate, thus rendering it useful in agriculture. Similar conclusions can be found in other publications [5,7,8,17]. In the case of treating the liquid fraction of digestate from a biogas plant, the receiving solution can be a concentrate of fertilizer substances [18–20]. The water from the digestate permeating the membrane can be used to dilute this concentrate, as its original concentration may be too high for it to be used directly in agriculture. Another example might be seawater or groundwater with increased salinity, which might be made suitable for watering plants after dilution with water extracted from the digest [21,22].

While many studies have been conducted on the agricultural biogas plant digestate liquid fraction (e.g., [23–25]), there are few reports on the treatment of liquid municipal waste biogas plant digestate. Furthermore, despite the examples of various applications of the FO process cited earlier in the literature, to our knowledge there are no reports on the application of the integrated struvite precipitation—forced osmosis process for treatment of the digestate liquid fraction. The digestate liquid fraction is a mixture of many organic and inorganic components; most of these can be utilized, however, they may negatively affect the FO water recovery process. Thus, we decided to conduct a study to determine the effect of struvite precipitation (as one of the potential products obtainable during digestate management) on the performance of the membrane process. The effect of the struvite precipitation–FO integrated process on the changes in the organic matter properties present in the treated solution (particle size distribution, ζ-potential) was investigated, as was its suitability for treatment of the liquid fraction of digestate from municipal waste biogas plants.

2. Materials and Methods

The study was conducted for the liquid fraction of digestate from a biogas plant processing the organic fraction of municipal waste. This fraction was from one of the

Polish waste management plants located in the Lower Silesia province (50°53′15.5″ N 17°23′28.0″ E). It was separated from digestate pulp with sedimentation centrifuges. The characteristics of the test solution are shown in Table 1. Physico-chemical analysis of the solution was carried out according to the Standard Methods for the Examination of Water and Wastewater, 23rd Edition [26].

Table 1. Properties of the liquid fraction of municipal digestate.

Index	Value
pH	7.23
Temperature, °C	21
Conductivity, mS/cm	20.3
Dry residue, mg/dm^3	55,820
Alkalinity, mmol/dm^3	150
Total hardness, mval/dm^3	753
Chemical oxygen demand (COD), mg O_2/dm^3	11,450
5-days biochemical oxygen demand (BOD_5), mg O_2/dm^3	3600
Dissolved organic carbon (DOC), mg C/dm^3	4210
N-NH_4^+, mg/dm^3	776
N-NO_2^-, mg/dm^3	5.9
N-NO_3^-, mg/dm^3	below level of detection
PO_4^{3-}, mg/dm^3	21.2
Mg, mg/dm^3	235
Ca, mg/dm^3	420
K, mg/dm^3	3220
Li, mg/dm^3	7
P, mg/dm^3	21.4

The molar ratio of N:Mg:P in the test solution was 40.1:14.2:1; hence, there was a need to dose the Mg and P compounds in order to ensure the optimal proportion of these components for struvite precipitation. In the tests conducted, two chemical reactants were added to the digestate solution: $MgCl_2$ as magnesium supplement, and NaH_2PO_4 as phosphorus supplement. Their characteristics are shown in Table 2.

Table 2. Characterization of the chemical reagents used in struvite precipitation [27,28].

	Magnesium Chloride	Monosodium Phosphate
manufacturer	Chempur	Chempur
chemical formula	$MgCl_2$	NaH_2PO_4
molar mass, g/mol	95.211	119.98
form	solid	solid
color	colorless to white	white to colorless
odour	odourless	n.a.
pH	5–6.5 (5%, 20 °C)	4–4.5 (5%, 20 °C)
density, g/cm^3	1.57 (20 °C)	1.91 (20 °C)
solubility in water, g/dm^3	2430 (20 °C)	n.a.

The FTS H_2O membrane from Sterlitech Corporation made of cellulose triacetate with 140 cm^2 of effective filtration area was used to carry out FO. The manufacturer provides that the membrane can operate in temperatures up to 50 °C and within a pH range of 3–7 [29]. According to our own measurements, its wetting angle is 32.2°.

The integrated process involved a combination of struvite precipitation and FO. Experiments related to the precipitation of struvite from the liquid fraction of the digestate were carried out with parameters determined as being the most effective in earlier studies [30]. The process was carried out at pH 9.0. The sample of the digestate, with a volume of 500 cm^3, was set on a Velp Scientifica FC6S mechanical stirrer. $MgCl_2$ and NaH_2PO_4 were then dosed into the sample at 1.318 g/dm^3 and 2.084 g/dm^3, respectively. The doses of the reactants were set such that after taking into account the concentrations of Mg^{2+}, $N\text{-}NH_4^+$, and PO_4^{3-} in the test solution, the molar ratio of N:Mg:P reached 5.13:1:1. The temperature of the solution was 21 °C. After adding in the appropriate amount of reagents, the pH was adjusted to 9.0 with 0.1 mol/dm^3 NaOH. According to the literature, these properties (temperature and pH) promote struvite precipitation [31]. The samples were then stirred for 5 min at a rate of 160 rpm, followed by 30 min of sedimentation. The properties of the solution obtained after struvite precipitation are shown in Table 3.

Table 3. Properties of the liquid fraction of municipal digestate after struvite precipitation.

Index	Value
Conductivity, mS/cm	19.7
pH	9.0
COD, mg O_2/dm^3	10,360
BOD_5, mg O_2/dm^3	3510
DOC, mg C/dm^3	3990
$N\text{-}NH_4^+$, mg/dm^3	772
Mg, mg/dm^3	23.1
Ca, mg/dm^3	42.3
K, mg/dm^3	2970
Li, mg/dm^3	6.9
P, mg/dm^3	1.9

The pretreated liquid (indicated in the text as solution A) was then subjected to FO for 180 min. Experiments were performed using a laboratory SEPA FO CELL plant from Sterlitech Corporation (Figure 1), with a 0.014 m^2 membrane with an active layer facing the feed solution. The installation allows for cross-flow separation tests, with a flow rate in the range of 10–100 dm^3/d and a maximum operating pressure of 6.9 MPa. Its main elements include a SEPA FO cell (active surface dimensions: 0.97 mm × 14.7 cm × 9.53 cm), a feed solution tank, a draw solution tank, two peristaltic pumps (Gear Pump Drive by Masterflex) with variable flow rate and a pressure of 7.6 MPa that allows fluid circulation in the system, and two digital balances (Isolab by Archem) for monitoring weight changes of the feed and draw solution. The minimum volume of each tank was equal to 3.7 dm^3. The cross-flow velocity of fluid in the system was 0.2 m/s. In the experiments, NaCl solutions of 0.5, 1, and 3 mol/dm^3 were used as receiving (draw) solutions. During the process, chloride concentration, conductivity, and DOC concentration were measured in the feed solution and in the receiving solution.

All treatment experiments were duplicated.

Particle size distribution was measured by a Mastersizer 2000 laser diffractometer (Malvern Panalytical Ltd., Malvern, UK) equipped with a HydroMu dispersion device (Malvern Panalytical Ltd., Malvern, UK) with a particle size measurement range of 0.1–2000 µm. During measurement, depending on the concentration of solid particles, about 3 cm^3 of the solution was added into a 700 cm^3 beaker filled with water circulating in the measuring cell. Particle size distribution was measured without ultrasound (the suspension was circulated through the measuring cell, with no ultrasound generated), followed by ultrasound sonication of the suspension (this took place in a beaker from which the suspension was pumped and circulated through the measuring cell) until the particle size distribution stabilized (i.e., until the disintegration of any agglomerates). Particle size distributions were determined by a Nicomp 380 DLS apparatus (Nicomp Particle Sizing Systems, Santa Barbara, CA, USA), in which the dynamic light scattering (DLS) method is used to study particle size distributions in samples with particle sizes of 1 nm–5 µm [32]. The measurement was carried out in about 3.5 cm^3 of diluted suspension placed in a chamber. Using the Nicomp analysis algorithm, complex multimodal distributions were analyzed with the highest resolution and repeatability.

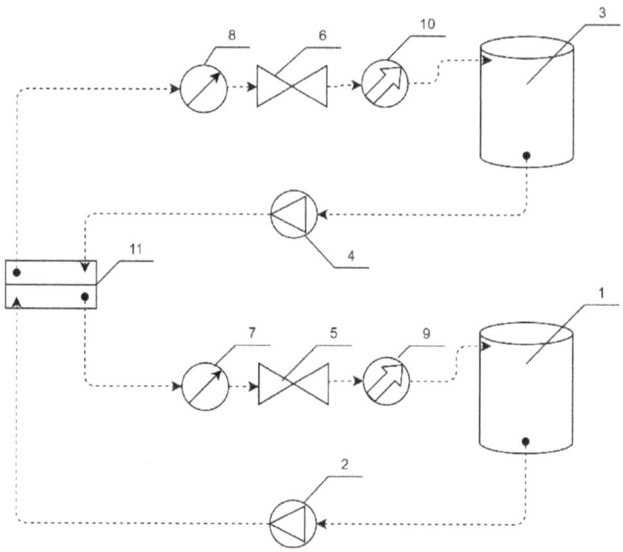

Figure 1. Diagram of the SEPA FO CELL installation (1—feed solution tank, 2—feed solution pump, 3—draw solution tank, 4—draw solution pump, 5—concentrate control valve, 6—draw solution control valve, 7—concentrate pressure gauge, 8—draw solution pressure gauge, 9—concentrate flow meter, 10—draw solution flow meter, 11—SEPA FO cell).

The ζ-potential was measured at 25 °C with a ζ-potential analyzer (Zetasizer 2000, Malvern Panalytical Ltd., Malvern, UK). The diluted test solution was conditioned in a beaker for 10 min at the specified pH. The pH was adjusted with sodium hydroxide or hydrochloric acid. Then, the test solution was placed in an electrophoretic chamber by means of a syringe. The value of the ζ-potential was determined as the average of five successive measurements.

3. Results

The preliminary studies were aimed at determining the feasibility of precipitating struvite from the liquid fraction of municipal-derived digestate. The research was intended to find a way to both improve the final quality of the test solution and reduce the fouling of membranes applied in the subsequent purification step. Figure 2 shows the removal efficiency of selected components from samples of the liquid fraction of the digestate in which struvite was precipitated by employing $MgCl_2$ salt (which is an external source of magnesium) and NaH_2PO_4 (which is an internal source of phosphorus). The adopted molar ratio of N:Mg:P of 5.13:1:1 brought about a low degree of organic compound removal. COD, BOD_5, and DOC concentrations in the pretreated digestated (supernatant) decreased by a maximum of 13% compared to the concentrations in the initial sample. The removal efficiency of N-NH_4^+ was 48%. The resulting effect may be partly due to the desorption of gaseous NH_3, which at pH 9 can account for about 30% of the ammonium nitrogen in solution. The efficiency of magnesium and calcium ion precipitation from the digestate was about 91%, while that of potassium was about 14% as compared to the concentration in the initial sample. Negligible removal of Li occurred, along with an increase in the concentration of Na ions, which is completely understandable due to the application of a phosphorus source in the form of NaH_2PO_4 to the treated digestate. On the other hand, the concentrations of phosphorus ions in the treated sample of digestate (84%) decreased significantly. The adopted conditions are likely to have favored the precipitation of this element in the crystalline form of potassium struvite ($MgKPO_4 \cdot 6\,H_2O$).

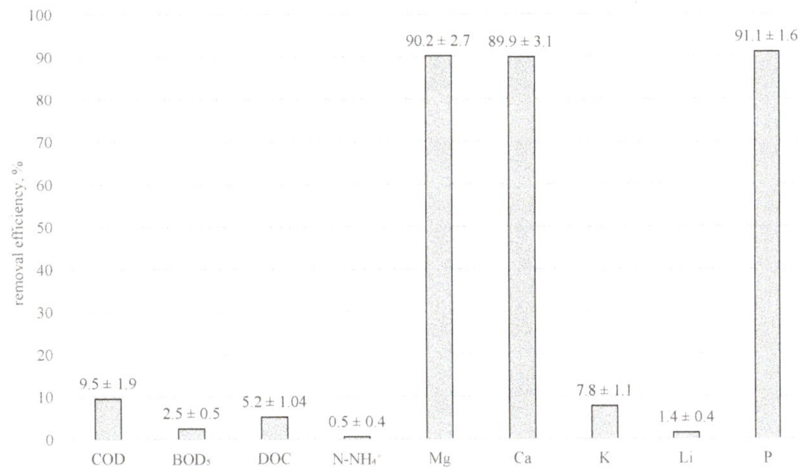

Figure 2. Removal efficiency of selected contaminants in the analyzed sample during struvite precipitation.

An analysis of the precipitates performed with a scanning electron microscope (Figure 3) confirmed that crystalline forms of struvite were separated from the solution of the liquid fraction of the digestate as a result of chemical precipitation. The pictures show a large amount of organic matter and a large number of microorganisms.

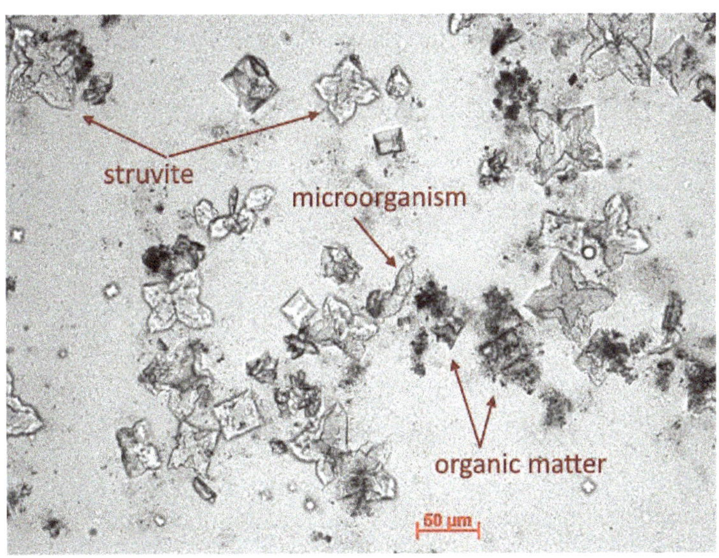

Figure 3. SEM images of sediment after struvite precipitation at a molar ratio of 5.13:1:1 from municipal liquid digestate.

Studies on the purification of the liquid fraction of digestate by the integrated process of struvite precipitation—FO were begun by determining the effect of the duration of the process and the draw solution osmotic pressure (Table 4) on the flux of water permeating the membrane from the solution of liquid digestate after struvite precipitation.

Table 4. Properties of draw solutions.

NaCl Concentration	Osmotic Pressure	
[mol/dm^3]	[MPa]	[atm]
0.5	2.478	24.46
1	4.985	48.93
3	14.87	146.8

Figures 4–6 present the results of experiments conducted with redistilled water as a feed solution. Redistilled water can serve as a reference point for the results, as it has negligibly low osmotic pressure and does not show the presence of any contaminants. It was observed (Figure 4) that as the concentration of NaCl in the draw solution decreased, the water flux declined. This was a result of the decreasing value of the driving force of the process. In the case of FO of liquid digestate, similar trends were observed as for redistilled water, with the flux values, however, being much lower than those measured for water. This was due to the fact that the difference in osmotic pressures on both sides of the membrane was smaller for the digestate solution/receiving solution combination than for the redistilled water/receiving solution system. Analyzing the effect of process duration on the change in water flux, it was noted that it successively decreases. This is due to the decreasing osmotic pressure of the receiving solution and the progressive blockage of the membrane. This effect was found regardless of the concentration of NaCl in the receiving solution. In this respect, the results are consistent with the literature reports [33–35]. This was due to the fact that the water diffused through the membrane, the receiving solution diluted, and the water flux decreased as a result of the decreasing value of the driving force of the process, which is the difference in osmotic pressures between the solutions on either side of the membrane.

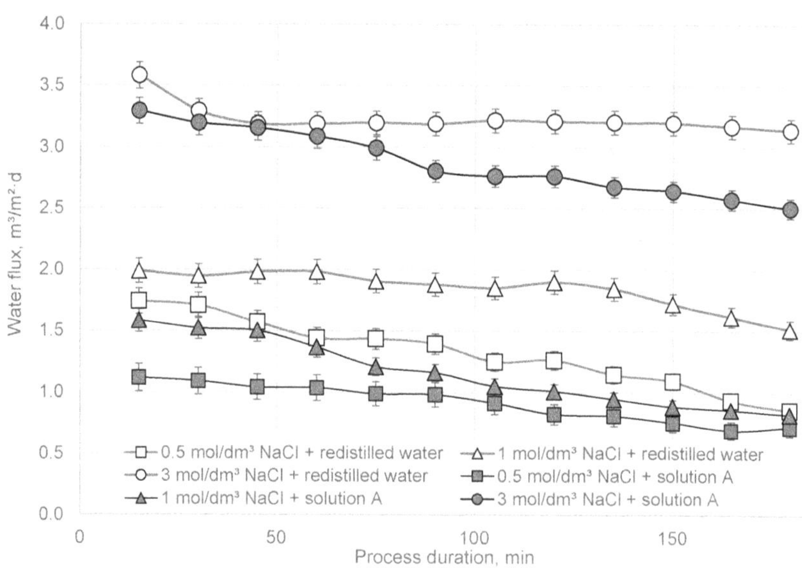

Figure 4. Water flux time dependence of FO for different concentrations of NaCl in the receiving solution for redistilled water and solution of the digestate after struvite precipitation.

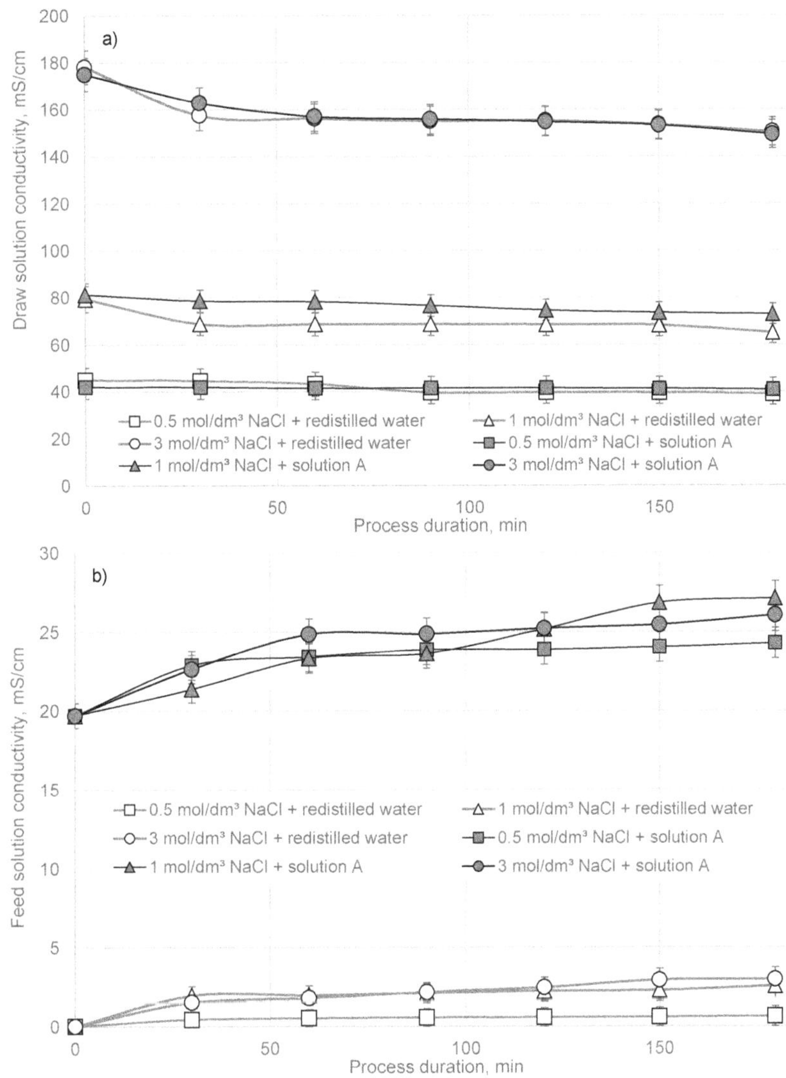

Figure 5. Changes of the conductivity in: (**a**) the draw solution and (**b**) the feed solution with the FO process duration.

Figures 5 and 6 show the changes in conductivity and chloride concentration in the feed and receiving solutions, respectively, as a function of time. The conductivity in the feed solution increased slightly with process time, from 28.9 mS/cm to 34.2 mS/cm on average (in the case of digestate solution after struvite precipitation) and from 0 to 1.9 mS/cm (in the case of redistilled water) due to the reverse permeation of salts through the membrane. The literature [36] indicates that the membrane used in FO with an average pore radius of 0.37 nm can retain most multivalent ions and organic pollutants. However, salt ions (a Na^+ ion with a hydration radius of 0.36 nm and a Cl^- ion of 0.33 nm) [37] were able to penetrate the FO membrane, a phenomenon known as reverse salt transport. In the draw solution, on the other hand, the conductivity decreased slightly over time as a result of dilution. Similar trends were observed when analyzing chloride concentrations in both solutions. Over time, the flow of water through the membrane caused dilution of the receiving solution.

Concurrently, a small amount of salt permeated through the membrane in the opposite direction (from the draw solution to the feed solution) due to diffusion (for the redistilled water) and concentration (for the digestion fluid), causing the salt concentration in the feed solution to increase on average down to 625 g/m^3 and from 4850 g/m^3 to 6820 g/m^3, respectively. Similar conclusions can be found in other works [38–41].

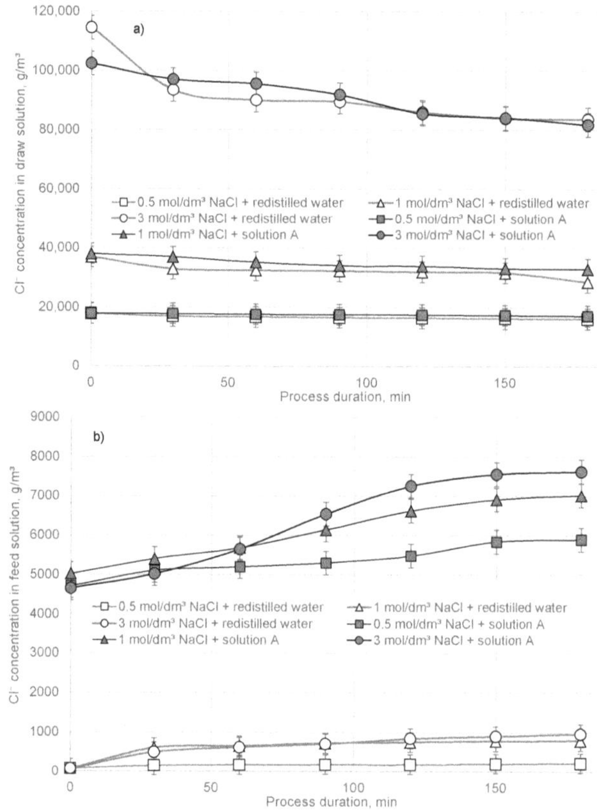

Figure 6. Changes of chloride concentration in: (**a**) the draw solution and (**b**) the feed solution with the FO process duration.

Due to the nature of FO processes, it was reasonable to check whether pollutants present in the liquid fraction of the digestate after struvite precipitation, which is the feed solution in the ongoing process, can permeate through the membrane into the receiving solution (NaCl solution). This could negatively affect the subsequent regeneration of the receiving solution and water recovery. The results presented in Figure 7 clearly show low penetration of organic substances (<0.1%). The amount of organic matter as determined by DOC concentration and colour increased to a non-significant degree with the duration of FO and as the initial concentration of NaCl in the draw solution increased. For example, after 180 min of the process with a brine solution of 3 mol/dm^3 NaCl as the receiving solution, the DOC concentration was 2.3 g C/m^3 due to the permeation of organic compounds and the colour intensity was 5 g Pt/m^3. COD and BOD_5 were determined in the draw solution; however, because the values were at a very low level they are not presented in the figure. The results show that the FO membrane is an effective barrier to the transport of organic compounds in the feed solution–receiving solution direction.

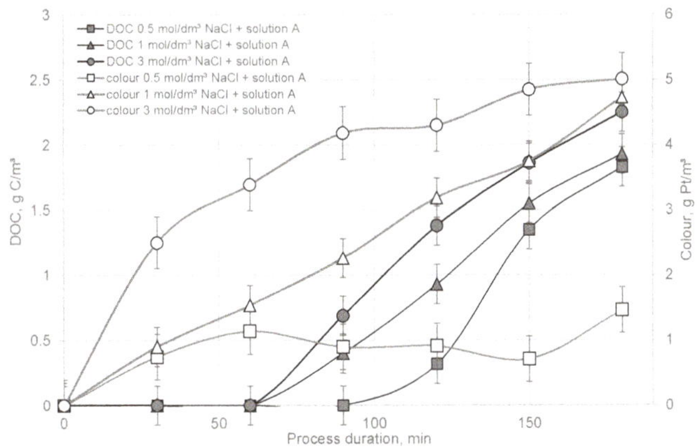

Figure 7. Changes in DOC concentration and colour intensity in the draw solution with the FO process duration.

Because the liquid fraction of the digestate in which struvite precipitation had previously been carried out was subjected to FO as a protection against excessive membrane fouling, and samples of the purified liquid (FO feed) were analyzed for particle size distribution and changes in ζ-potential. The visual effect of the purification of municipal digestion liquid at various stages, including by FO after struvite precipitation, can be observed in Figure 8. As can be seen, the precipitation of struvite, with which part of the organic matter (including compounds responsible for colour intensity), was separated from the digestate solution, resulting in a decrease in colour intensity. FO used in the next step to recover water affected the colour changes of the feed solution. The solution obtained by applying 3 mol/dm^3 NaCl as a receiving solution had the highest colour intensity after FO. As shown in Figure 4, the use of 3 mol/dm^3 NaCl as the draw solution produces the highest flux of water and retrieves the largest amount of water, which results in the greatest concentration of coloured substances in the feed solution. The lower colour intensity as compared to the solution of the digestate treated with struvite precipitation of the liquid after FO carried out with 0.5 and 1 mol/dm^3 NaCl as the draw solution could be the result, among other things, of a change in the chromophoric properties of the organic matter due to the penetration of salt into the feed solution.

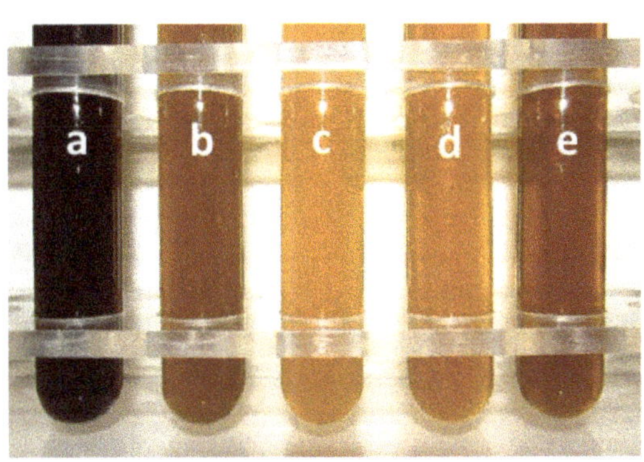

Figure 8. Visual changes in the quality of the liquid fraction of the digestate after successive stages of purification (a—liquid raw digestate; b—solution A (after struvite precipitation); c—solution A after FO with 0.5 mol/dm^3 NaCl; d—solution A after FO with 1 mol/dm^3 NaCl; e—solution A after FO with 3 mol/dm^3 NaCl).

In addition, particle size changes were analyzed during integrated struvite precipitation/FO. As shown in Figure 9, pretreatment of the digest by chemical precipitation of struvite resulted in a significant reduction in the mean diameter of the particles present in the tested solution. This confirms early observations that macromolecules and colloids of the largest size are separated from the solution along with struvite. Measurements of particle size distribution taken by means of the DLS method show that the finest particles (about 675 nm) are present in the sample after FO, when a 0.5 mol/dm^3 NaCl solution was applied as the receiving solution. The increase in the driving force of the process (higher concentration of NaCl in the receiving solution) resulted in an increase in the particle size in the feed solution after FO (to about 807 nm and 1113 nm for 1 and 3 mol/dm^3 NaCl, respectively). This could have been a result of the concentration of the feed solution indicated earlier and the formation of larger particles due to their bridging by salts permeating from the feed solution.

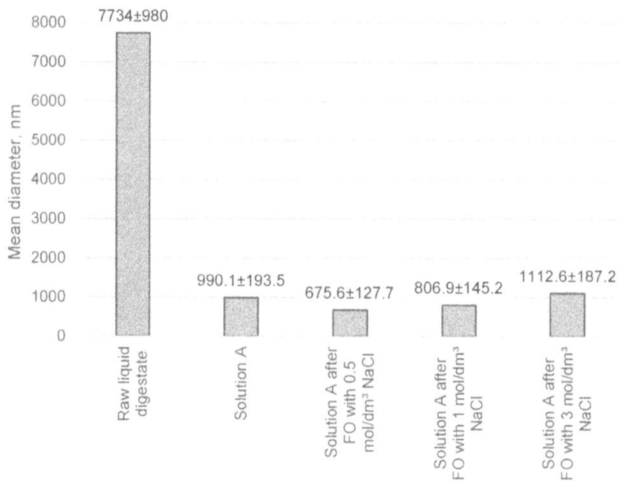

Figure 9. Average particle diameter in the tested solutions after struvite precipitation and FO.

The results of particle size distribution analyses based on the laser diffraction method are shown in Figures 10 and 11. This method is employed under conditions of continuous mixing and pumping of the dilute dispersion through the measuring cell. In this case, there is no possibility of sedimentation of larger particles, which is otherwise common in dynamic light scattering measurements, where the sample is placed in a cuvette and larger particles settle on the bottom. For the raw digestate, lower and upper deciles of about 1 μm and 34 μm were obtained, respectively, and half of the particle population had diameters of up to about 7.7 μm. Sonication of the sample during measurement of the particle size distribution of the raw digestate allowed the individual diameters (d_{10}, d_{50} and d_{90}) to be reduced to 0.8, 4.6 and 23 μm, respectively (Figure 10). The analysis of particle size distribution in solution A (Figure 11) revealed that the diameters of d_{50} and d_{90} decreased to 2 μm and 19 μm, respectively, compared to the values obtained for the raw digestate. In contrast, the lower decile remained virtually unchanged (about 0.9 μm). The use of an integrated process combining struvite precipitation with FO did not practically change the particle size distribution in terms of the finest and medium fractions. The only noticeable differences were in the share of the largest particles. The results confirm that the digestate particles form agglomerates that can be easily broken up.

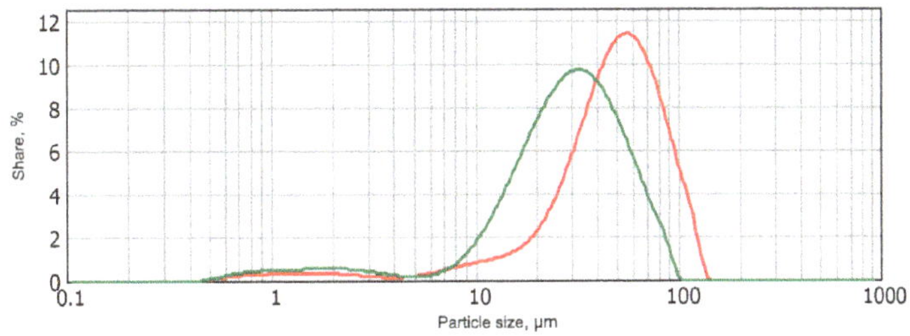

Figure 10. Raw digestate liquid fraction particle size distribution (red line—without sonication, green line—with sample sonication).

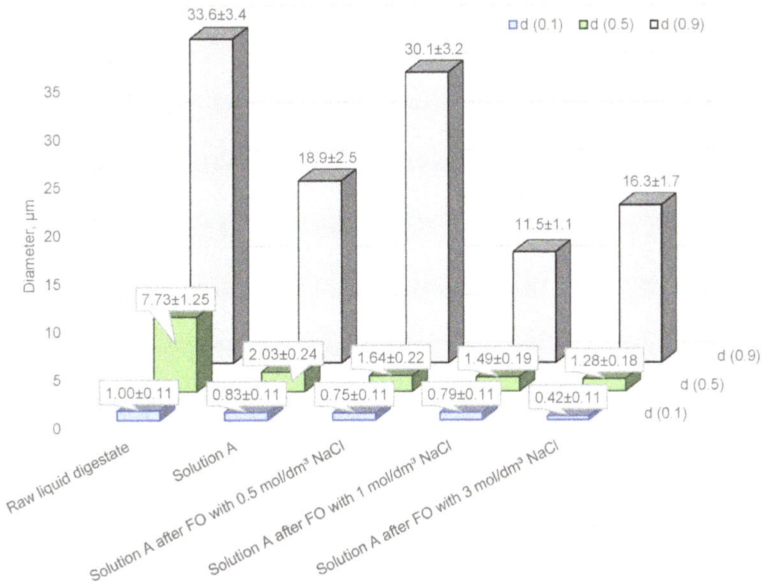

Figure 11. Particle size distribution for the liquid raw digestate, solution A (digestate after struvite precipitation) and the feed solution (solution A after FO).

In the course of the study, the isoelectric point values of the particles were determined in the liquid fraction of digestate, in the digestate after struvite precipitation, and in the samples after integrated struvite precipitation/FO. This may be important from the point of view of conducting the integrated process, as at the isoelectric point (pH_{IEP}) the particles have the lowest solubility, the smallest viscosity, and the smallest osmotic pressure. Measurements of the ζ potential in the pH range from 1.2 to 12.7 allowed the pH_{IEP} to be determined. For all samples tested, the value was about 1.9 (Figure 12). Above pH_{IEP}, the value of the ζ potential was negative, reaching the lowest value of −33 mV for samples of the digestate solution subjected to struvite precipitation and FO using NaCl solutions of 1 mol/dm^3 and 3 mol/dm^3. In general, the course of the correlation of ζ potential with pH did not differ significantly between samples; only the raw digestate and the sample after struvite precipitation showed lower values of ζ potential at the same pH. This may indicate that despite the higher ionic strength in the samples after FO with electrolyte (NaCl), the potential at the slip boundary is higher.

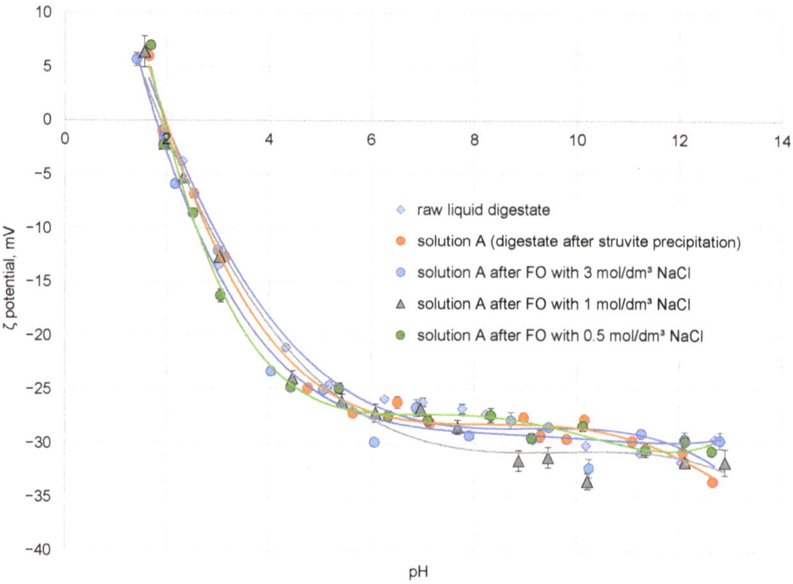

Figure 12. ζ potential of raw liquid digestate, digestate after struvite precipitation, and after FO with various NaCl concentrations in draw solution.

The results obtained with the struvite precipitation—FO integrated process indicate that it can be effectively used to treat the municipal digestate liquid fraction. The water recovered from the digestate (after removal of excess salt or dilution) can be used agriculturally, e.g., for crop watering or fertiliser preparation. However, it should be kept in mind that the efficiency of water recovery from the liquid fraction of the digestate preceded by struvite precipitation is determined by the difference in electrochemical potential between the feed solution and the receiving solution. As shown earlier, an increase in the salt concentration in the draw solution causes a more intense reverse flow of salt into the liquid to be concentrated, which adversely affects the properties of the liquid via increased salinity. This may be undesirable, for example, when trying to use the concentrated digestate as an organic fertilizer.

The proposed technological solution for the management of digestate, i.e., preceding the water recovery stage using forward osmosis with struvite precipitation, fits the trend of research aimed at promoting a closed-loop economy in the waste management sector. Such an approach to the management of digestate has been presented, for example, by Jurgutis et al. [42], who showed that digestate used as biofertilizer can affect soil properties and plant growth. However, the work carried out thus far on incorporating the management of digestate into the idea of a circular economy [43] has been limited mainly to the use of its fertilizing properties (source of organic matter and certain elements needed for plant growth). The proposed approach to processing the liquid fraction of the digestate makes it possible to take advantage of the properties of this waste stream and allows for the recovery of water, which is an extremely valuable and limited resource these days. Another advantage of the proposed technology is that it requires much less energy consumption than alternative processes used to produce fertilizers (i.e., thermal or vacuum processes [44]) or treat the liquid fraction of the digestate (i.e., pressure-driven membrane processes [45]. It should be borne in mind that in previous work, in an effort to close the circularity of matter, the precipitation of struvite, for use as a mineral fertilizer was carried out during the management/treatment of wastewater [46,47], while in our study [30] this fertilizer was obtained from the liquid fraction of digestate, which shows the potential of this liquid in terms of promoting a circular economy.

4. Conclusions

The present study showed that the liquid fraction of the digestate generated during the treatment of the organic fraction of municipal waste, when further processed using an integrated process involving struvite precipitation and forward osmosis, can allow for the obtaining of very valuable products such as struvite, concentrated organic matter, and water. Any of these products can be used in agriculture; with natural, mineral and organic materials in decline, as well as limited water resources, all sectors of the economy, including waste management, are seeking to introduce the concept of a circular economy, in which that existing waste products becomes a source of valuable raw materials. The precipitated struvite and concentrated organic matter can be used as fertilizers, while the water can be utilized to irrigate plant crops.

Studies on the effectiveness of digestate purification carried out by struvite precipitation and FO have demonstrated the usefulness of this process for recovering water and valuable components from digestate obtained from municipal waste processing. The research presented in this thesis leads to the following conclusions:

- Struvite precipitation allowed for organic compounds removal of 2.5–9.5% and removal of Mg/Ca/P up to ca. 90%.
- A novel concept of struvite precipitation from liquid digestate applied at the solution pretreatment stage before FO allowed us to obtain a final solution of significant quality (DOC removed by 99.9%) and protected the membranes from excessive blockage.
- FO performed on a stream of digestate pretreated by chemical precipitation of struvite yielded higher values of water flux, as the concentration of salt in the receiving solution increased and there was practically no infiltration of organic substances into the draw solution.
- The removal of the macromolecular fraction of organic compounds from the digestate took place mainly simultaneously with the chemical precipitation of struvite.
- An increase in the salt concentration of the draw solution, which allowed for a greater water recovery, resulted in the aggregation of the concentrated organic molecules in the feed solution.
- The correlation of the ζ potential with pH did not change when comparing raw digestate and concentrated samples (after process performance); pH_{IEP} was about 1.9.

Author Contributions: Conceptualization, A.U.; methodology, A.U.; validation, A.U. and M.K.-K.; formal analysis, A.U.; investigation, A.U. and I.P.; resources, A.U. and M.K.-K.; data curation, A.U.; writing—original draft preparation, A.U. and I.P.; writing—review and editing, M.K.-K.; visualization, A.U. and I.P.; supervision, M.K.-K.; project administration, A.U.; funding acquisition, A.U. and M.K.-K. All authors have read and agreed to the published version of the manuscript.

Funding: This research was funded by the National Centre for Research and Development (Poland), WATERWORKS2017/I/RECOWATDIG/01/2019.

Data Availability Statement: The data that support the findings of this study are available from the corresponding author, upon reasonable request.

Acknowledgments: Research was conducted as part of the interdisciplinary project: Sustainable technology for the staged recovery of an agricultural water from high moisture fermentation products (RECOWATDIG) carried out under the Water Joint Programming Initiative co-financed by 20 Funding Partner Organizations from 18 countries and with the support of the European Commission (full list: http://www.waterjpi.eu/images/documents/JC2018/waterjpi2018joint_call_participating_countries.pdf, accessed on 14 December 2022). I.P. is grateful for the financial support given by the Polish Ministry of Science and Higher Education by subvention activity for the Faculty of Chemistry at Wrocław University of Science and Technology.

Conflicts of Interest: The authors declare no conflict of interest. The funders had no role in the design of the study; in the collection, analyses, or interpretation of data; in the writing of the manuscript, or in the decision to publish the results.

References

1. Fuchs, W.; Drosg, B. Assessment of the State of the Art of Technologies for the Processing of Digestate Residue from Anaerobic Digesters. *Water Sci. Technol.* **2013**, *67*, 1984–1993. [CrossRef] [PubMed]
2. Kabsch-Korbutowicz, M. Zastosowanie procesu wymuszonej osmozy do odsalania i odnowy wody. *Ochr. Srodowiska* **2016**, *38*, 9–14.
3. Kucera, J. *Desalination*; Kucera, J., Ed.; John Wiley & Sons, Inc.: Hoboken, NJ, USA, 2014; Volume 9781118208526, ISBN 9781118904855.
4. ForwardOsmosisTech Guide to Forward Osmosis Membranes. Available online: www.forwardosmosistech.com/forward-osmosis-membranes (accessed on 17 July 2021).
5. Li, L.; Shi, W.; Yu, S. Research on Forward Osmosis Membrane Technology Still Needs Improvement in Water Recovery and Wastewater Treatment. *Water* **2019**, *12*, 107. [CrossRef]
6. Korenak, J.; Hélix-Nielsen, C.; Bukšek, H.; Petrinić, I. Efficiency and Economic Feasibility of Forward Osmosis in Textile Wastewater Treatment. *J. Clean. Prod.* **2019**, *210*, 1483–1495. [CrossRef]
7. Jafarinejad, S. Forward Osmosis Membrane Technology for Nutrient Removal/Recovery from Wastewater: Recent Advances, Proposed Designs, and Future Directions. *Chemosphere* **2021**, *263*, 128116. [CrossRef]
8. Corzo, B.; de la Torre, T.; Sans, C.; Escorihuela, R.; Navea, S.; Malfeito, J.J. Long-Term Evaluation of a Forward Osmosis-Nanofiltration Demonstration Plant for Wastewater Reuse in Agriculture. *Chem. Eng. J.* **2018**, *338*, 383–391. [CrossRef]
9. Ang, W.L.; Mohammad, A.W.; Johnson, D.; Hilal, N. Unlocking the Application Potential of Forward Osmosis through Integrated/Hybrid Process. *Sci. Total Environ.* **2020**, *706*, 136047. [CrossRef]
10. Wu, Z.; Zou, S.; Zhang, B.; Wang, L.; He, Z. Forward Osmosis Promoted In-Situ Formation of Struvite with Simultaneous Water Recovery from Digested Swine Wastewater. *Chem. Eng. J.* **2018**, *342*, 274–280. [CrossRef]
11. Li, Y.; Xu, Z.; Xie, M.; Zhang, B.; Li, G.; Luo, W. Resource Recovery from Digested Manure Centrate: Comparison between Conventional and Aquaporin Thin-Film Composite Forward Osmosis Membranes. *J. Membr. Sci.* **2020**, *593*, 117436. [CrossRef]
12. Ezugbe, E.; Kweinor Tetteh, E.; Rathilal, S.; Asante-Sackey, D.; Amo-Duodu, G. Desalination of Municipal Wastewater Using Forward Osmosis. *Membranes* **2021**, *11*, 119. [CrossRef]
13. Jashrapuria, K.; Singh, S.P. Forward Osmosis in Desalination and Wastewater Treatment. In *Pollution Control Technologies. Energy, Environment, and Sustainability*; Singh, S.P., Rathinam, K., Gupta, T., Agarwal, A.K., Eds.; Springer: Singapore, 2021; pp. 157–175.
14. Minier-Matar, J.; Hussain, A.; Janson, A.; Wang, R.; Fane, A.G.; Adham, S. Application of Forward Osmosis for Reducing Volume of Produced/Process Water from Oil and Gas Operations. *Desalination* **2015**, *376*, 1–8. [CrossRef]
15. Ang, W.L.; Wahab Mohammad, A.; Johnson, D.; Hilal, N. Forward Osmosis Research Trends in Desalination and Wastewater Treatment: A Review of Research Trends over the Past Decade. *J. Water Process. Eng.* **2019**, *31*, 100886. [CrossRef]
16. Camilleri-Rumbau, M.S.; Soler-Cabezas, J.L.; Christensen, K.V.; Norddahl, B.; Mendoza-Roca, J.A.; Vincent-Vela, M.C. Application of Aquaporin-Based Forward Osmosis Membranes for Processing of Digestate Liquid Fractions. *Chem. Eng. J.* **2019**, *371*, 583–592. [CrossRef]
17. Shi, L.; Simplicio, W.S.; Wu, G.; Hu, Z.; Hu, H.; Zhan, X. Nutrient Recovery from Digestate of Anaerobic Digestion of Livestock Manure: A Review. *Curr. Pollut. Rep.* **2018**, *4*, 74–83. [CrossRef]
18. Awad, A.M.; Jalab, R.; Minier-Matar, J.; Adham, S.; Nasser, M.S.; Judd, S.J. The Status of Forward Osmosis Technology Implementation. *Desalination* **2019**, *461*, 10–21. [CrossRef]
19. Chekli, L.; Kim, Y.; Phuntsho, S.; Li, S.; Ghaffour, N.; Leiknes, T.; Shon, H.K. Evaluation of Fertilizer-Drawn Forward Osmosis for Sustainable Agriculture and Water Reuse in Arid Regions. *J. Environ. Manag.* **2017**, *187*, 137–145. [CrossRef]
20. Abdul Wahid, R.; Ang, W.L.; Mohammad, A.W.; Johnson, D.J.; Hilal, N. Evaluating Fertilizer-Drawn Forward Osmosis Performance in Treating Anaerobic Palm Oil Mill Effluent. *Membranes* **2021**, *11*, 566. [CrossRef]
21. Nasr, P.; Sewilam, H. The Potential of Groundwater Desalination Using Forward Osmosis for Irrigation in Egypt. *Clean Technol. Environ. Policy* **2015**, *17*, 1883–1895. [CrossRef]
22. Cath, T.Y.; Hancock, N.T.; Lundin, C.D.; Hoppe-Jones, C.; Drewes, J.E. A Multi-Barrier Osmotic Dilution Process for Simultaneous Desalination and Purification of Impaired Water. *J. Membr. Sci.* **2010**, *362*, 417–426. [CrossRef]
23. Durdević, D.; Hulenić, I. Anaerobic Digestate Treatment Selection Model for Biogas Plant Costs and Emissions Reduction. *Processes* **2020**, *8*, 142. [CrossRef]
24. Monfet, E.; Aubry, G.; Ramirez, A.A. Nutrient Removal and Recovery from Digestate: A Review of the Technology. *Biofuels* **2018**, *9*, 247–262. [CrossRef]
25. Szymańska, M.; Szara, E.; Sosulski, T.; Wąs, A.; Van Pruissen, G.W.P.; Cornelissen, R.L.; Borowik, M.; Konkol, M. A Bio-Refinery Concept for n and p Recovery—A Chance for Biogas Plant Development. *Energies* **2019**, *12*, 155. [CrossRef]
26. Baird, R.; Eaton, A.D.; Rice, E.W.; Bridgewater, L.; Federation, W.E. *Standard Methods for the Examination of Water and Wastewater*, 23rd ed.; American Public Health Association: Washington, DC, USA, 2017; ISBN 9780875532875.
27. Chempur Karta Charakterystyki Substancji Chemicznej—Sodu Diwodorofosforan Bezwodny. Available online: http://chempur.pl/pliki/karty_charakterystyk/sodu_fosforan_I_bezwodny.pdf (accessed on 16 August 2021).
28. Chempur Karta Charakterystyki Substancji Chemicznej—Chlorek Magnezu. Available online: http://chempur.pl/pliki/karty_charakterystyk/magnezu_chlorek_6h.pdf (accessed on 16 August 2021).

29. Sterlitech Forward Osmosis (FO) Membranes. Available online: https://www.sterlitech.com/forward-osmosis-membranes.html (accessed on 22 June 2022).
30. Urbanowska, A.; Polowczyk, I.; Kabsch-Korbutowicz, M. Struvite Precipitation from the Liquid Fraction of the Digestate of a Municipal Waste Biogas Plant. *Environ. Prot. Eng.* **2021**, *47*. [CrossRef]
31. Cerrillo, M.; Palatsi, J.; Comas, J.; Vicens, J.; Bonmatí, A. Struvite Precipitation as a Technology to Be Integrated in a Manure Anaerobic Digestion Treatment Plant—Removal Efficiency, Crystal Characterization and Agricultural Assessment. *J. Chem. Technol. Biotechnol.* **2015**, *90*, 1135–1143. [CrossRef]
32. Wang, Z.-H.; Liu, X.-Y.; Zhang, H.-Q.; Wang, Y.; Xu, Y.-F.; Peng, B.-L.; Liu, Y. Modeling of Kinetic Characteristics of Alkaline-Surfactant-Polymer-Strengthened Foams Decay under Ultrasonic Standing Wave. *Pet. Sci.* **2022**, *19*, 1825–1839. [CrossRef]
33. Holloway, R.; Childress, A.; Dennett, K.; Cath, T. Forward Osmosis for Concentration of Anaerobic Digester Centrate. *Water Res.* **2007**, *41*, 4005–4014. [CrossRef] [PubMed]
34. Wang, L.; Li, T.; Chu, H.; Zhang, W.; Huang, W.; Dong, B.; Wu, D.; Chen, F. Natural Organic Matter Separation by Forward Osmosis: Performance and Mechanisms. *Water Res.* **2021**, *191*, 116829. [CrossRef]
35. Almoalimi, K.; Liu, Y.-Q. Fouling and Cleaning of Thin Film Composite Forward Osmosis Membrane Treating Municipal Wastewater for Resource Recovery. *Chemosphere* **2022**, *288*, 132507. [CrossRef]
36. Xie, M.; Nghiem, L.D.; Price, W.E.; Elimelech, M. Comparison of the Removal of Hydrophobic Trace Organic Contaminants by Forward Osmosis and Reverse Osmosis. *Water Res.* **2012**, *46*, 2683–2692. [CrossRef]
37. Israelachvili, J.N. *Intermolecular and Surface Forces*; Elsevier: Amsterdam, The Netherlands, 2011; ISBN 9780123919274.
38. Liu, X.; Wu, J.; Liu, C.; Wang, J. Removal of Cobalt Ions from Aqueous Solution by Forward Osmosis. *Sep. Purif. Technol.* **2017**, *177*, 8–20. [CrossRef]
39. Eddouibi, J.; Abderafi, S.; Vaudreuil, S.; Bounahmidi, T. Water Desalination by Forward Osmosis: Dynamic Performance Assessment and Experimental Validation Using MgCl2 and NaCl as Draw Solutes. *Comput. Chem. Eng.* **2021**, *149*, 107313. [CrossRef]
40. Phillip, W.A.; Yong, J.S.; Elimelech, M. Reverse Draw Solute Permeation in Forward Osmosis: Modeling and Experiments. *Environ. Sci. Technol.* **2010**, *44*, 5170–5176. [CrossRef]
41. Hickenbottom, K.L.; Hancock, N.T.; Hutchings, N.R.; Appleton, E.W.; Beaudry, E.G.; Xu, P.; Cath, T.Y. Forward Osmosis Treatment of Drilling Mud and Fracturing Wastewater from Oil and Gas Operations. *Desalination* **2013**, *312*, 60–66. [CrossRef]
42. Jurgutis, L.; Šlepetienė, A.; Šlepetys, J.; Cesevičienė, J. Towards a Full Circular Economy in Biogas Plants: Sustainable Management of Digestate for Growing Biomass Feedstocks and Use as Biofertilizer. *Energies* **2021**, *14*, 4272. [CrossRef]
43. Peng, W.; Pivato, A. Sustainable Management of Digestate from the Organic Fraction of Municipal Solid Waste and Food Waste Under the Concepts of Back to Earth Alternatives and Circular Economy. *Waste Biomass Valorization* **2019**, *10*, 465–481. [CrossRef]
44. Chiumenti, A.; da Borso, F.; Chiumenti, R.; Teri, F.; Segantin, P. Treatment of Digestate from a Co-Digestion Biogas Plant by Means of Vacuum Evaporation: Tests for Process Optimization and Environmental Sustainability. *Waste Manag.* **2013**, *33*, 1339–1344. [CrossRef]
45. Gienau, T.; Brüß, U.; Kraume, M.; Rosenberger, S. Nutrient Recovery from Biogas Digestate by Optimised Membrane Treatment. *Waste Biomass Valorization* **2018**, *9*, 2337–2347. [CrossRef]
46. Ramprasad, C.; Alekhya, D.; Bhishmitha, C.; Deepika, C.S. Precipitation of Struvite by Sustainable Waste Materials and Use as Slow Release Fertilizer—A Circular Economy Approach. *IOP Conf. Ser. Mater. Sci. Eng.* **2020**, *955*, 012088. [CrossRef]
47. Achilleos, P.; Roberts, K.R.; Williams, I.D. Struvite Precipitation within Wastewater Treatment: A Problem or a Circular Economy Opportunity? *Heliyon* **2022**, *8*, e09862. [CrossRef]

energies

MDPI

Review

Recent Findings on Fly Ash-Derived Zeolites Synthesis and Utilization According to the Circular Economy Concept

Agata Mlonka-Mędrala

Department of Thermal and Fluid Flow Machines, Faculty of Energy and Fuels, AGH University of Science and Technology, Al. Mickiewicza 30, 40-053 Kraków, Poland; amlonka@agh.edu.pl

Abstract: The synthesis and utilization of zeolites derived from fly ash (FA) gained significant attention years ago due to their potential to address environmental challenges and promote sustainable practices subscribing to the circular economy concept. This paper highlights the recent findings regarding the synthesis and utilization of zeolites derived from FA. It begins with a discussion about the recent challenges regarding industrial waste management and statistics regarding its availability on the global market with a special insight into the situation in Poland. The characteristics of FA obtained from various fuels were presented and the main differences were highlighted. Then, different methods used for the synthesis of zeolites from FA were discussed in small and pilot scales taking into consideration the main challenges and problems. The analytical methods used in porous materials synthesis verification and properties determination were described. The sorption properties of FA-derived zeolites were presented and discussed. Finally, the paper emphasizes the potential applications of fly ash-derived zeolites in different fields. Their importance as sustainable alternatives to conventional materials in industry, construction, agriculture, power, medicine, and other industrial sectors was analyzed.

Keywords: fly ash; waste management; zeolite; circular economy

Citation: Mlonka-Mędrala, A. Recent Findings on Fly Ash-Derived Zeolites Synthesis and Utilization According to the Circular Economy Concept. *Energies* **2023**, *16*, 6593. https://doi.org/10.3390/en16186593

Academic Editor: Lidia Lombardi

Received: 31 July 2023
Revised: 31 August 2023
Accepted: 4 September 2023
Published: 13 September 2023

1. Introduction

The environment has been negatively affected by the exponential growth of our economy as well as our heavy dependence on conventional energy sources. The degradation of the natural environment leads to severe climate changes. Another significant global concern is the proper management of waste, especially industrial ones, like fly ash from coal [1] and biomass-firing power plants [2,3].

Fly ash (FA) is a solid residue of fuel combustion in power boilers. It has a complex composition depending upon the raw feedstock properties [4]. FA is mostly composed of aluminosilicates [5] forming fine particulate matter captured using dust removal equipment, ranging in size from 0.5 μm to 300 μm [6]. FA might be used directly as a waste material or after preprocessing, like for example in zeolite synthesis. Zeolites are well-defined crystalline aluminosilicate porous materials, built up with a framework of alumina and silica tetrahedrals [7]. These porous materials are composed not only of aluminum and silica, but also oxygen, water, and cations such as Na^+, K^+, Mg^{2+}, and Ca^{2+} as counterions [8,9]. Their general chemical formula is $M_{y/n}[(AlO_2)_y(SiO_2)_m]_zH_2O$ [10], but water within their structure can be removed with the application of heat [11]. In the tetrahedral structure, zeolite includes silica (Si^{4+}) and aluminum (Al^{3+}) cations, which creates large negatively charged channels. Therefore, the mechanism of adsorption using zeolites was identified to be ion-exchange [12]. Additionally, their unique micro- and mesoporous structure, high ion-exchange capacity, substantial specific surface area, and thermal stability contribute to their prominent position in the industrial market [7].

Zeolites can be obtained from natural sources or synthesized using various methods [9]. There is a diverse array of natural zeolites, with more than 40 different varieties known.

Among these, the most frequently encountered types are analcime, chabazite, heulandite, phillipsite, and stilbite [13].

The evaluation of synthetic zeolite production using various materials like clay minerals, low-carbon materials, and FA has created new value-added products that find applications in various branches of industry, agriculture, biochemical, and chemical sectors such as catalysis [14,15], water treatment [16,17], and gas storage [18,19]. Utilizing FA as a feedstock for zeolite synthesis aligns with waste minimization strategies and the circular economy concept, but it requires specific procedures [20–22]. Commonly used and well-known synthesis methods include hydrothermal synthesis [23], the molten salt method, alkali activation, and microwave synthesis [24,25].

This study focuses on analyzing FA-derived zeolites for their potential use as adsorption materials in various industries. Recent findings in zeolite synthesis and utilization are analyzed and presented.

2. Industrial Waste Management in Poland

2.1. Industrial Waste Generation Globally and in Poland—FA Generation and Utilization

The largest global coal fly ash (CFA) producers are India, China, USA, Russia, and the EU [4]. China's annual production of FA is equal to almost 500 million tons, but only about 40% of this quantity is effectively utilized [26]. In India more than 230 million tons of FA is generated but the waste is properly managed and 92% of FA is utilized [27]. In the USA, FA generation has been decreasing since 2014, whereas the amount used in other sectors is increasing and in 2021 more than 60% of coal combustion products were further utilized [28]. A very low CFA utilization rate is observed in Russia, only 8% of CFA is used [29].

Poland is one of the countries with the relatively lowest production of municipal solid waste in the European Union [30]. Unfortunately, due to our heavy dependence on fossil fuels in the production and energy sectors, there is a noticeable generation of industrial waste. According to the Environment 2022 report published by Statistics Poland in 2021, a total of 107.7 million tons of waste (excluding municipal waste) was produced, with over 61.9% coming from the mining and extractive industries. As much as 48% of the waste was recovered, but only 7% was disposed using methods other than landfilling, and a significant 44% of waste was landfilled (Figure 1). It is essential to remember that there are also over 1811.4 million tons of waste currently being stored in Poland [31]. However, only 10% of the total amount of FA generated each year is landfilled in Poland. Whereas in China only 30% of CFA is used, mostly in building materials and the large amount of 3 billion tons of CFA is landfilled [32].

In Poland, the largest amounts of industrial waste are produced by industrial processing plants (46.9%), as well as the mining (31.8%) and energy (18.6%) industries. Consequently, the regions with prominent industrial activities generate the most significant quantity of waste, mainly from metallurgical, mining, and energy sectors. The storage of industrial waste leads to health hazards due to potential exposure to toxic and harmful substances present in the waste. Industrial waste often contains heavy metals, as well as aliphatic and aromatic hydrocarbons, and pesticides, all of which have physicochemical and toxic properties that can endanger human health and the environment [33].

The improper location, construction, and operation of landfills, along with uncontrolled storage of industrial waste, create opportunities for the release of chemical substances from the accumulated waste, leading to potential threats to the environment [33].

The efficiency of the recovery process depends on the waste's properties, its consistent composition, and the level of contaminants present. In the case of waste generated during the rinsing and purification of minerals, slag from smelting processes, soil and stones waste, as well as mixtures of FA and solid waste from wet flue gas desulfurization and FA from coal combustion, the recovery process is quite effective. However, for waste resulting from the flotation enrichment of non-ferrous metal ores and mixtures of ash-slag from the wet disposal of combustion residues, the recovery process is conducted for less than 20% of the generated waste (Figure 1).

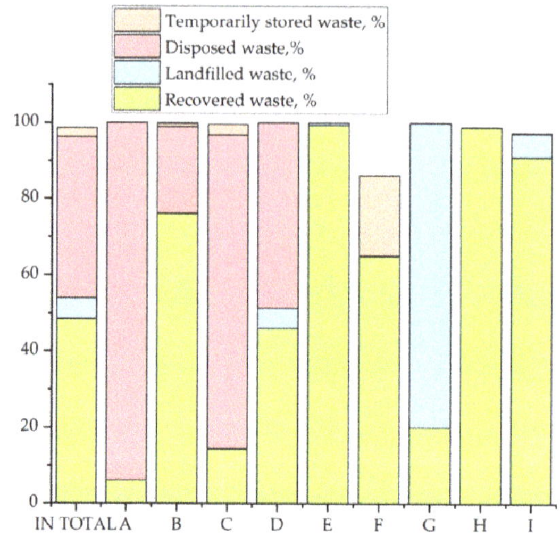

A	Waste generated during the rinsing and purification of minerals
B	Mixtures of ash and slag from wet disposal of combustion residues
C	Waste from the extraction of non-metallic minerals other than metal ores
D	Slag from smelting processes
E	Soil and earth, including stones
F	Soil and earth, including stones
G	Clarified water sludges
H	Mixtures of fly ashes and solid waste from calcium-based flue gas desulfurization methods
I	Coal fly ashes

Figure 1. Industrial waste generated and landfilled in Poland, data for 2021 [31].

The high utilization of FA in the industry is due to changes made in Poland since 1965; new regulations introduced the possibility of FA-doping cement [34]. This addition improves the properties of concrete and increases the utilization of this waste in the industry [35]. However, the use of fly ash as a substitute for cement is limited by the amount of free lime in the ash [36]. The chemical and physical properties of silica FA used as an additive in concrete production have to be standardized [4]. It was observed that the addition of FA and zeolite to the cement slightly decreases the compressive strength of the material. However, it was also noted that incorporation of FA and zeolites improves the flowability of concrete [37].

FA fulfilling the requirements described in the standard serves as a valuable additive in the production of commodity concretes, precast concrete products, dry mixes, and mineral adhesives [38]. CFA can also be used in the production of concrete-ash mixtures used in road construction. However, due to the increasing co-firing of conventional fuels with other fuels (such as biomass or alternative fuels) in power plants, as well as the use of flue gas treatment systems, such as sorbents dosed directly into the combustion chamber or flue gas ducts, which directly affect the composition of FA, new opportunities for using this waste in the industry are constantly growing. FA can be directly used in agriculture as fertilizer [39], or if it contains large amounts of carbon and phosphorus, FA can be used as a modified material in the production of mineral-organic fertilizers [5,19,40] or as material for the synthesis of potassium zeolites [41]. Methods for synthesizing advanced porous materials from FA are also being developed, such as commonly used zeolites [19,42,43] and metal-organic frameworks (MOFs) [44]. Suitable treatment also allows for the use of FA from municipal waste incineration in phosphorus recovery processes [45].

FA which is rich in carbon is also a valuable adsorbent for capturing heavy metals, such as mercury, from flue gas streams [46], or heavy metals like copper and nickel in water treatment systems [47]. The disposal of wastewater containing dyes into aquatic streams poses a significant challenge due to its adverse impact on the water ecosystem. Direct dyes with their complex aromatic structure make their separation from industrial wastewater a difficult task. However, a promising solution was found with the use of chemically modified CFA, which demonstrated a noticeable removal efficiency for these types of contaminants [48]. Additionally, after hydrothermal treatment, FA can be used as a hydrophilic material with a high potential for use in solar air-conditioning systems [49].

In recent years, there has been a growing trend of using ashes from biomass combustion to enhance soil quality, both in agricultural practices and in remediation and reclamation techniques. There is also consideration for using ash as a binding additive in the production of fertilizers from other waste materials, such as sewage sludge [50].

The method of treating FA for its utilization in industry is largely dependent on the combusted fuel or fuel mixtures, combustion technology, and applied flue gas cleaning methods [51]. Co-firing biomass and high unburned carbon content create possibilities for using FA in agriculture. However, a significant barrier is the contamination of FA with heavy metals when co-firing fossil fuels with alternative fuels. In such cases, the FA needs to be classified as hazardous waste if it contains high levels of heavy metals. Monitoring the stability of specific contaminants is also necessary due to the potential for uncontrolled environmental pollution.

2.2. FA Types and Properties

During the combustion process of fuels in power boilers, two types of ashes are obtained: bottom ash (constituting 10 to 20% of the total ash mass) and FA (70 to 90%). The proportion of each type depends primarily on the combustion technology and the type of fuel being burned. Besides ashes, other solid by-products of combustion include desulfurization products, slag, gypsum, and other residues from flue gas cleaning processes [52].

The standard ash analysis generally determines the mineral composition of the solid residue remaining after the combustion process. Typically, elements such as silica, aluminum, iron, calcium, magnesium, sodium, potassium, manganese, phosphorus, and titanium are considered, along with any specific trace elements in certain cases [53–55]. For example, in ashes from the incineration or co-incineration of waste, the content of heavy metals is of key importance [56]. The main components of the mineral phase are expressed as the highest oxides of these elements, assuming that all bonds and compounds present in the raw fuel sample are destroyed during the combustion process, converting into oxide form. Table 1 presents major FA constituents, mainly oxides depending on the combusted fuel. The classification of FA is based on the sum of selected elements in oxide form, leading to three main groups:

1. Silica-based, determined as a sum of the concentrations of selected oxides: SiO_2 + Al_2O_3 + TiO_2
2. Calcium-based, determined as a sum of the concentrations of selected oxides: CaO + MgO + Na_2O + K_2O
3. Iron-based, determined as a sum of the concentrations of selected oxides: Fe_2O_3 + MnO + SO_3 + P_2O_5,

Intermediate groups are also distinguished based on the concentration of individual elements in the mineral phase [57].

Table 1. Major FA constituents depending on the combusted fuel.

FA Component	PL-L1	PL-HC 2	PL-L3	PL-B4	PL-B5	PL-W6	PL-W7
SiO_2	27.37	48.43	47.3	36.03	44.41	14.02	5.55
Al_2O_3	6.63	28.72	31.4	8.33	10.80	22.73	18.19
CaO	34.48	4.63	1.7	27.41	23.84	20.22	32.49
MgO	8.23	2.60	1.9	3.56	3.76	6.53	2.29
Na_2O	1.08	1.77	-	0.87	1.27	5.83	5.55
K_2O	0.41	2.81	-	4.92	3.99	3.08	3.22
Fe_2O_3	3.75	6.35	7.7	4.12	3.63	0.56	0.62
TiO_2	0.96	1.17	1.6	0.94	1.05	0.76	0.76
P_2O_5	-	-	-	3.21	2.02	0.82	1.23
Heavy metals (sum)	-	-	-	-	-	10.49	7.48
Reference	[58]	[59]	[60]	[3]	[61]	[62]	[62]

PL-L1—FA obtained from lignite combusted in a pulverized coal boiler (Poland).

PL-HC2—FA obtained from hard coal combusted in a pulverized coal boiler (Poland).

PL-L3—FA obtained from lignite combusted in a fluidized bed boiler (Poland).

PL-B4—FA obtained from a pulp and paper mill, from a boiler fueled with waste forest biomass (Portugal).

PL-B5—FA obtained from biomass combusted in a fluidized bed boiler (China).

PL-W6—FA obtained from a grate boiler in a waste incineration plant (Italy).

PL-W7—FA obtained from waste combusted in a fluidized bed boiler (Italy).

Major constituents of FA are usually silica and aluminum, but in some cases calcium is also one of the main components of the mineral matter, especially in the case of biomass and waste. Additionally, in the case of biomass a noticeable content of alkali metals, mostly potassium, might be noted. FA collected from the waste incineration process is usually contaminated with heavy metals which might affect its potential use in industry.

3. Circular Economy (CE) Concept

In 2015, a circular economy (CE) package was published by the European Commission to enhance the transition from a linear to a circular economy, where resource efficiency is increased and the value of products and materials is maintained as long and as productively possible (Figure 2) [63]. In the CE concept, when a product cannot be further used, options like remanufacturing and recycling are promoted to create additional value. The economic benefits associated with the CE concept are connected to additional job creation and the promotion of innovation and environmental benefits [64].

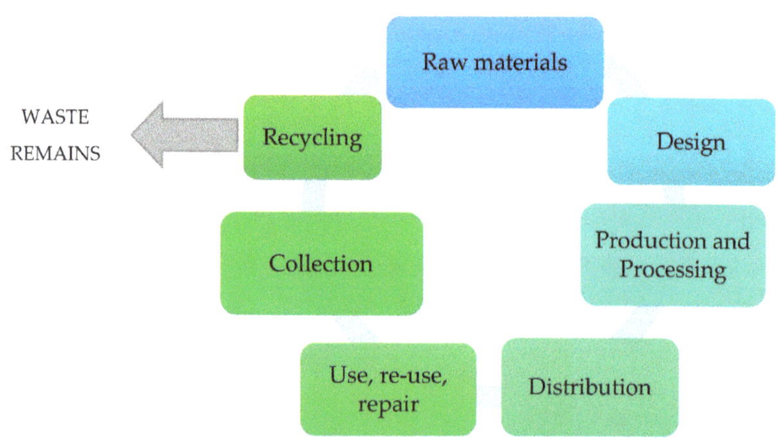

Figure 2. General idea of the circular economy concept.

Society has been aware for the last several years that the increasing consumption of non-renewable material resources cannot sustain human development anymore; resources must be utilized at their maximum. Until now, many intensive environmental problems have arisen, and some boundaries that defined the safe environment for humans have already been transgressed. Very soon, we will be challenged by the running out of global fossil fuel resources. However, pollution and climate change are even more influential factors forcing our society to change its habits.

The awareness of this situation led to changes in waste processing from a linear approach to a circular, which is depicted by sequential circles. Earlier, linear waste processing was based on the traditional take–make–consume–dispose approach (Figure 3).

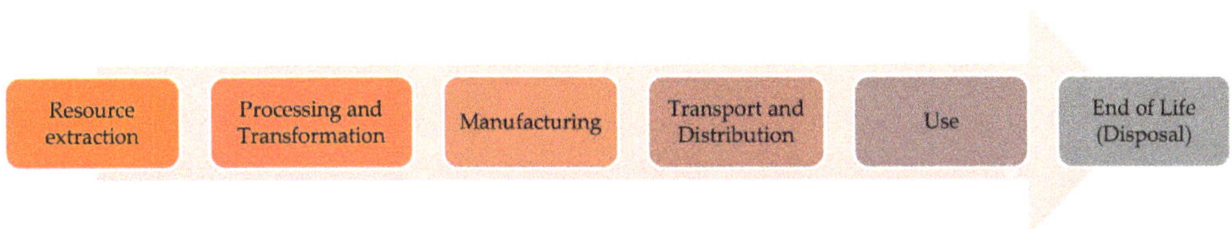

Figure 3. General idea of the linear economy.

The CE is an economic model that aims to maximize resource efficiency and minimize waste by keeping products, materials, and resources in circulation for as long as possible. The concept of the circular economy is based on the idea of closing the loop of production and consumption, where products are designed to be durable, repairable, and recyclable, and where waste and emissions are reduced to a minimum [65–67].

The CE concept has gained attraction globally as a promising solution to address environmental challenges and create a more sustainable and resilient economic system. It encourages businesses, governments, and consumers to adopt more circular practices, which can lead to reduced resource depletion, less waste generation, and a more inclusive and sustainable economy.

CE strategies are widely described in the literature, but the official definitions of the strategies promoting a CE are not defined. The most developed R-framework uses ten strategies to increase circularity: refuse, rethink, reduce, reuse, repair, refurbish, remanufacture, repurpose, recycle, and recover. CE strategies were designed to preserve raw materials and decrease their consumption [68].

The seven pillars of the CE supporting the evolution of economic rules and incentive structures that actually fulfill the end results were proposed by Metabolic [69]. The most efficient and successful technologies and business models support all seven pillars at the same time, but not all of them are equally important. Some of the most important aspects of the pillars are:

- long-term consequences and irreversibility,
- making the ability of our planet to provide a safe space for us weaker,
- high uncertainty of outcomes.

The seven pillars of the CE help us approach problems in a systemic manner. They might be translated into quantitative tools: metrics and indicators. These are useful tools for the evaluation of the circularity of products, projects, businesses, etc.

The CE concept is already implemented in some big companies like Canon, Renault, Philips, or Tata Motors Limited, which are collecting used equipment to remanufacture, recycle, or reuse it and resell with the same guarantee as brand new products [70]. Such actions lead to the minimization of waste generation and the preservation of natural resources. The circularity is also enhanced by extended guarantee of the products and encouraging customers to repair broken and used goods instead of buying new ones.

4. FA-Derived Zeolites

4.1. FA as a Potential Precursor for Zeolites Synthesis

Zeolites are a group of aluminosilicate minerals containing oxides of alkali metals and alkaline earth metals in their composition. They belong to the class of so-called molecular sieves. Depending on their origin, they can be classified as natural or synthetic. Natural zeolites are formed as a result of the weathering of volcanic rocks. Synthetic zeolites, on the other hand, form a much larger group of materials, with over 150 types. Many synthetic zeolites have a similar structure and geometric arrangement as natural zeolites [5]. However, synthetic zeolites have better properties and a lower contamination content than natural ones [71].

The most commonly used zeolites in industry are of type A, X, and Y. Figure 4 shows the morphology of commercial molecular sieves 4A and 13X. Type A zeolites are characterized by relatively low silica content, with a molar Si to Al ratio below 2. Type X and Y zeolites, unlike type A zeolites, exhibit increased resistance to acids and high temperatures. For type X zeolites, the silicon module does not exceed 2, while for type Y zeolites, it ranges from 2 to 5 [72]. The artificial synthesis of these materials has improved their heat conduction properties. Satisfactory results have been obtained in the synthesis of zeolites from natural aluminosilicates like halloysite [73,74] and kaolin [75], synthetic materials like kaolinite [76], as well as combustion by-products like FA [77–79]. To achieve specific properties, zeolite materials undergo additional modifications. Table 2 presents the properties of reference materials—selected natural and commercial zeolites described in the literature.

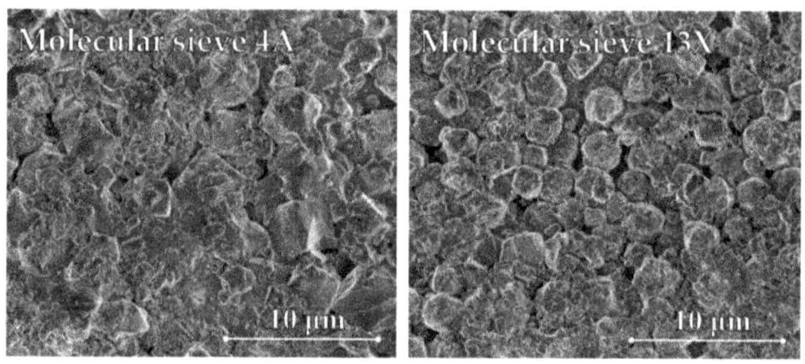

Figure 4. Scanning electron microscope images of molecular sieves 4A and 13X.

Table 2. Comparison of specific surface area and pore diameter values for natural and commercial zeolite materials.

Adsorbent	Specific Surface Area, m^2/g	Pore Size, nm	Reference
Natural zeolite	9–15	9.2–14.6	[80,81]
Zeolite 4A	7–559	22	[82–85]
Zeolite 13X	310–730	1–2.5	[84,86–88]
ZSM-5	304–417	-	[89–92]
TS-1/2zeolite	500	-	[93]
Zeolite Y	527	2.5–3	[92,94]
Zeolite HY	620–625	7.4	[95,96]

Natural zeolites are characterized by very low specific surface area resulting in weak and moderate sorption capacities. However, synthetic zeolites which are very frequently used in the industry might be characterized by a specific surface area several times higher, mostly due to microporous structure development.

4.2. Methods of Synthesis for FA-Derived Zeolites

One of the most innovative applications of FA is the synthesis of sorption materials such as zeolites. FA-derived porous materials might increase the amount of this industrial waste management according to the CE concept. FA has a similar chemical composition to zeolites, although the combustion and co-combustion of biomass and waste fuels affect the composition of the mineral phase, which may affect the properties of the obtained porous materials. The production of new porous materials from heterogeneous materials with variable composition, such as FA, forced the adaptation of existing and the development of new special procedures dedicated to the synthesis of zeolites from FA. There are many synthetic methods, which vary depending on the temperature and time of the process (hours/days), environment, Si to Al ratio, reaction components ratio, process steps, etc. The most commonly used and well-known methods for the synthesis of zeolite materials are hydrothermal synthesis, the molten salt method, alkaline activation, microwave synthesis [97,98], and ultrasonic treatment [99]. Hydrothermal crystallization consists of three stages of synthesis: dissolution, condensation, and crystallization in an alkaline solution with a $pH > 8.5$. Typically, the process is carried out in autoclaves with increased pressure [100]. Zeolite samples prepared via the hydrothermal method using CFA as a pre-cursor were analyzed in [101]. It was noted that an increase in NaOH concentration led to an increase in open volumes (or open porosity) or defects in the zeolite structure. On the other hand, increasing the temperature during sample preparation tended to reduce or diminish the open volumes, indicating a change in the zeolite's structural characteristics [101]. The direct hydrothermal method is a straightforward process, but it has drawbacks in terms of time and energy consumption, leading to low product purity [102]. Also, the presence of contaminations in FA like, for example calcium oxides, limits the purity of the final product [103]. To overcome these limitations, alternative techniques like alkaline fusion-assisted and microwave-assisted hydrothermal methods have been introduced to enhance the yield and quality of zeolites. The advantage of using microwave-assisted over alkaline fusion-assisted hydrothermal methods lies in the increased crystallization rate, leading to reduced crystallization time [9]. In turn, the use of microwave synthesis allows for high control of the temperature and time of crystallization, which is essential in obtaining specific crystalline phases. Additional extraction of silicon from FA allows us to achieve zeolites of high quality and purity. The desired Si/Al molar ratio might be obtained by adding aluminum ions [52]. The more frequently used method is also hydrothermal alkaline fusion, in which a solid mixture of NaOH with FA is used, which is then heated to a temperature of about 500–600 °C [104]. However, due to high energy consumption the fusion method is rather expensive and the production line on the technical or industrial scale is not economically justified [105]. By employing ultrasonic energy, the aging time for the alkaline fusion followed by the ultrasonic-assisted synthesis method is significantly reduced from 24 to 2 h. This ultrasonic aging process allows for an approximately 90% reduction in aging time and related costs, primarily attributed to the decrease in electricity consumption [106].

More frequently, combinations of methods are also used to enhance the quality and properties of the final product. Just recently, for the first time, researchers have successfully synthesized a highly crystalline zeolite from CFA using a unique approach that combines in situ microwave and ultrasound activation. This method is known as microwave and ultrasound collaborative activation (MUCA) and it gives better results than individual activation methods using microwave or ultrasound alone. MUCA exhibits higher efficiency in dissolving silica-aluminum species from CFA, while preserving other impurities and avoiding their digestion. As a result, MUCA significantly reduces energy consumption during the zeolite synthesis process [107].

In addition to the potential reduction in energy consumption during zeolite synthesis, a high-purity zeolite has been successfully produced from FA using a green synthesis approach. This environmentally friendly method proposed in [108] involves the minimal use of water, with only trace amounts utilized, and it achieves zero discharge of wastewater during the synthesis process [108]. A CFA-derived zeolite might also be produced using another new and convenient solvent-free method, ensuring minimal energy consumption due to its short processing time as described in [109].

Recent findings also prove that the particle size distribution of utilized FA significantly influences the quality and yield of the obtained zeolites. By incorporating fine FA particles, a larger reaction area might be achieved, resulting in the formation of zeolite phases with a higher degree of crystal formation, increased purity, and enhanced crystallinity. These improvements in the physicochemical properties directly correlate with a considerable improvement in their potential applications. Additionally, this can be achieved with only a relatively small financial investment required for the FA grinding process [110].

In [111], an additional step of wet grinding was proposed to enhance the synthesis process of zeolites from FA. During grinding, the raw material undergoes destructive forces such as friction, shear force, and impact, leading to a decrease in particle size. The process generates heat, which promotes the activation process of FA. The key parameters affecting the particle size change and system temperature increase are grinding time, grinding speed, ball–powder ratio, and solid–liquid ratio [111]. Grinding has the advantageous effect of reducing particle size, increasing the surface area, and enhancing the surface structure of coal-based solid wastes. These improvements are beneficial as they facilitate better exposure and subsequent removal of impurities [112]. It was noted that, when the wet-grinding time increases, the particle size gradually decreases until reaching stability after 2 h. During wet-grinding, the raw material develops structural defects, but with prolonged grinding, these defects decrease, and particle strength increases. Consequently, the crushing difficulty increases, making 2 h the optimal grinding time. Using FA as a raw material for zeolite synthesis through the wet-grinding activation hydrothermal method reduces energy consumption and the number of process steps. This approach holds the potential for the large-scale use of FA to create an inexpensive adsorbent, making it a promising option for practical application [111].

The incorporation of mesopores into microporous zeolites, known as hierarchical zeolite synthesis, represents a highly promising method to enhance the surface properties and porous structure of zeolites. When the diffusion of ions and molecules to an active site is limited, the internal surface area of a zeolite remains empty, leading to reduced zeolite activity. However, this challenge can be addressed by synthesizing hierarchical zeolites that possess enhanced diffusion capabilities [99]. This approach combines the benefits of shape selectivity from the microporous framework with the facilitated mass transfer achieved through the mesoporous channels [113]. Hierarchical zeolites, possessing a dual-porous structure, have emerged as a significant and novel category within the field of zeolites. Their potential lies in their ability to enhance mass transfer and molecular accessibility, crucial for overcoming obstacles such as steric hindrance, diffusional constraints, and coke formation in catalytic reactions [114].

Numerous techniques have been developed to synthesize hierarchical zeolites, including organosilane-based procedures, surfactant-assisted methods, framework atom removal, zeolite synthesis of preformed solids, templating, and the modification of porous materials like activated carbon and diatomite. However, these methods are expensive due to their multi-step processes involving templates and acid-base treatment after synthesis. Alternatively, hierarchical zeolites can be synthesized in a cost-effective and straightforward manner using the microwave heating method [99].

The technological production line described in [115] generates very good results regarding high-purity Na-P1 zeolite synthesis. Most pilot-scale facilities are conducting the conversion process of FA with NaOH at a moderate temperature at atmospheric [115] or elevated pressure [116]. The obtained zeolites were tested as water treatment adsorbents for capturing heavy metals and ammonium [116]. Whereas zeolites obtained via fusion followed by a bench-scale hydrothermal reaction and hydrothermal pilot-scale reaction were analyzed as potential fertilizer carriers [117].

4.3. Standard Characterization Methods of Zeolite Materials

Zeolite materials can be characterized using various analytical methods to gain insights into their structure, composition, porosity, surface properties, and other relevant parameters, which depend upon their intended function.

Some of the common methods of characterization for zeolites include:

1. X-ray Diffraction (XRD) is used to determine the crystalline structure and phase composition of zeolites. It provides information about the arrangement of atoms in the zeolite framework and helps identify different zeolite phases [118]. It is a valuable tool in the determination of the efficiency of synthetic zeolite synthesis methods [7].

X-ray diffraction (XRD) analysis can be utilized to estimate the crystallinity degree of a specific zeolite. This calculation involves comparing the intensities of the characteristic zeolite profile with those of the reference materials, as expressed in the following equation:

$$Cristallinity(I/I_0) = \frac{\sum relative\ intensities\ of\ fly\ ash - derived\ zeolite}{\sum relative\ intensities\ of\ a\ reference\ material} \times 100 \quad (1)$$

2. Scanning Electron Microscopy allows us to visualize the surface morphology of zeolite materials at a high resolution. It provides information about particle size, shape, and distribution. Additional EDS (Energy-Dispersive X-ray Spectroscopy) gives information about the semiquantitative chemical composition of the material. Observation of the sample morphology enables an estimation of its porous structure development through synthesis procedures. SEM has been used to investigate zeolites' microstructure, particle morphology, and crystal growth [115,119,120].
3. Transmission Electron Microscopy (TEM) offers a higher resolution than SEM and allows us to examine the internal structure of zeolite crystals at atomic resolution, including the arrangement of atoms and crystal defects [121–123].
4. Low-Temperature Gas Adsorption using nitrogen or carbon dioxide (BET Analysis) is used to determine the specific surface area and porosity of zeolites [9,18,124]. It provides information about the structure of the material and its potential sorption capacity. To more precisely determine the sorption properties of the material toward a given adsorbate or contamination more advanced methods are needed, like, for example the dynamic vapor sorption method (DVS) [125].
5. Fourier Transform Infrared Spectroscopy (FTIR) is used to identify the chemical bonds and functional groups present in zeolite materials, providing information about their composition and surface properties [118,126].
6. Thermogravimetric Analysis (TGA) with Differential Scanning Calorimetry (DSC) together with Simultaneous Thermal Analysis (STA) is used to study the thermal stability of zeolites and to determine the water content and their desorption behavior. TGA and DSC are used to determine the thermal properties of zeolites, including phase transitions and thermal stability. These properties are crucial when zeolites are dedicated to high-temperature purposes like flue gas treatment. A unique method based on TGA was also proposed in [127], dedicated to determining the degree of fly ash conversion efficiency.
7. Nuclear Magnetic Resonance (NMR) Spectroscopy can provide information about the local environment of certain atoms in the zeolite structure, giving insights into their connectivity and coordination [126]. Due to its remarkable sensitivity to the atomic-scale environment, this method is well-suited for investigating local structure, disorder, and chemical reactivity in solid-state materials [128].
8. X-ray Photoelectron Spectroscopy (XPS) is used in some cases to study the chemical composition and oxidation states of the surface of zeolite materials [129].

By combining these characterization methods, research groups working in the field of zeolite synthesis can gain a comprehensive understanding of zeolite materials and optimize their synthesis methods and use in various applications.

4.4. FA-Derived Zeolites as Value-Added Products in Different Industrial Branches According to the CE Concept

In general, zeolites are characterized by a very large specific surface area, due to which they can be used as sorbents, catalysts, and ion exchange materials [130–132]. In nature, zeolites are represented as minerals (e.g., clinoptilolite, chabazite, and mordenite). The type of zeolites depends on the SiO_2 content (SiO_2/Al_2O_3 molar ratio), therefore they can be classified as low-, medium-, and high-silica zeolites. The Si/Al ratio determines the properties of the zeolite. Low-silica zeolites are characterized by higher acid resistance, stability at higher temperatures and hydrophilicity. On the other hand, high-silica zeolites are more hydrophobic and have high ion-exchange properties. The properties of natural zeolites in aqueous solution depend on the pH; a very low pH value can cause damage to the structure of the zeolite. Zeolites of the NaP1 type, A type, X type, Y type, and ZSM-5 zeolite are the most frequently studied and used in industry [89,133,134].

Currently, synthetic zeolites are successfully used in a wide range in industry and environmental protection, for example as soil-improving additives and as adsorbents in water and gas purification systems (Figure 5). FA-derived zeolites X and A might be employed to capture elemental mercury from a gas stream. The proper capture and immobilization of mercury is especially important as large EU combustion plants (LCP) are obliged to monitor and minimize the emission of this contaminant from the flue gas stream [135]. A unique hybrid synthesis approach was developed, incorporating the crystallization stage of zeolites, along with silver or iron modification. This method successfully yielded pure and high-crystalline zeolites. Notably, it was observed that silver-modified zeolites exhibited significant potential in adsorbing mercury [136]. The zeolites synthesized from coal FA (CFA) were also effectively employed to remove methylene blue [119] and Cu ions from aqueous solutions [137].

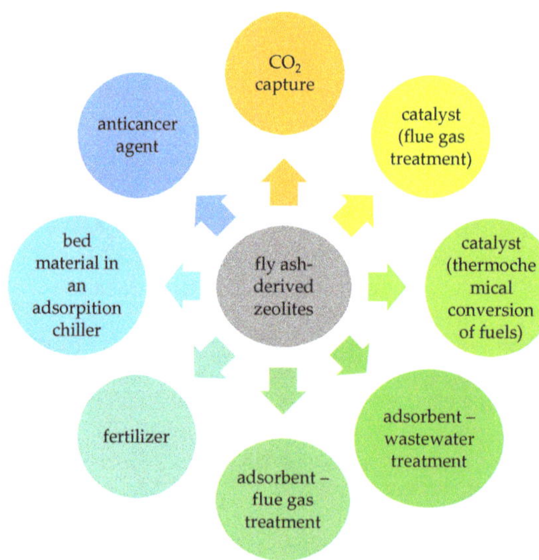

Figure 5. FA-derived zeolites' utilization in industry, agriculture, and power sectors.

As for the more specific use of zeolites synthesized from FA, they can also be used in adsorption systems for CO_2 [138,139]. It has been observed that a higher calcium content in a FA positively affects CO_2 retention under equilibrium conditions. However, this effect is not as significant in dynamic conditions, likely due to the presence of accompanying chemisorption processes [140]. Zeolites also serve as great adsorbents in gas treatment facilities, capturing contaminations [97] like nickel (Ni) [141], arsenic (As) [142], zinc (Zn), copper (Cu) [143], and lead (Pb) [111,144]. Zeolites are also used as a bed material in adsorption refrigerators [145–147], because they are sorption materials characterized by excellent sorption properties: high adsorption capacity, molecular shape selectivity, and a very large surface area. A synthesized zeolite may be characterized by superior properties compared to a commercial zeolite, making it a more appealing choice for gas separation applications. Notably, in some cases synthesized zeolite displayed a higher surface area, smaller crystallite size, and improved thermal stability, all of which contribute to its enhanced performance in this specific context [148,149]. Zeolites might be also used as catalysts in gas treatment systems. ZSM-5 zeolite obtained from CFA was used as a catalytic converter for the reduction of NOx in gasoline-powered engines. By incorporating copper and cobalt-doped zeolite on a ceramic monolith, the catalytic converter demonstrates superior activity in rapidly reducing NOx emissions compared to conventional catalytic converters [150]. Another CFA-based ZSM-5 zeolite characterized by a flexible textural design and obtained using a solvent-free method gave a good result for toluene adsorption elimination [151].

Additionally, zeolites are characterized by a three times greater water adsorption capacity compared to silica gel, and the artificial synthesis of these materials allows for improvement of the efficiency of heat exchange, making them a good adsorbent for adsorption chillers [152,153]. A certain limitation to the use of zeolites in refrigeration is the fact that the bed regeneration process is carried out at temperatures higher than in the case of silica gels, i.e., above 120 °C [154]. In addition, as in the case of other porous materials, zeolites are characterized by a low coefficient of thermal conductivity [155–157]. However, the high sorption capacity and durability of these materials mean that commercial solutions based on a zeolite bed are available on the market.

The recycling of coal FA offers a promising opportunity to create highly efficient Fenton-like catalysts capable of removing organic contaminants from wastewater effectively [118]. An additional product of synthesis might include a magnetic type which could be used as a substrate in nano-iron synthesis [109].

FA-derived zeolites hold potential not only as adsorbers in various industrial processes but also as valuable new products or components. For instance, the treatment of FAs significantly influenced both the synthesis yield and properties of the resulting zeolites. A well-designed zeolite synthesis method effectively reduced the viscosity of modified asphalt cement during heating, greatly improving its workability. The observed excellent performance of zeolite A, derived from FA, in reducing asphalt cement viscosity makes it well-suited for this specific purpose [158].

Fertilizing zeolites have demonstrated significant potential in providing fertilizer elements over extended periods of time, making them a suitable solution for more demanding crops' nutrient requirements. As a result, these synthesized materials effectively serve their purpose as slow-release nutrient suppliers, showing promising applications as K fertilizers. Moreover, a cost-effective method of FA utilization in producing a zeolitic product with a well-structured framework capable of efficiently retaining K ions was proposed. This product, called MER zeolite, exhibits improved performance in releasing K nutrients, thus minimizing nutrient wastage and reducing the risk of environmental pollution [159].

Prior to disposal in landfills, municipal solid waste incineration fly ash (MSWIFA) with elevated heavy metal concentrations requires consolidation. The alkali-activated technique plays a crucial role in enhancing the degree of heavy metal solidification and the utilization ratio on MSWIFA material. Incorporating zeolite into the process shows promising results, as it increases the efficiency of heavy metal solidification within the FA matrix, making it a potentially valuable application of FA [160].

Self-supporting adsorbents composed of zeolite-geopolymer composites, synthesized using industrial solid waste, exhibit remarkable mechanical strength, a high specific surface area, and cation exchange capability. These composites serve as effective substitutes for natural heulandite zeolite in removing metal ions from water. The simple two-step method allows for the synthesis of zeolite–geopolymer composites with a high zeolite content, making them highly suitable as self-supporting adsorbents for water treatment applications [81].

An effective method for defluorination of water involved the use of aluminum hydroxide-coated zeolite (AHZ), which was synthesized from coal FA. AHZ proves to be efficient in removing fluoride from water and offers a promising solution for addressing fluoride contamination issues in wastewater treatment [161]. The synthesized zeolite can be then effectively employed as an adsorbent to eliminate organic dye from aqueous solutions as well [162].

The zeolites obtained from the coal FA were also subjected to testing in tgas-phase phenol alkylation using diethyl carbonate (DEC) as a novel alkylating agent. The results from the experiments demonstrate the practicality and suitability of obtained innovative synthetic zeolites for catalytic applications as heterogeneous basic systems. The method employed yielded high conversion rates, was environmentally friendly, cost-effective, and presented a clean approach. Consequently, the synthetic zeolites proved to be highly efficient catalyst alternatives to the commonly used commercial materials in industrial applications [163].

In Ref. [164], an investigation was conducted on the mineral and chemical compositions of FA to determine its suitability as a source material for synthesizing zeolites, which serve as useful matrices for nuclear waste vitrification. Additionally, the findings in [2] indicate that K-zeolites synthesized from biomass FA are well-suited for the removal of radioactive cesium. Biomass-origin FA is characterized by a low content of impurities making it possible to exclude additional purification from the zeolite synthesis procedure [165]. Additionally, this process offers a potential solution for recycling FA from biomass power plants, thus avoiding the high economic costs associated with its disposal.

Zeolites might be also used as catalysts in thermochemical fuel conversion processes [95,132,166]. FA-derived zeolite X was used as a suitable catalyst for the production of linear aliphatic hydrocarbons (HC), although some concerns have arisen due to the deactivation caused by coke formation. The data obtained from research conducted in [167] can be applied to enhance the quality of bio-oil produced from local biomass resources, pyrolyzing palm press fiber (PPW), through catalytic fast pyrolysis.

The groundbreaking experiments conducted on zeolite X synthesized from coal fly ash have revealed its potential as a significant anticancer agent. However, further research is still necessary to understand the specific mode of action [149].

To sum up the recent findings regarding FA-derived zeolites, the properties of the obtained materials are presented in Table 3.

Table 3. Comparison of specific surface area and pore diameter values for fly ash-derived zeolite materials.

Adsorbent	Specific Surface Area	Pore Size	Reference
Unit	[m^2/g]	[nm]	-
Na-X zeolite obtained from FA from oil shale combustion (China)	252	3.817	[167]
Na-P1 zeolite obtained from FA from lignite combustion (Greece)	64.5	0.8	[168]
NaA zeolite obtained from CFA (South Africa)	33	-	[139]
NaX zeolite obtained from CFA (South Africa)	228	-	[139]
Na-P1 zeolite obtained from CFA (India)	63	-	[169]
Zeolite ZSM-22 obtained from CFA (Taiwan)	30.2	-	[170]
Zeolite X obtained from CFA (Poland)	629	1.7	[171]
Zeolite NaP1 obtained from FA from the waste incineration process	23.86	-	[172]

Table 3. *Cont.*

Adsorbent	Specific Surface Area	Pore Size	Reference
Unit	[m^2/g]	[nm]	-
FA-derived zeolite NaP1 from class F FA, obtained from a coal power plant (Poland)	60	11	[143]
Na-P1 zeolite obtained from CFA (Poland)	88	-	[115]
Sodalite zeolites obtained from CFA (South Africa)	366–399	-	[109]
Zeolite-geopolymer composites obtained from FA and metakaolin (China)	52–100	4–7	[81]
Zeolites-calcium silicate hydrate composite obtained from CFA (China)	96.5	-	[103]
Zeolites obtained from CFA collected from the electrostatic precipitators (Bulgaria)	284–486	13.2–61.1	[140]
Zeolite X prepared from FA (China)	473.56	1.9	[108]
ZSM5 obtained from CFA (Thailand)	329	6.0–27.8	[173]
X zeolite obtained from CFA (Thailand)	722	5.7–24.2	[173]
Zeolite 4A obtained from CFA (China)	18.33	11.3	[137]
Zeolite 4A obtained from CFA (Pakistan)	122	-	[148]

Zeolites synthesized from fly ashes have a significantly lower specific surface area compared to natural and commercial zeolites (Table 2), which is due, among other things, to the fact that fly ashes are not a pure mixture of silicon and aluminum compounds. They also have other impurities that affect the content of the zeolite phase obtained in the final product. In addition, the final product depends to a large extent on the method of synthesis and its parameters. FA preparation in order to obtain native material more susceptible to the synthesis of zeolite material, such as silicon extraction, significantly increases the cost of synthesis.

However, the synthesis of zeolites with coal-based solid waste as a raw material can reduce the environmental pollution of industrial wastes, minimize natural resource utilization, and realize their valuable utilization in line with the CE concept.

5. Conclusions and Future Directions

Recent research findings present fly ash-derived zeolites as promising materials that subscribe to the principles of CE by turning a waste stream into a valuable resource. Their diverse applications in environmental remediation, catalysis, material substitution, and medicine demonstrate their potential to implement sustainable practices and contribute to a more circular and resource-efficient future. The main challenges are associated with synthesis by-products, especially additional waste generation and energy consumption. Furthermore, in not all methods can high purity of the final product be achieved. FA contaminated with other elements like heavy metals and calcium might not be suitable for zeolite synthesis. High energy consumption in the case of hydrothermal methods at a high temperature makes them economically unjustified in a larger scale.

Further research and collaboration between researchers, industrial stakeholders, and policymakers are essential to fully realize the benefits of these materials and their integration into CE practices. However, utilization of FA-derived zeolites requires the precisely defined chemical composition and properties of raw FA, as synthesis methods are adjusted to the given feedstock.

Funding: This research was funded by the Ministry of Science and Higher Education, Poland, grant AGH no 16.16.210.476 and partly supported by the program "Excellence initiative—research university" of the AGH University of Science and Technology.

Data Availability Statement: Not applicable.

Conflicts of Interest: The author declares no conflict of interest.

References

1. Tirva, D.; Palkar, R.R.; Agheda, K. An Optimization of Synthesis Technique for Na-X Zeolite from Coal-Fly Ash Using Taguchi Experimental Design. *Chem. Pap.* **2023**, *77*, 5143–5153. [CrossRef]
2. Fukasawa, T.; Horigome, A.; Tsu, T.; Karisma, A.D.; Maeda, N.; Huang, A.N.; Fukui, K. Utilization of Incineration Fly Ash from Biomass Power Plants for Zeolite Synthesis from Coal Fly Ash by Hydrothermal Treatment. *Fuel Process. Technol.* **2017**, *167*, 92–98. [CrossRef]
3. Teixeira, E.R.; Camões, A.; Branco, F.G.; Aguiar, J.B.; Fangueiro, R. Recycling of Biomass and Coal Fly Ash as Cement Replacement Material and Its Effect on Hydration and Carbonation of Concrete. *Waste Manag.* **2019**, *94*, 39–48. [CrossRef] [PubMed]
4. Gollakota, A.R.K.; Volli, V.; Shu, C.M. Progressive Utilisation Prospects of Coal Fly Ash: A Review. *Sci. Total Environ.* **2019**, *672*, 951–989. [CrossRef] [PubMed]
5. Szerement, J.; Szatanik-Kloc, A.; Jarosz, R.; Bajda, T.; Mierzwa-Hersztek, M. Contemporary Applications of Natural and Synthetic Zeolites from Fly Ash in Agriculture and Environmental Protection. *J. Clean. Prod.* **2021**, *311*, 127461. [CrossRef]
6. Bhatt, A.; Priyadarshini, S.; Acharath Mohanakrishnan, A.; Abri, A.; Sattler, M.; Techapaphawit, S. Physical, Chemical, and Geotechnical Properties of Coal Fly Ash: A Global Review. *Case Stud. Constr. Mater.* **2019**, *11*, e00263. [CrossRef]
7. Ait Baha, A.; Tabit, K.; Idouhli, R.; Khadiri, M.; Zakir, O.; Dikici, B.; Abouelfida, A. Zeolitization of Fumed Silica and Coal Fly Ash Using the Taguchi Method toward Organic Pollutant Removal. *Silicon* **2023**. [CrossRef]
8. Zeolite Structure and Types—Lenntech. Available online: https://www.lenntech.pl/zeolites-structure-types.htm (accessed on 23 July 2023).
9. Truttim, P.; Asavapisit, S.; Piyaphanuwat, R. Microwave Synthesis of Zeolite X from Bituminous Fly Ash and Its Characterization. *Mater. Today Proc.* 2023; *in press*. [CrossRef]
10. Wang, R.; Wang, L.; Wu, J. *Adsorption Refrigeration Technology, Theory and Application*; Wiley: Singapore, 2014.
11. Moshoeshoe, M.; Nadiye-Tabbiruka, M.S.; Obuseng, V. A Review of the Chemistry, Structure, Properties and Applications of Zeolites. *Am. J. Mater. Sci.* **2017**, *7*, 196–221. [CrossRef]
12. Chojnacki, A.; Chojnacka, K.; Hoffmann, J.; Górecki, H. The Application of Natural Zeolites for Mercury Removal: From Laboratory Tests to Industrial Scale. *Min. Eng.* **2004**, *17*, 933–937. [CrossRef]
13. Gottardi, G.; Galli, E. *Natural Zeolites*; Minerals and Rocks; Springer: Berlin/Heidelberg, Germany, 1985; Volume 18, ISBN 978-3-642-46520-8.
14. Cao, Z.; Niu, J.; Gu, Y.; Zhang, R.; Liu, Y.; Luo, L. Catalytic Pyrolysis of Rice Straw: Screening of Various Metal Salts, Metal Basic Oxide, Acidic Metal Oxide and Zeolite Catalyst on Products Yield and Characterization. *J. Clean. Prod.* **2020**, *269*, 122079. [CrossRef]
15. Khalil, U.; Vongsvivut, J.; Shahabuddin, M.; Samudrala, S.P.; Srivatsa, S.C.; Bhattacharya, S. A Study on the Performance of Coke Resistive Cerium Modified Zeolite Y Catalyst for the Pyrolysis of Scrap Tyres in a Two-Stage Fixed Bed Reactor. *Waste Manag.* **2020**, *102*, 139–148. [CrossRef] [PubMed]
16. Algieri, C.; Drioli, E. Zeolite Membranes: Synthesis and Applications. *Sep. Purif. Technol.* **2022**, *278*, 119295. [CrossRef]
17. Michalev, T.; Petrov, I. The Removal of Heavy Metal Ions by Synthetic Zeolites: A Review. *Proc. Univ. Ruse* **2012**, *51*, 79–84.
18. Hai, T.; Alenizi, F.A.; Mohammed, A.H.; Chauhan, B.S.; Al-Qargholi, B.; Metwally, A.S.M.; Ullah, M. Machine Learning-Aided Modeling of the Hydrogen Storage in Zeolite-Based Porous Media. *Int. Commun. Heat Mass Transf.* **2023**, *145*, 106848. [CrossRef]
19. Czuma, N.; Zarębska, K.; Motak, M.; Gálvez, M.E.; Da Costa, P. Ni/Zeolite X Derived from Fly Ash as Catalysts for CO_2 Methanation. *Fuel* **2020**, *267*, 117139. [CrossRef]
20. Strzałkowska, E. Fly Ash—A Valuable Material for the Circular Economy. *Gospod. Surowcami Miner. Miner. Resour. Manag.* **2021**, *37*, 49–62. [CrossRef]
21. Boycheva, S.; Szegedi, Á.; Lázár, K.; Popov, C.; Popova, M. Advanced High-Iron Coal Fly Ash Zeolites for Low-Carbon Emission Catalytic Combustion of VOCs. *Catal. Today* **2023**, *418*, 114109. [CrossRef]
22. Saldarriaga, J.F.; Gaviria, X.; Gene, J.M.; Aguado, R. Improving Circular Economy by Assessing the Use of Fly Ash as a Replacement of Lime Pastes Reducing Its Environmental Impact. *Process Saf. Environ. Prot.* **2022**, *159*, 1008–1018. [CrossRef]
23. Ameh, A.E.; Fatoba, O.O.; Musyoka, N.M.; Petrik, L.F. Influence of Aluminium Source on the Crystal Structure and Framework Coordination of Al and Si in Fly Ash-Based Zeolite NaA. *Powder Technol.* **2017**, *306*, 17–25. [CrossRef]
24. Indira, V.; Abhitha, K. A Review on Recent Developments in Zeolite A Synthesis for Improved Carbon Dioxide Capture: Implications for the Water-Energy Nexus. *Energy Nexus* **2022**, *7*, 100095. [CrossRef]
25. Ma, B.; Lothenbach, B. Synthesis, Characterization, and Thermodynamic Study of Selected Na-Based Zeolites. *Cem. Concr. Res.* **2020**, *135*, 106111. [CrossRef]
26. Luo, Y.; Wu, Y.; Ma, S.; Zheng, S.; Zhang, Y.; Chu, P.K. Utilization of Coal Fly Ash in China: A Mini-Review on Challenges and Future Directions. *Environ. Sci. Pollut. Res.* **2021**, *28*, 18727–18740. [CrossRef] [PubMed]

27. Central Electricity Authority Thermal Civil Design Division, New Delhi. Report on Fly Ash Generation at Coal/Lignite Based Thermal Power Stations and Its Utilization in the Country for the Year 2020–2021. 2021. Available online: https://cea.nic.in/wp-content/uploads/tcd/2021/09/Report_Ash_Yearly_2020_21.pdf (accessed on 23 July 2023).
28. Update on Fly Ash in the US, April 2023—Cement Industry News from Global Cement. Available online: https://www.globalcement.com/news/item/15657-update-on-fly-ash-in-the-us-april-2023 (accessed on 31 August 2023).
29. Valeev, D.; Bobylev, P.; Osokin, N.; Zolotova, I.; Rodionov, I.; Salazar-Concha, C.; Verichev, K. A Review of the Alumina Production from Coal Fly Ash, with a Focus in Russia. *J. Clean. Prod.* **2022**, *363*, 132360. [CrossRef]
30. Smol, M.; Duda, J.; Czaplicka-Kotas, A.; Szołdrowska, D. Transformation towards Circular Economy (CE) in Municipal Waste Management System: Model Solutions for Poland. *Sustainability* **2020**, *12*, 4561. [CrossRef]
31. Statistics Poland. *Environment 2022*; Statistics Poland: Warsaw, Poland, 2022.
32. Zhao, X.; Zeng, L.; Guo, J.; Zhu, Q.; Huang, Z.; Lin, L.; Chen, X.; Cao, J.; Zhou, Z. Efficient Separation and Comprehensive Extraction of Aluminum, Silicon, and Iron from Coal Fly Ash by a Cascade Extraction Method. *J. Clean. Prod.* **2023**, *406*, 137090. [CrossRef]
33. Brzeszczak, A. Analysis of Industrial Waste Management in Poland. *World Sci. News* **2019**, *127*, 337–348.
34. Igliński, B.; Buczkowski, R. Development of Cement Industry in Poland—History, Current State, Ecological Aspects. A Review. *J. Clean. Prod.* **2017**, *141*, 702–720. [CrossRef]
35. Kowsalya, M.; Sindhu Nachiar, S.; Anandh, S. A Review on Fly Ash Cenosphere as a Solid Waste in Concrete Application. *Mater. Today Proc.* **2022**, *68*, 2072–2078. [CrossRef]
36. Kumar Nayak, D.; Abhilash, P.P.; Singh, R.; Kumar, R.; Kumar, V. Fly Ash for Sustainable Construction: A Review of Fly Ash Concrete and Its Beneficial Use Case Studies. *Clean. Mater.* **2022**, *6*, 100143. [CrossRef]
37. Das, M.; Adhikary, S.K.; Rudzionis, Z. Effectiveness of Fly Ash, Zeolite, and Unburnt Rice Husk as a Substitute of Cement in Concrete. *Mater. Today Proc.* **2022**, *61*, 237–242. [CrossRef]
38. Lafarge. Popiół Lotny—Dodatki Mineralne Do Betonów. Available online: https://www.lafarge.pl/popiol-lotny-certyfikowany-dodatek-mineralny-do-betonow (accessed on 23 July 2023).
39. Su, H.F.; Lin, J.F.; Chen, H.; Wang, Q.Y. Production of a Novel Slow-Release Coal Fly Ash Microbial Fertilizer for Restoration of Mine Vegetation. *Waste Manag.* **2021**, *124*, 185–194. [CrossRef] [PubMed]
40. Popova, M.; Boycheva, S.; Lazarova, H.; Zgureva, D.; Lázár, K.; Szegedi, Á. VOC Oxidation and CO_2 Adsorption on Dual Adsorption/Catalytic System Based on Fly Ash Zeolites. *Catal. Today* **2020**, *357*, 518–525. [CrossRef]
41. Flores, C.G.; Schneider, H.; Marcilio, N.R.; Ferret, L.; Oliveira, J.C.P. Potassic Zeolites from Brazilian Coal Ash for Use as a Fertilizer in Agriculture. *Waste Manag.* **2017**, *70*, 263–271. [CrossRef]
42. Majchrzak-Kucęba, I.; Nowak, W. Thermal Analysis of Fly Ash-Based Zeolites. *J. Therm. Anal. Calorim.* **2004**, *77*, 125–131. [CrossRef]
43. Yao, Z.T.; Xia, M.S.; Ye, Y.; Zhang, L. Synthesis of Zeolite Li-ABW from Fly Ash by Fusion Method. *J. Hazard. Mater.* **2009**, *170*, 639–644. [CrossRef]
44. Rambau, K.M.; Musyoka, N.M.; Panek, R.; Franus, W.; Wdowin, M.; Manyala, N. Preparation of Coal Fly Ash Derived Metal Organic Frameworks and Their Carbon Derivatives. *Mater. Today Commun.* **2021**, *27*, 102433. [CrossRef]
45. Kalmykova, Y.; Karlfeldt Fedje, K. Phosphorus Recovery from Municipal Solid Waste Incineration Fly Ash. *Waste Manag.* **2013**, *33*, 1403–1410. [CrossRef]
46. Jia, L.; Fan, B.; Zheng, X.; Qiao, X.; Yao, Y.; Zhao, R.; Guo, J.; Jin, Y. Mercury Emission and Adsorption Characteristics of Fly Ash in PC and CFB Boilers. *Front. Energy* **2020**, *15*, 112–123. [CrossRef]
47. Lekgoba, T.; Ntuli, F.; Falayi, T. Application of Coal Fly Ash for Treatment of Wastewater Containing a Binary Mixture of Copper and Nickel. *J. Water Process Eng.* **2021**, *40*, 101822. [CrossRef]
48. Eteba, A.; Bassyouni, M.; Saleh, M. Utilization of Chemically Modified Coal Fly Ash as Cost-Effective Adsorbent for Removal of Hazardous Organic Wastes. *Int. J. Environ. Sci. Technol.* **2023**, *20*, 7589–7602. [CrossRef]
49. Karamanis, D.; Vardoulakis, E. Application of Zeolitic Materials Prepared from Fly Ash to Water Vapor Adsorption for Solar Cooling. *Appl. Energy* **2012**, *97*, 334–339. [CrossRef]
50. Odzijewicz, J.I.; Wołejko, E.; Wydro, U.; Wasil, M.; Jabłońska-Trypuć, A. Utilization of Ashes from Biomass Combustion. *Energies* **2022**, *15*, 9653. [CrossRef]
51. Ahmaruzzaman, M. A Review on the Utilization of Fly Ash. *Prog. Energy Combust. Sci.* **2010**, *36*, 327–363. [CrossRef]
52. Kar, K.K. *Handbook of Fly Ash*; Elsevier: Amsterdam, The Netherlands, 2022.
53. Jaworek, A.; Czech, T.; Sobczyk, A.T.; Krupa, A. Properties of Biomass vs. Coal Fly Ashes Deposited in Electrostatic Precipitator. *J. Electrost.* **2013**, *71*, 165–175. [CrossRef]
54. Vassilev, S.V.; Baxter, D.; Andersen, L.K.; Vassileva, C.G. An Overview of the Composition and Application of Biomass Ash. Part 1. Phase-Mineral and Chemical Composition and Classification. *Fuel* **2013**, *105*, 40–76. [CrossRef]
55. Vassilev, S.V.; Vassileva, C.G.; Song, Y.C.; Li, W.Y.; Feng, J. Ash Contents and Ash-Forming Elements of Biomass and Their Significance for Solid Biofuel Combustion. *Fuel* **2017**, *208*, 377–409. [CrossRef]
56. Mlonka-Mędrala, A.; Dziok, T.; Magdziarz, A.; Nowak, W. Composition and Properties of Fly Ash Collected from a Multifuel Fluidized Bed Boiler Co-Firing Refuse Derived Fuel (RDF) and Hard Coal. *Energy* **2021**, *234*, 121229. [CrossRef]

57. Vassilev, S.V.; Vassileva, C.G. A New Approach for the Classification of Coal Fly Ashes Based on Their Origin, Composition, Properties, and Behaviour. *Fuel* **2007**, *86*, 1490–1512. [CrossRef]
58. Sarbak, Z.; Kramer-Wachowiak, M. Porous Structure of Waste Fly Ashes and Their Chemical Modifications. *Powder Technol.* **2002**, *123*, 53–58. [CrossRef]
59. Wons, W.; Rzepa, K.; Reben, M.; Murzyn, P.; Sitarz, M.; Olejniczak, Z. Effect of Thermal Processing on the Structural Characteristics of Fly Ashes. *J. Mol. Struct.* **2018**, *1165*, 299–304. [CrossRef]
60. Sarbak, Z.; Stańczyk, A.; Kramer-Wachowiak, M. Characterisation of Surface Properties of Various Fly Ashes. *Powder Technol.* **2004**, *145*, 82–87. [CrossRef]
61. Liang, G.; Li, Y.; Yang, C.; Zi, C.; Zhang, Y.; Hu, X.; Zhao, W. Production of Biosilica Nanoparticles from Biomass Power Plant Fly Ash. *Waste Manag.* **2020**, *105*, 8–17. [CrossRef] [PubMed]
62. Sharifikolouei, E.; Baino, F.; Salvo, M.; Tommasi, T.; Pirone, R.; Fino, D.; Ferraris, M. Vitrification of Municipal Solid Waste Incineration Fly Ash: An Approach to Find the Successful Batch Compositions. *Ceram. Int.* **2021**, *47*, 7738–7744. [CrossRef]
63. Zając, P.; Avdiushchenko, A. The Impact of Converting Waste into Resources on the Regional Economy, Evidence from Poland. *Ecol. Model.* **2020**, *437*, 109299. [CrossRef]
64. Morseletto, P. Targets for a Circular Economy. *Resour. Conserv. Recycl.* **2020**, *153*, 104553. [CrossRef]
65. Kirchherr, J.; Reike, D.; Hekkert, M. Conceptualizing the Circular Economy: An Analysis of 114 Definitions. *Resour. Conserv. Recycl.* **2017**, *127*, 221–232. [CrossRef]
66. Stahel, W.R. The Circular Economy. *Nature* **2016**, *531*, 435–438. [CrossRef]
67. Geissdoerfer, M.; Savaget, P.; Bocken, N.M.P.; Hultink, E.J. The Circular Economy—A New Sustainability Paradigm? *J. Clean. Prod.* **2017**, *143*, 757–768. [CrossRef]
68. Moraga, G.; Huysveld, S.; Mathieux, F.; Blengini, G.A.; Alaerts, L.; Van Acker, K.; de Meester, S.; Dewulf, J. Circular Economy Indicators: What Do They Measure? *Resour. Conserv. Recycl.* **2019**, *146*, 452–461. [CrossRef]
69. The Seven Pillars of the Circular Economy. Available online: https://www.metabolic.nl/news/the-seven-pillars-of-the-circular-economy (accessed on 22 July 2023).
70. World Business Council for Sustainable Development. *8 Bussines Cases for the Circular Economy*; World Business Council for Sustainable Development: Geneva, Switzerland, 2017.
71. Shichalin, O.O.; Papynov, E.K.; Nepomnyushchaya, V.A.; Ivanets, A.I.; Belov, A.A.; Dran'kov, A.N.; Yarusova, S.B.; Buravlev, I.Y.; Tarabanova, A.E.; Fedorets, A.N.; et al. Hydrothermal Synthesis and Spark Plasma Sintering of NaY Zeolite as Solid-State Matrices for Cesium-137 Immobilization. *J. Eur. Ceram. Soc.* **2022**, *42*, 3004–3014. [CrossRef]
72. Yoldi, M.; Fuentes-Ordoñez, E.G.; Korili, S.A.; Gil, A. Zeolite Synthesis from Industrial Wastes. *Microporous Mesoporous Mater.* **2019**, *287*, 183–191. [CrossRef]
73. Gualtieri, A.F. Synthesis of Sodium Zeolites from a Natural Halloysite. *Phys. Chem. Min.* **2001**, *28*, 719–728. [CrossRef]
74. Novembre, D.; di Sabatino, B.; Gimeno, D. Synthesis of Na-A Zeolite from 10 Å Halloysite and a New Crystallization Kinetic Model for the Transformation of Na-A into HS Zeolite. *Clays Clay Min.* **2005**, *53*, 28–36. [CrossRef]
75. Kirdeciler, S.K.; Akata, B. One Pot Fusion Route for the Synthesis of Zeolite 4A Using Kaolin. *Adv. Powder Technol.* **2020**, *31*, 4336–4343. [CrossRef]
76. Belviso, C.; Cavalcante, F.; Lettino, A.; Fiore, S. A and X-Type Zeolites Synthesised from Kaolinite at Low Temperature. *Appl. Clay Sci.* **2013**, *80–81*, 162–168. [CrossRef]
77. Belviso, C.; Cavalcante, F.; Fiore, S. Synthesis of Zeolite from Italian Coal Fly Ash: Differences in Crystallization Temperature Using Seawater Instead of Distilled Water. *Waste Manag.* **2010**, *30*, 839–847. [CrossRef]
78. Ju, T.; Meng, Y.; Han, S.; Lin, L.; Jiang, J. On the State of the Art of Crystalline Structure Reconstruction of Coal Fly Ash: A Focus on Zeolites. *Chemosphere* **2021**, *283*, 131010. [CrossRef]
79. Yang, T.; Han, C.; Liu, H.; Yang, L.; Liu, D.; Tang, J.; Luo, Y. Synthesis of Na-X Zeolite from Low Aluminum Coal Fly Ash: Characterization and High Efficient As(V) Removal. *Adv. Powder Technol.* **2019**, *30*, 199–206. [CrossRef]
80. Florez, C.; Restrepo-Baena, O.; Tobon, J.I. Effects of Calcination and Milling Pre-Treatments on Natural Zeolites as a Supplementary Cementitious Material. *Constr. Build. Mater.* **2021**, *310*, 125220. [CrossRef]
81. Yang, S.; Yang, L.; Gao, M.; Bai, H.; Nagasaka, T. Synthesis of Zeolite-Geopolymer Composites with High Zeolite Content for Pb(II) Removal by a Simple Two-Step Method Using Fly Ash and Metakaolin. *J. Clean. Prod.* **2022**, *378*, 134528. [CrossRef]
82. Le, T.M.; Nguyen, G.T.; Dat, N.D.; Tran, N.T. An Innovative Approach Based on Microwave Radiation for Synthesis of Zeolite 4A and Porosity Enhancement. *Results Eng.* **2023**, *19*, 101235. [CrossRef]
83. Cheng, M.; Luo, Y.; Geng, J.; Cui, R.; Qu, Y.; Sun, L.; Dou, Q.; Fu, H. Adsorption Behavior of Iodide Ion by Silver-Doped Zeolite 4A in LiCl-KCl Molten Salt. *Adv. Powder Technol.* **2022**, *33*, 103415. [CrossRef]
84. Sowunmi, A.R.; Folayan, C.O.; Anafi, F.O.; Ajayi, O.A.; Omisanya, N.O.; Obada, D.O.; Dodoo-Arhin, D. Dataset on the Comparison of Synthesized and Commercial Zeolites for Potential Solar Adsorption Refrigerating System. *Data Brief.* **2018**, *20*, 90–95. [CrossRef] [PubMed]
85. Wang, P.; Sun, Q.; Zhang, Y.; Cao, J. Synthesis of Zeolite 4A from Kaolin and Its Adsorption Equilibrium of Carbon Dioxide. *Materials* **2019**, *12*, 1536. [CrossRef]
86. Cortés, F.B.; Chejne, F.; Carrasco-Marín, F.; Pérez-Cadenas, A.F.; Moreno-Castilla, C. Water Sorption on Silica- and Zeolite-Supported Hygroscopic Salts for Cooling System Applications. *Energy Convers. Manag.* **2012**, *53*, 219–223. [CrossRef]

87. Jacobs, J.H.; Deering, C.E.; Sui, R.; Cann, A.P.; Lesage, K.L.; Marriott, R.A. The Role of Carbon Dioxide and Water in the Degradation of Zeolite 4A, Zeolite 13X and Silica Gels. *New J. Chem.* **2023**, *47*, 5249–5261. [CrossRef]
88. Ojha, K.; Pradhan, N.C.; Nath Samanta, A. Zeolite from Fly Ash: Synthesis and Characterization. *Bull. Mater. Sci.* **2004**, *27*, 555–564. [CrossRef]
89. Hoang, P.H.; Dat, N.M. Study on Using Cellulose Derivatives as Pore Directing Agent for Preparation of Hierarchical ZSM-5 Zeolite Catalyst. *Adv. Powder Technol.* **2021**, *32*, 3927–3933. [CrossRef]
90. Gevert, B.; Eriksson, L.; Törncrona, A. Preparation of Discrete Colloidal ZSM-5 Crystals with High Al-Content. *J. Porous Mater.* **2011**, *18*, 723–728. [CrossRef]
91. Hoff, T.C.; Gardner, D.W.; Thilakaratne, R.; Wang, K.; Hansen, T.W.; Brown, R.C.; Tessonnier, J.P. Tailoring ZSM-5 Zeolites for the Fast Pyrolysis of Biomass to Aromatic Hydrocarbons. *ChemSusChem* **2016**, *9*, 1473–1482. [CrossRef]
92. Jerzak, W.; Gao, N.; Kalemba-Rec, I.; Magdziarz, A. Catalytic Intermediate Pyrolysis of Post-Extraction Rapeseed Meal by Reusing ZSM-5 and Zeolite Y Catalysts. *Catal. Today* **2022**, *404*, 63–77. [CrossRef]
93. Zhu, M.-H.; Chen, L.; Ding, W.; Zou, L.; Wu, T.; Li, Y.; Hu, N.; Chen, X.; Kita, H. Preparation and Catalytic Performance of TS-2 Zeolite Membrane. *Microporous Mesoporous Mater.* **2022**, *331*, 111660. [CrossRef]
94. Hosseini Hashemi, M.S.; Eslami, F.; Karimzadeh, R. Organic Contaminants Removal from Industrial Wastewater by CTAB Treated Synthetic Zeolite Y. *J. Environ. Manag.* **2019**, *233*, 785–792. [CrossRef] [PubMed]
95. Zou, Q.; Lin, W.; Xu, D.; Wu, S.; Mondal, A.K.; Huang, F. Study the Effect of Zeolite Pore Size and Acidity on the Catalytic Pyrolysis of Kraft Lignin. *Fuel Process. Technol.* **2022**, *237*, 107467. [CrossRef]
96. Doyle, A.M.; Albayati, T.M.; Abbas, A.S.; Alismaeel, Z.T. Biodiesel Production by Esterification of Oleic Acid over Zeolite Y Prepared from Kaolin. *Renew. Energy* **2016**, *97*, 19–23. [CrossRef]
97. Querol, X.; Moreno, N.; Umaña, J.C.; Alastuey, A.; Hernández, E.; López-Soler, A.; Plana, F. Synthesis of Zeolites from Coal Fly Ash: An Overview. *Int. J. Coal Geol.* **2002**, *50*, 413–423. [CrossRef]
98. Abdullahi, T.; Harun, Z.; Othman, M.H.D. A Review on Sustainable Synthesis of Zeolite from Kaolinite Resources via Hydrothermal Process. *Adv. Powder Technol.* **2017**, *28*, 1827–1840. [CrossRef]
99. Murakami, T.; Otsuka, K.; Fukasawa, T.; Ishigami, T.; Fukui, K. Hierarchical Porous Zeolite Synthesis from Coal Fly Ash via Microwave Heating. *Colloids Surf. A Physicochem. Eng. Asp.* **2023**, *661*, 130941. [CrossRef]
100. Król, M. Natural vs. Synthetic Zeolites. *Crystals* **2020**, *10*, 622. [CrossRef]
101. Supelano, G.I.; Palacio, C.A.; Weber, M.; McCoy, J.; Jennings, J.; Mejía Gómez, J.A.; Parra Vargas, C.A.; Ortiz, C.; Lynn, K. Study of Fly Ash Based Zeolite by Using Slow Positrons. *Mater. Sci. Eng. B Solid. State Mater. Adv. Technol.* **2021**, *263*, 114890. [CrossRef]
102. Koshlak, H. Synthesis of Zeolites from Coal Fly Ash Using Alkaline Fusion and Its Applications in Removing Heavy Metals. *Materials* **2023**, *16*, 4837. [CrossRef] [PubMed]
103. Li, G.; Li, M.; Zhang, X.; Cao, P.; Jiang, H.; Luo, J.; Jiang, T. Hydrothermal Synthesis of Zeolites-Calcium Silicate Hydrate Composite from Coal Fly Ash with Co-Activation of $Ca(OH)_2$-NaOH for Aqueous Heavy Metals Removal. *Int. J. Min. Sci. Technol.* **2022**, *32*, 563–573. [CrossRef]
104. Verrecchia, G.; Cafiero, L.; de Caprariis, B.; Dell'Era, A.; Pettiti, I.; Tuffi, R.; Scarsella, M. Study of the Parameters of Zeolites Synthesis from Coal Fly Ash in Order to Optimize Their CO_2 Adsorption. *Fuel* **2020**, *276*, 118041. [CrossRef]
105. Panek, R.; Madej, J.; Bandura, L.; Słowik, G. Recycling of Waste Solution after Hydrothermal Conversion of Fly Ash on a Semi-Technical Scale for Zeolite Synthesis. *Materials* **2021**, *14*, 1413. [CrossRef]
106. Dere Ozdemir, O.; Piskin, S. A Novel Synthesis Method of Zeolite X From Coal Fly Ash: Alkaline Fusion Followed by Ultrasonic-Assisted Synthesis Method. *Waste Biomass Valorization* **2019**, *10*, 143–154. [CrossRef]
107. Chen, W.; Song, G.; Lin, Y.; Qiao, J.; Wu, T.; Yi, X.; Kawi, S. Synthesis and Catalytic Performance of Linde-Type A Zeolite (LTA) from Coal Fly Ash Utilizing Microwave and Ultrasound Collaborative Activation Method. *Catal. Today* **2022**, *397–399*, 407–418. [CrossRef]
108. Shi, B.; Zhao, J.; Chang, Q. Green Synthesize of Fly Ash-Based Zeolite X: A Potential Microwave Absorbent. *J. Mater. Sci. Mater. Electron.* **2021**, *32*, 26097–26104. [CrossRef]
109. Ndlovu, N.Z.N.; Ameh, A.E.; Petrik, L.F.; Ojumu, T.V. Synthesis and Characterisation of Pure Phase ZSM-5 and Sodalite Zeolites from Coal Fly Ash. *Mater. Today Commun.* **2023**, *34*, 105436. [CrossRef]
110. Kunecki, P.; Panek, R.; Wdowin, M.; Bień, T.; Franus, W. Influence of the Fly Ash Fraction after Grinding Process on the Hydrothermal Synthesis Efficiency of Na-A, Na-P1, Na-X and Sodalite Zeolite Types. *Int. J. Coal Sci. Technol.* **2021**, *8*, 291–311. [CrossRef]
111. Yang, L.; Jiang, T.; Xiong, P.; Yang, S.; Gao, M.; Nagasaka, T. Green Activating Silica-Alumina Insoluble Phase of Fly Ash to Synthesize Zeolite P with High Adsorption Capacity for Pb(II) in Solution. *Adv. Powder Technol.* **2023**, *34*, 103938. [CrossRef]
112. Zhang, X.; Li, C.; Zheng, S.; Di, Y.; Sun, Z. A Review of the Synthesis and Application of Zeolites from Coal-Based Solid Wastes. *Int. J. Miner. Metall. Mater.* **2022**, *29*, 1–21. [CrossRef]
113. Dabbawala, A.A.; Ismail, I.; Vaithilingam, B.V.; Polychronopoulou, K.; Singaravel, G.; Morin, S.; Berthod, M.; Al Wahedi, Y. Synthesis of Hierarchical Porous Zeolite-Y for Enhanced CO_2 Capture. *Microporous Mesoporous Mater.* **2020**, *303*, 110261. [CrossRef]
114. Bai, R.; Song, Y.; Li, Y.; Yu, J. Creating Hierarchical Pores in Zeolite Catalysts. *Trends Chem.* **2019**, *1*, 601–611. [CrossRef]
115. Wdowin, M.; Franus, M.; Panek, R.; Badura, L.; Franus, W. The Conversion Technology of Fly Ash into Zeolites. *Clean. Technol. Environ. Policy* **2014**, *16*, 1217–1223. [CrossRef]

116. Querol, X.; Umaña, J.C.; Plana, F.; Alastuey, A.; Lopez-Soler, A.; Medinaceli, A.; Valero, A.; Domingo, M.J.; Garcia-Rojo, E. Synthesis of Zeolites from Fly Ash at Pilot Plant Scale. Examples of Potential Applications. *Fuel* **2001**, *80*, 857–865. [CrossRef]
117. Bonetti, B.; Ferrarini, S.F.; Hammercshmitt, M.E.; De Aquino, T.F.; Pezente, D.P.; Dos Santos Zavarize, M.; Pires, M.J.R. Use of Pilot Scale Produced Zeolites as Fertilizer. *J. Environ. Chem. Eng.* **2022**, *10*, 107907. [CrossRef]
118. Xu, H.Y.; Yuan, X.Q.; Dai, L.Y.; Wang, W.S.; Li, Y.; Dong, L.M. Fenton-like Degradation of Rhodamine B over Fe_2O_3–Al_2O_3-Zeolite Hybrids Derived from Coal Fly Ash. *Int. J. Environ. Sci. Technol.* **2023**. [CrossRef]
119. Supelano, G.I.; Gómez Cuaspud, J.A.; Moreno-Aldana, L.C.; Ortiz, C.; Trujillo, C.A.; Palacio, C.A.; Parra Vargas, C.A.; Mejía Gómez, J.A. Synthesis of Magnetic Zeolites from Recycled Fly Ash for Adsorption of Methylene Blue. *Fuel* **2020**, *263*, 116800. [CrossRef]
120. Basaldella, E.I.; Torres Sánchez, R.M.; Conconi, M.S. Conversion of Exhausted Fluid Cracking Catalysts into Zeolites by Alkaline Fusion. *Appl. Clay Sci.* **2009**, *42*, 611–614. [CrossRef]
121. Tian, X.; Shan, Y.; Zhang, J.; Yan, Z.; Sun, Y.; Ding, W.; Yu, Y. The Study of Pt/Zeolites for CO Oxidation: Effects of Skeleton Structure and Si/Al Ratio. *Catal. Commun.* **2023**, *178*, 106679. [CrossRef]
122. Yu, L.; Xu, C.; Zhou, Q.; Fu, X.; Liang, Y.; Wang, W. Facile Synthesis of Hierarchical Porous ZSM-5 Zeolite with Tunable Mesostructure and Its Application in Catalytic Cracking of LDPE. *J. Alloys Compd.* **2023**, *965*, 171454. [CrossRef]
123. Diaz, I.; Mayoral, A. TEM Studies of Zeolites and Ordered Mesoporous Materials. *Micron* **2011**, *42*, 512–527. [CrossRef] [PubMed]
124. Valbuena, H.M.G.; Medina, A.F.; Vargas, J.C.; Fandiño, O.H. Synthesis of Zeolites Na-A, Na-X, and Analcime from Crushed Stone Waste and Their Applications in Heavy Metal Removal in Aqueous Media. *Chem. Eng. Res. Des.* **2023**, *197*, 159–172. [CrossRef]
125. Mlonka-Mędrala, A.; Hasan, T.; Kalawa, W.; Sowa, M.; Sztekler, K.; Pinto, M.L.; Mika, Ł. Possibilities of Using Zeolites Synthesized from Fly Ash in Adsorption Chillers. *Energies* **2022**, *15*, 7444. [CrossRef]
126. Baran, R.; Millot, Y.; Onfroy, T.; Krafft, J.M.; Dzwigaj, S. Influence of the Nitric Acid Treatment on Al Removal, Framework Composition and Acidity of BEA Zeolite Investigated by XRD, FTIR and NMR. *Microporous Mesoporous Mater.* **2012**, *163*, 122–130. [CrossRef]
127. Majchrzak-Kucęba, I. A Simple Thermogravimetric Method for the Evaluation of the Degree of Fly Ash Conversion into Zeolite Material. *J. Porous Mater.* **2013**, *20*, 407–415. [CrossRef]
128. Ashbrook, S.E.; Morris, R.; Rice, C.M. Understanding the Synthesis and Reactivity of ADORable Zeolites Using NMR Spectroscopy. *Curr. Opin. Colloid Interface Sci.* **2022**, *61*, 101634. [CrossRef]
129. Shigemoto, N.; Sugiyama, S.; Hayashi, H.; Miyaura, K. Characterization of Na-X, Na-A, and Coal Fly Ash Zeolites and Their Amorphous Precursors by IR, MAS NMR and XPS. *J. Mater. Sci.* **1995**, *30*, 5777–5783. [CrossRef]
130. Panagiotou, T.; Morency, J.R.; Senior, C.L. Zeolite-Based Mercury Sorbent-Laboratory Testing and Modeling. *ACS Div. Fuel Chem. Prepr.* **2000**, *45*, 426–430.
131. Ghorbannezhad, P.; Park, S.; Onwudili, J.A. Co-Pyrolysis of Biomass and Plastic Waste over Zeolite- and Sodium-Based Catalysts for Enhanced Yields of Hydrocarbon Products. *Waste Manag.* **2020**, *102*, 909–918. [CrossRef]
132. Bellat, J.P.; Weber, G.; Bezverkhyy, I.; Lamonier, J.F. Selective Adsorption of Formaldehyde and Water Vapors in NaY and NaX Zeolites. *Microporous Mesoporous Mater.* **2019**, *288*, 109563. [CrossRef]
133. Dabbawala, A.A.; Suresh Kumar Reddy, K.; Mittal, H.; Al Wahedi, Y.; Vaithilingam, B.V.; Karanikolos, G.N.; Singaravel, G.; Morin, S.; Berthod, M.; Alhassan, S.M. Water Vapor Adsorption on Metal-Exchanged Hierarchical Porous Zeolite-Y. *Microporous Mesoporous Mater.* **2021**, *326*, 111380. [CrossRef]
134. Maia, A.Á.B.; Neves, R.F.; Angélica, R.S.; Pöllmann, H. Synthesis, Optimisation and Characterisation of the Zeolite NaA Using Kaolin Waste from the Amazon Region. Production of Zeolites KA, MgA and CaA. *Appl. Clay Sci.* **2015**, *108*, 55–60. [CrossRef]
135. European Comission. Commission Implementing Decision (EU) 2021/2326 of 30 November 2021 Establishing Best Available Techniques (BAT) Conclusions, under Directive 2010/75/EU of the European Parliament and of the Council, for Large Combustion Plants (Notified under Document C (2021) 8580). Official Journal of the European Union. 2021. Available online: https://eur-lex.europa.eu/legal-content/EN/TXT/?uri=CELEX%3A32021D2326 (accessed on 23 July 2023).
136. Kunecki, P.; Wdowin, M.; Hanc, E. Fly Ash-Derived Zeolites and Their Sorption Abilities in Relation to Elemental Mercury in a Simulated Gas Stream. *J. Clean. Prod.* **2023**, *391*, 136181. [CrossRef]
137. Yang, L.; Qian, X.; Yuan, P.; Bai, H.; Miki, T.; Men, F.; Li, H.; Nagasaka, T. Green Synthesis of Zeolite 4A Using Fly Ash Fused with Synergism of NaOH and Na_2CO_3. *J. Clean. Prod.* **2019**, *212*, 250–260. [CrossRef]
138. Lee, K.M.; Jo, Y.M. Synthesis of Zeolite from Waste Fly Ash for Adsorption of CO_2. *J. Mater. Cycles Waste Manag.* **2010**, *12*, 212–219. [CrossRef]
139. Muriithi, G.N.; Petrik, L.F.; Doucet, F.J. Synthesis, Characterisation and CO_2 Adsorption Potential of NaA and NaX Zeolites and Hydrotalcite Obtained from the Same Coal Fly Ash. *J. CO2 Util.* **2020**, *36*, 220–230. [CrossRef]
140. Boycheva, S.; Zgureva, D.; Lazarova, H.; Popova, M. Comparative Studies of Carbon Capture onto Coal Fly Ash Zeolites Na-X and Na–Ca-X. *Chemosphere* **2021**, *271*, 129505. [CrossRef]
141. He, X.; Yao, B.; Xia, Y.; Huang, H.; Gan, Y.; Zhang, W. Coal Fly Ash Derived Zeolite for Highly Efficient Removal of Ni^{2+} Inwaste Water. *Powder Technol.* **2020**, *367*, 40–46. [CrossRef]
142. Yang, L.; Gao, M.; Wei, T.; Nagasaka, T. Synergistic Removal of As(V) from Aqueous Solution by Nanozero Valent Iron Loaded with Zeolite 5A Synthesized from Fly Ash. *J. Hazard. Mater.* **2022**, *424*, 127428. [CrossRef]

143. Ankrah, A.F.; Tokay, B.; Snape, C.E. Heavy Metal Removal from Aqueous Solutions Using Fly-Ash Derived Zeolite NaP1. *Int. J. Environ. Res.* **2022**, *16*, 17. [CrossRef]
144. Han, L.; Wang, J.; Liu, Z.; Zhang, Y.; Jin, Y.; Li, J.; Wang, D. Synthesis of Fly Ash-Based Self-Supported Zeolites Foam Geopolymer via Saturated Steam Treatment. *J. Hazard. Mater.* **2020**, *393*, 122468. [CrossRef] [PubMed]
145. Myat, A.; Kim Choon, N.; Thu, K.; Kim, Y.D. Experimental Investigation on the Optimal Performance of Zeolite-Water Adsorption Chiller. *Appl. Energy* **2013**, *102*, 582–590. [CrossRef]
146. Liu, Y.M.; Yuan, Z.X.; Wen, X.; Du, C.X. Evaluation on Performance of Solar Adsorption Cooling of Silica Gel and SAPO-34 Zeolite. *Appl. Therm. Eng.* **2021**, *182*, 116019. [CrossRef]
147. Li, A.; Ismail, A.B.; Thu, K.; Ng, K.C.; Loh, W.S. Performance Evaluation of a Zeolite—Water Adsorption Chiller with Entropy Analysis of Thermodynamic Insight. *Appl. Energy* **2014**, *130*, 702–711. [CrossRef]
148. Iqbal, A.; Sattar, H.; Haider, R.; Munir, S. Synthesis and Characterization of Pure Phase Zeolite 4A from Coal Fly Ash. *J. Clean. Prod.* **2019**, *219*, 258–267. [CrossRef]
149. Subhapriya, S.; Gomathipriya, P. Zeolite X from Coal Fly Ash Inhibits Proliferation of Human Breast Cancer Cell Lines (MCF-7) via Induction of S Phase Arrest and Apoptosis. *Mol. Biol. Rep.* **2018**, *45*, 2063–2074. [CrossRef]
150. Rajakrishnamoorthy, P.; Karthikeyan, D.; Saravanan, C.G. Emission Reduction Technique Applied in SI Engines Exhaust by Using Zsm5 Zeolite as Catalysts Synthesized from Coal Fly Ash. *Mater. Today Proc.* **2020**, *22*, 499–506. [CrossRef]
151. Wu, Y.; Liang, G.; Zhao, X.; Wang, H.; Qu, Z. Flexible Textural Design of ZSM-5 Zeolite Adsorbent from Coal Fly Ash via Solvent-Free Method for Toluene Elimination. *J. Environ. Chem. Eng.* **2023**, *11*, 109589. [CrossRef]
152. Krzywanski, J.; Grabowska, K.; Sosnowski, M.; Zylka, A.; Kulakowska, A.; Czakiert, T.; Sztekler, K.; Wesolowska, M.; Nowak, W. Heat Transfer in Adsorption Chillers with Fluidized Beds of Silica Gel, Zeolite, and Carbon Nanotubes. *Heat Transf. Eng.* **2021**, *43*, 172–182. [CrossRef]
153. Kayal, S.; Baichuan, S.; Saha, B.B. Adsorption Characteristics of AQSOA Zeolites and Water for Adsorption Chillers. *Int. J. Heat Mass Transf.* **2016**, *92*, 1120–1127. [CrossRef]
154. Wei, H.; Teo, B.; Chakraborty, A.; Han, B. Water Adsorption on CHA and AFI Types Zeolites: Modelling and Investigation of Adsorption Chiller under Static and Dynamic Conditions. *Appl. Therm. Eng.* **2017**, *127*, 35–45. [CrossRef]
155. Choudhury, B.; Saha, B.B.; Chatterjee, P.K.; Sarkar, J.P. An Overview of Developments in Adsorption Refrigeration Systems towards a Sustainable Way of Cooling. *Appl. Energy* **2013**, *104*, 554–567. [CrossRef]
156. Grabowska, K.; Krzywanski, J.; Nowak, W.; Wesolowska, M. Construction of an Innovative Adsorbent Bed Configuration in the Adsorption Chiller—Selection Criteria for Effective Sorbent-Glue Pair. *Energy* **2018**, *151*, 317–323. [CrossRef]
157. Fernandes, M.S.; Brites, G.J.V.N.; Costa, J.J.; Gaspar, A.R.; Costa, V.A.F. Review and Future Trends of Solar Adsorption Refrigeration Systems. *Renew. Sustain. Energy Rev.* **2014**, *39*, 102–123. [CrossRef]
158. Amoni, B.C.; Freitas, A.D.L.; Bessa, R.A.; Oliveira, C.P.; Bastos-Neto, M.; Azevedo, D.C.S.; Lucena, S.M.P.; Sasaki, J.M.; Soares, J.B.; Soares, S.A.; et al. Effect of Coal Fly Ash Treatments on Synthesis of High-Quality Zeolite A as a Potential Additive for Warm Mix Asphalt. *Mater. Chem. Phys.* **2022**, *275*, 125197. [CrossRef]
159. Estevam, S.T.; de Aquino, T.F.; da Silva, T.D.; da Cruz, R.; Bonetti, B.; Riella, H.G.; Soares, C. Synthesis of K-Merlinoite Zeolite from Coal Fly Ash for Fertilizer Application. *Braz. J. Chem. Eng.* **2022**, *39*, 631–643. [CrossRef]
160. Zhang, X.; Wang, B.; Chang, J.; Fan, C.; Liu, Z. Effect of Zeolite Contents on Mineral Evolution and Heavy Metal Solidification in Alkali-Activated MSWI Fly Ash Specimens. *Constr. Build. Mater.* **2022**, *345*, 128309. [CrossRef]
161. Chen, J.; Yang, R.; Zhang, Z.; Wu, D. Removal of Fluoride from Water Using Aluminum Hydroxide-Loaded Zeolite Synthesized from Coal Fly Ash. *J. Hazard. Mater.* **2022**, *421*, 126817. [CrossRef]
162. Teng, L.; Jin, X.; Bu, Y.; Ma, J.; Liu, Q.; Yang, J.; Liu, W.; Yao, L. Facile and Fast Synthesis of Cancrinite-Type Zeolite from Coal Fly Ash by a Novel Hot Stuffy Route. *J. Environ. Chem. Eng.* **2022**, *10*, 108369. [CrossRef]
163. Gjyli, S.; Korpa, A. High Catalytic Activity of a Seawater Fly Ash Based Zeolite for Phenol Alkylation. *Kinet. Catal.* **2021**, *62*, 270–278. [CrossRef]
164. Mahima Kumar, M.; Senthilvadivu, R.; Brahmaji Rao, J.S.; Neelamegam, M.; Ashok Kumar, G.V.S.; Kumar, R.; Jena, H. Characterization of Fly Ash by ED-XRF and INAA for the Synthesis of Low Silica Zeolites. *J. Radioanal. Nucl. Chem.* **2020**, *325*, 941–947. [CrossRef]
165. Panasenko, A.E.; Shichalin, O.O.; Yarusova, S.B.; Ivanets, A.I.; Belov, A.A.; Dran'kov, A.N.; Azon, S.A.; Fedorets, A.N.; Buravlev, I.Y.; Mayorov, V.Y.; et al. A Novel Approach for Rice Straw Agricultural Waste Utilization: Synthesis of Solid Aluminosilicate Matrices for Cesium Immobilization. *Nucl. Eng. Technol.* **2022**, *54*, 3250–3259. [CrossRef]
166. Lin, X.; Zhang, Z.; Wang, Q. Evaluation of Zeolite Catalysts on Product Distribution and Synergy during Wood-Plastic Composite Catalytic Pyrolysis. *Energy* **2019**, *189*, 116174. [CrossRef]
167. Bai, S.; Zhou, L.; Chang, Z.; Zhang, C.; Chu, M. Synthesis of Na-X Zeolite from Longkou Oil Shale Ash by Alkaline Fusion Hydrothermal Method. *Carbon. Resour. Convers.* **2018**, *1*, 245–250. [CrossRef]
168. Roulia, M.; Koukouza, K.; Stamatakis, M.; Vasilatos, C. Fly-Ash Derived Na-P1, Natural Zeolite Tuffs and Diatomite in Motor Oil Retention. *Clean. Mater.* **2022**, *4*, 100063. [CrossRef]
169. Kumar, M.M.; Jena, H. Direct Single-Step Synthesis of Phase Pure Zeolite Na–P1, Hydroxy Sodalite and Analcime from Coal Fly Ash and Assessment of Their Cs^{+} and Sr^{2+} Removal Efficiencies. *Microporous Mesoporous Mater.* **2022**, *333*, 111738. [CrossRef]

170. Gollakota, A.R.K.; Volli, V.; Munagapati, V.S.; Wen, J.C.; Shu, C.M. Synthesis of Novel ZSM-22 Zeolite from Taiwanese Coal Fly Ash for the Selective Separation of Rhodamine 6G. *J. Mater. Res. Technol.* **2020**, *9*, 15381–15393. [CrossRef]
171. Mokrzycki, J.; Fedyna, M.; Marzec, M.; Szerement, J.; Panek, R.; Klimek, A.; Bajda, T.; Mierzwa-Hersztek, M. Copper Ion-Exchanged Zeolite X from Fly Ash as an Efficient Adsorbent of Phosphate Ions from Aqueous Solutions. *J. Environ. Chem. Eng.* **2022**, *10*, 108567. [CrossRef]
172. Chen, Q.; Zhao, Y.; Qiu, Q.; Long, L.; Liu, X.; Lin, S.; Jiang, X. Zeolite NaP1 Synthesized from Municipal Solid Waste Incineration Fly Ash for Photocatalytic Degradation of Methylene Blue. *Environ. Res.* **2023**, *218*, 114873. [CrossRef]
173. Soongprasit, K.; Vichaphund, S.; Sricharoenchaikul, V.; Atong, D. Activity of Fly Ash-Derived ZSM-5 and Zeolite X on Fast Pyrolysis of *Millettia (Pongamia) Pinnata* Waste. *Waste Biomass Valorization* **2020**, *11*, 715–724. [CrossRef]

energies

MDPI

Article

Possibilities of RDF Pyrolysis Products Utilization in the Face of the Energy Crisis

Magdalena Skrzyniarz [1], Marcin Sajdak [2,*], Monika Zajemska [1], Anna Biniek-Poskart [3], Józef Iwaszko [1] and Andrzej Skibiński [3]

1 Faculty of Production Engineering and Materials Technology, Czestochowa University of Technology, 19 Armii Krajowej Ave., 42-200 Czestochowa, Poland; monika.zajemska@pcz.pl (M.Z.)
2 Center of New Technologies, Department of Air Protection, Faculty of Energy and Environmental Engineering, Silesian University of Technology, 22 B Konarskiego Ave., 44-100 Gliwice, Poland
3 Faculty of Management, Czestochowa University of Technology, 19 B Armii Krajowej Ave., 42-200 Czestochowa, Poland; anna.poskart@pcz.pl (A.B.-P.)
* Correspondence: marcin.sajdak@polsl.pl; Tel.: +48-32-237-15-19

Abstract: The main goal of the study was to assess the possibility of practical use of products of pyrolysis of refuse-derived fuel (RDF), i.e., pyrolysis gas, biochar and pyrolysis oil, as an alternative to standard fossil fuels. The subject matter of the paper reaches out to the challenges faced by the global economy, not only in the context of the energy crisis, but also in the context of the energy transformation currently beginning in Europe. The increase in fuel and energy prices prompts countries to look for alternative solutions to Russian minerals. At the same time, the growing amount of municipal waste forces the implementation of solutions based on energy recovery (the amount of municipal waste per EU inhabitant in 2021 is 530 kg). One such solution is pyrolysis of RDF, i.e., fuels produced from the over-sieve fraction of municipal waste. In Poland, insufficient processing capacity of thermal waste conversion plants has led to significant surpluses of RDF (1.2 million Mg of undeveloped RDF in Poland in 2021). RDF, due to their high calorific value, can be a valuable energy resource (16–18 MJ/k). This issue is analyzed in this study.

Keywords: RDF; pyrolysis; steel and iron industry; sustainable production; energy crisis

Citation: Skrzyniarz, M.; Sajdak, M.; Zajemska, M.; Biniek-Poskart, A.; Iwaszko, J.; Skibiński, A. Possibilities of RDF Pyrolysis Products Utilization in the Face of the Energy Crisis. *Energies* **2023**, *16*, 6695. https://doi.org/10.3390/en16186695

Academic Editor: Shusheng Pang

Received: 26 July 2023
Revised: 1 September 2023
Accepted: 3 September 2023
Published: 19 September 2023

1. Introduction

The energy crisis that is currently affecting many countries around the world is largely due to the global economy's dependence on non-renewable fossil fuels such as coal, oil and natural gas [1]. Rising energy prices have contributed to the rising inflation, forcing some industrial plants to reduce or even stop production. Energy costs have also slowed down economic growth to such an extent that some countries are now heading for a deep recession. Production in Europe in particular, which consumes large amounts of gas in production processes, has suffered the most. These plants were forced to reduce their output because they could not afford to continue operating.

In addition to the energy crisis, the management of municipal solid waste is currently one of the biggest challenges, especially with regards to the environment [2]. The amount of municipal solid waste is increasing day by day all over the world [3] due to the continuous growth of industrialization, urbanization, consumerism, population growth and other factors [4]. Improper waste management has a negative impact on the environment [5], contributes to an increase in the amount of methane emitted into the atmosphere (which is one of the main factors responsible for global warming), causes deterioration of water quality and soil properties, and adversely affects human health [6]. Literature data show that since 2004 the world's largest producer of municipal solid waste (MSW) is China, which in one year collected six billion tons of unprocessed MSW in about 200 out of the 660 metropolises of the country. Improper management of MSW not only has a negative impact on the environment, but also poses a threat to public health and raises other socio-economic problems. Waste-to-energy (WtE) technologies such as pyrolysis, gasification, incineration and biomethanization [7] can convert MSW into usable energy (electricity and heat) in a safe and environmentally friendly way [8].

The results of literature research clearly prove that pyrolysis of municipal waste is a valuable source of secondary fuels, the wider use of which will contribute to further reducing the amount of landfilled waste, while increasing the diversification of fuel and energy sources in Poland [9,10].

On a national scale, it is not possible to convert all waste into RDF and, consequently, it is not possible to use them to the maximum extent in terms of energy [11]. An alternative to an incineration plant may be pyrolysis and waste gasification technology producing gas, the composition of which depends not only on the conditions of the process, but above all on the type of waste incinerated [12]. At present, the most problematic group of waste in terms of thermal conversion is plastic [13], rubber [14] and textile [14] waste. Spent railroad sleepers are also noteworthy, impregnated with creosote oil containing, among others, anthracene, fluorene, phenanthrene, benzo(a)anthracene and benzo(a)pyrene [15]. In addition to hydrogen, carbon monoxide and a small amount of methane, polycyclic aromatic hydrocarbons (PAHs) (Table 1) are also present in the composition of gas obtained as a result of thermal conversion of waste. The chemical composition of both pyrolysis gas and gas from waste gasification can be highly diverse, which translates into the impact of their combustion on the environment [16].

Table 1. PAH contents in the gaseous product after thermal conversion of municipal solid waste (MSW) and municipal solid waste co-incinerated with coal (MSW/Coal) in μg/g [16].

PAHs	**Process Temperature**									
	500 °C		600 °C		700 °C		800 °C		900 °C	
	Type of thermally converted fuel									
	MSW	MSW/Coal	MSW	MSW/Coal	MSW	MSW/Coal	MSW	MSW/Coal	MSW	MSW/Coal
2-ring	57.3	21.3	184.0	41.2	288.1	44.4	43.42	22.73	400.0	25.2
3-ring	671.8	67.26	1221.6	108.1	4283.7	257.9	5619.3	285.2	3078.9	884.5
4-ring	225.5	5.30	783.8	9.6	4040.6	15.2	1110.2	9.2	2002.3	1132.5
5-ring	14.4	0.52	98.9	0.2	835.4	4.9	100.1	0.1	235.8	70.9
6-ring	3.36	0.06	30.6	0.0	201.2	2.0	0.4	0.1	119.1	37.5
Sum	972.4	94.60	2319.0	159.2	9649.1	324.5	6873.4	317.5	5836.2	2150.8

Due to the energy crisis, many industries, including the iron and steel sectors, are facing economic problems and a loss of competitive advantage [17]. In connection with the above, numerous efforts are being made to increase energy efficiency of the metallurgical industry and thus reduce production costs, through, inter alia, the use of waste energy and alternative fuels (Figure 1) [18,19].

As revealed by literature sources, the iron and steel industries have great potential and opportunities to increase their energy efficiency, reduce carbon dioxide emissions and, as a result, promote sustainable development by replacing conventional fuels used in metallurgical processes with alternative fuels, including biomass (Figure 2) [20].

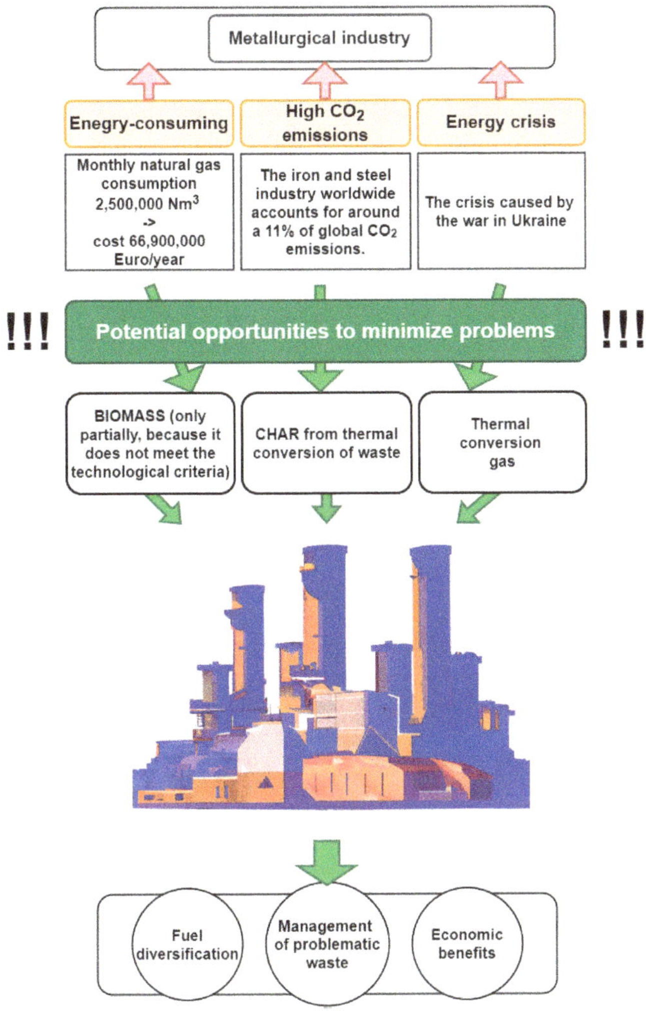

Figure 1. Potential opportunities to improve energy efficiency of the metallurgical industry [20].

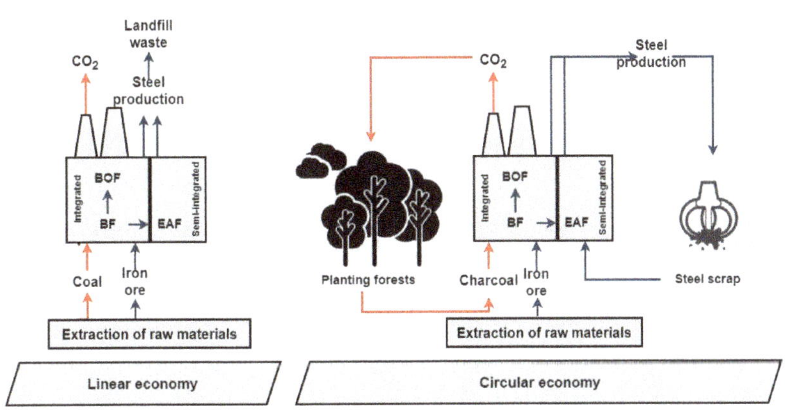

Figure 2. The idea of linear and circular economy in the metallurgical sector [8].

2. Theoretical Background

2.1. Energy Potential of RDF

RDF is a type of alternative fuel produced from MSW. Their calorific value is estimated at about 16–18 MJ/kg [21]. The process of producing RDF consists of separating combustible substances (rubber, paper, plastics, wood, etc.) from the municipal waste stream by sorting [10]. Thanks to pelletization, RDF is characterized by a homogeneous particle size, but the diverse composition of these fuels may cause high variability of their physicochemical properties. It is worth noting, however, that the Polish waste segregation system primarily uses mechanical-biological waste treatment plants, thanks to which adverse variability of the composition is reduced and a good input material for the production of RDF is obtained. The fraction selectively separated from the solid waste stream is characterized by increased stability of the material composition, and the impact of the waste collection site is negligible (Figure 3) [22,23].

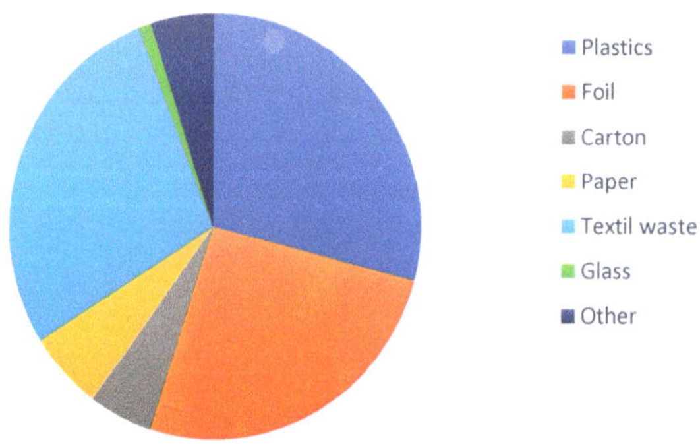

Figure 3. Typical composition of alternative fuels produced from mixed municipal waste in Poland [24].

Research by the authors of the publication [24] showed that the composition of 12 samples of alternative fuels was dominated by combustible components, such as plastics, wood, textiles, paper and cardboard, which together accounted for 94% of the mass of RDF. Other components found in the fuels were non-flammable and their presence negatively affected the calorific value of RDF. Producers and consumers of alternative fuels want the non-flammable fraction to constitute the lowest possible percentage of the fuel mass. This fraction should not be present in the alternative fuel, but often contaminates it as a result of bonding with other waste [24]. For comparison, Figure 4 shows the composition of solid recovered fuel (SRF)/refuse-derived fuel (RDF) in selected European countries. Composition studies were conducted between 2009 and 2019 and the results were averaged for all the countries [25]. Differences in the composition of fuels produced from waste may result from different waste management models in the individual countries. The culture and the ways of disposing of waste can also have an impact, as well as, for example, the way products are packed and the type of material used for this purpose.

2.2. Legal Background of Energy Recovery from RDF

Thanks to the departure in the EU from the linear economy model to the circular economy (CE) and the implementation of the "zero waste" idea, all of the EU countries are striving to stop landfilling municipal waste [26]. Changes in waste management are expected to have a beneficial impact on the economy, the environment and the health of residents of EU countries. One of the main documents introducing changes in waste management is the "Circular Economy Package" (2015), which implements long-term measures to reduce landfilling and to increase the reuse and recycling of waste [27].

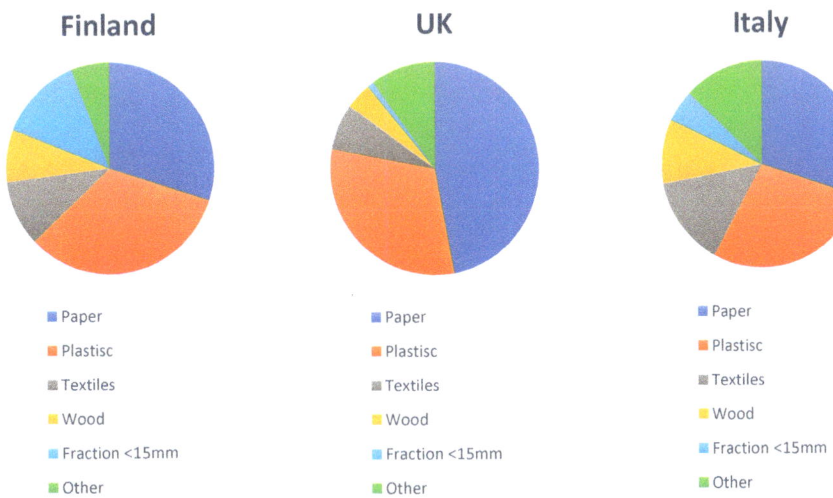

Figure 4. Typical compositions of samples of SRF/RDF commercially produced from municipal solid waste (Finland, UK, Italy) [25].

In Poland, appropriate legislative measures were also taken, convergent with the EU legislation and with the Regulation of the Minister of Economy (in force since 1 January 2016) which prohibit, among other things, the landfilling of high-calorific waste, i.e., waste with a combustion heat above 6 MJ/kg of dry matter (Figure 5) [28]. Although legal solutions to minimize the problem of waste storage have already been introduced and are in force, RDF have not obtained the status of "fuels" yet. These materials are still categorized as waste identified with code "19 12 10". Additionally, until now, neither the EU nor the Polish legislator has clearly defined what parameters RDF should have [29]. The lack of top-down unification of RDF parameters results in great freedom in determining the requirements of various industries or enterprises regarding these fuels. For example, the main clients for RDF (cement plants) approach these requirements very rigorously, setting fuel specifications at a very high level (Table 2).

Figure 5. Waste management in terms of calorific value [24,28,29].

Table 2. Required specifications of RDF (19 December 2010), ODRA S.A. cement plant.

Parameters	Units	Acceptable Limits Values
Total moisture content	%	≤18.0
Total sulfur content (operating condition)	%	≤0.5
Chlorine content (dry state)	%	≤0.7
LHV (operating condition)	GJ/Mg	≥20.0

Taking into account the amount of municipal waste produced each year in Poland (about 13.7 million tons, year 2021), the part suitable for thermal conversion accounts for 21.5% (2.7 million tons) (GUS data source). Cement plants burn only about 1 million tons per year (very restrictive calorific value standards > 18–20 MJ/kg), which means that other consumers, for example heating plants, are unable to consume all the fuel that is produced [24]. The constant increase in the amount of municipal waste also causes an increase in the production of fuel from waste. The need to manage the newly created waste fuel may be noticed in the extraction of fossil fuels, which in turn will have an impact on the climate and energy economy [30].

2.3. Processes of Thermal Conversion of Municipal Waste

In order to prioritise environmental concerns, European legislation mandates Member States to adhere to progressively rigorous rules pertaining to waste management and the mitigation of pollutant emissions. The objective of this initiative is to reduce the amount of trash that is disposed in landfills by promoting the utilisation of thermal waste conversion techniques for the recovery of materials and energy [31]. Thermochemical conversion methods like as pyrolysis and gasification have been found to result in reduced emissions of pollutants into the atmosphere as compared to combustion [32]. Moreover, these methods are considered economically feasible for the treatment of municipal or industrial waste. In recent years, there has been a growing recognition of the significance of pyrolysis and gasification techniques in mitigating the environmental consequences associated with combustion processes and the generation of power from waste materials. The employment of these methodologies facilitates significant adaptability in the utilisation of primary resources and also facilitates the acquisition of chemical energy in the shape of hydrocarbon products, alongside electrical power. The widespread adoption of waste-to-energy technology has the potential to reduce reliance on fossil fuels. According to a study conducted in the United States [33], the substitution of 0.4 tons of coal used for electricity production can be achieved by recovering each ton of municipal waste. It is anticipated that the widespread use of these technologies in the United States will result in a significant decrease in coal consumption, estimated to be approximately 100 tons year [33].

The circular economy model incorporates the utilisation of energy derived from trash, resulting in the neutralisation of the energy balance for several waste materials. The utilisation of the latent energy present in waste materials should be harnessed in order to fulfil energy demands, specifically for the generation of heat and electricity. The economic and environmental justification for harnessing the potential of garbage has been documented [34].

It is important to acknowledge that, in accordance with the relevant waste management hierarchy, the process of energy recovery takes precedence above recycling and composting. Hence, it is imperative to take into account the categorization of trash that is neither economically viable or technically unfeasible for recycling and composting (as seen in Figure 6) [26,35,36].

To release energy stored in waste, various methods of thermal conversion are used: incineration, pyrolysis, torrefaction and gasification (Figure 7) [37].

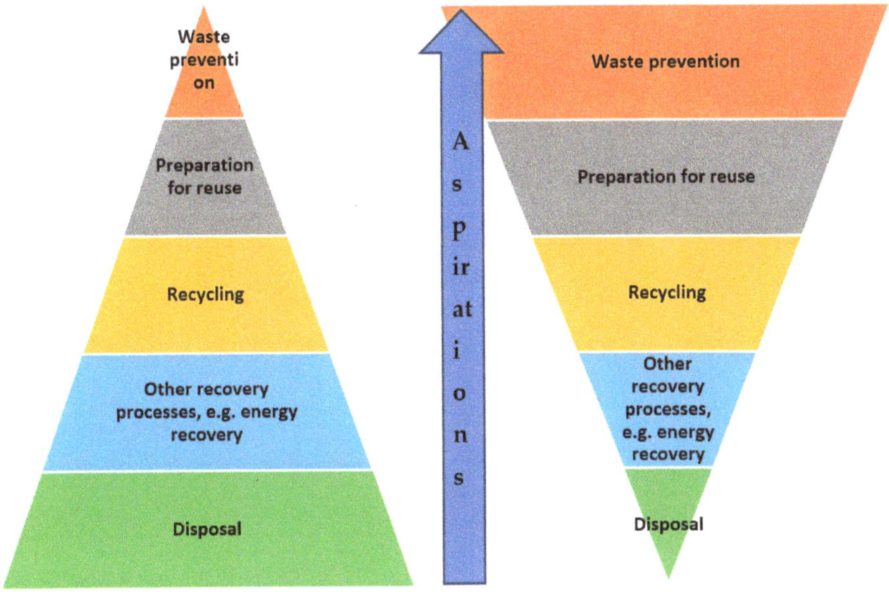

Figure 6. Waste management hierarchy [28].

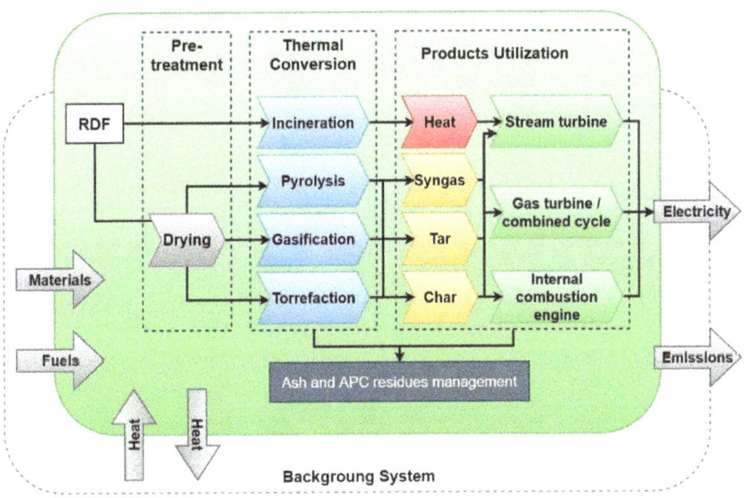

Figure 7. Methods of thermal conversion of waste [38].

The thermal conversion techniques depicted in Figure 7 are employed globally in diverse combined heat and power facilities to generate both electricity and heat [27]. However, it is important to note that each of these processes exhibits variations in parameters, including atmospheric conditions (namely the presence of oxygen), operating temperature, reactor system, and other factors. The assessment of end products and intermediate products is conducted to ascertain their quality. The temperature of a thermal process is mostly influenced by the appropriate design of the plant, the kind of reactor, and the input materials [39,40]. In underdeveloped nations, the practise of pre-treating trash is typically not observed in the context of incineration, as untreated municipal waste is commonly utilised as a primary input. Tables 3–6 in the cited literature [9,41–44] provide a comprehensive analysis and comparison of the various approaches employed for the thermal conversion of waste. Without a doubt, combustion is the prevailing mechanism of thermal conversion [45]. The technique described is well recognised and may be easily initiated and implemented in various regions, thanks to the prevalence of technology and the existence of ubiquitous infrastructure [27]. Nevertheless, combustion has numerous drawbacks in comparison to pyrolysis or gasification techniques [46].

Tables 3–6 present a comparative analysis of the primary benefits and drawbacks associated with pyrolysis and gasification methods, as indicated by reference [47].

Table 3. Comparison of advantages and disadvantages of the pyrolysis process [48–50].

PYROLYSIS	
✔	✘
• The sole thermochemical process capable of producing liquid, solid, and gaseous fractions suitable for immediate utilisation in diverse power plants; • A process with high flexibility, ability to control parameters (temperature, pressure, dwell time) so as to obtain the desired product; • Complete destruction of toxins (furans and dioxins); • The process is suitable for the disposal of many types of plastics and other waste groups; • Environmentally friendly—low NO_X and CO_2 emissions; • Less exhaust gas; • [Higher efficiency of energy recovery from] waste compared to other methods of disposal; • Energy recovery rate of up to 80%.	× High initial investment costs; × The need for fuel with a high calorific value is driven by the necessity for processed waste, particularly plastics, to possess a high calorific value in order to ensure the quality of the secondary fuels derived from them; × High water content in liquid oily products; × Technology little known to the "average person", which raises concerns about its use.

Table 4. Comparison of advantages and disadvantages of the gasification process [41,48,49,51].

GASIFICATION	
✔	✘
✓ High energy efficiency of the process and reduction of pollutant emissions; ✓ The resulting syngas is easy to purify compared to flue gas treatment after conventional waste incineration; ✓ Waste gasification technologies using cogeneration systems based on reciprocating engines enable their use in existing waste management systems; ✓ Reduction of waste volume by up to 95%; ✓ The possibility of using the method to manage various groups of waste.	× Tar production; × High initial investment costs; × Poor availability of infrastructure with reactors for waste gasification; × Corrosion of metal pipes during reaction; × Higher power consumption; × The amount and composition of syngas from gasification depending on the type of waste, gasifying agent, temperature, pressure and method of gasification.

Table 5. Comparison of the benefits and drawbacks of the torrefaction process [52–54].

TORREFACTION	
✓	✗
✓ Torrefaction products are biochar and torgas, ✓ Torrefied raw material has very low moisture (1–6%), higher calorific value and energy density, better abrasion and more uniform characteristics, lowering waste volume by 30% and keeping up to 90% of the basic energy in biochar., ✓ The possibility of using the method to manage various groups of waste, ✓ Torgas produced can be used to dry torrefied waste/fuel (torgas contains around 10% of the energy present in the initial material), ✓ Financial analyses encourage the construction of heating plants adapted to torrefied RDF combustion, showing high return on investment.	× The torrefaction technology has not yet reached commercial status and is still at the stage of research and testing in pilot installations, × High ash content in the fuel, × High initial investment outlays,

Table 6. Comparison of advantages and disadvantages of the combustion process [55–57].

COMBUSTION	
✓	✗
✓ Technology widely known and used on a massive scale, high availability of boilers and infrastructure, ✓ The technological process does not raise objections from society due to its universality, ✓ Combustion reduces up to 90% of the volume of the burned material.	× Combustion releases carcinogenic compounds such as dioxins and furans, as well as increases the emissivity of greenhouse gases and other air and atmosphere pollutants., × The need to adapt an expensive exhaust gas treatment system, × The greatest source of low emission is heat production (the combustion of low-quality fuels in boilers). × Combustion wastes resources whose re-use saves more energy and reduces the amount of pollution generated, × High initial investment outlays, × Highly energy-intensive process.

The concept underlying the pyrolysis process is the heat decomposition of organic matter [58], specifically in the case of refuse-derived fuel (RDF), in the absence of oxygen. This process yields gaseous, liquid, and solid byproducts (Figure 8) [59].

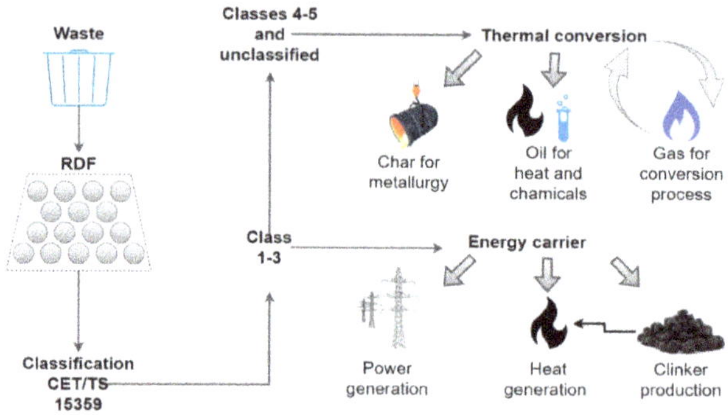

Figure 8. Diagram of the RDF pyrolysis process [60].

3. Possibilities of Using RDF Pyrolysis Products

Today, both Polish and global heavy industry (including the power generation sector) is characterized by high energy intensity, in addition to high pollutant emissions [42]. Given that improving energy efficiency is one of the priorities of the EU's energy policy, particular emphasis is placed on all forms of fuel and energy saving [61]. When it comes to Poland, there are many opportunities to save energy with relatively small financial resources. One solution is to use pyrolysis gas from the thermal conversion of RDF to fire metallurgical furnaces without the need for their pre-treatment, which in the case of polycyclic aromatic hydrocarbons (PAHs) is extremely problematic and expensive [16,62,63].

The issue of management of RDF pyrolysis products deserves attention, as evidenced by the interest of both waste management and metallurgical companies, among others, due to the following [64,65]:

- The possibility of managing the surplus RDF;
- The possibility of managing environmentally harmful plastic waste, such as polyethylene (PE), polypropylene (PP), polyvinyl chloride (PVC), polystyrene (PS) [66], textile and rubber wastes, as well as gaseous products of thermal conversion of these wastes;
- The possibility of reducing the consumption of natural gas for heating furnaces, which will translate into economic benefits [20,67].

The products of RDF pyrolysis are currently used mainly for energy generation purposes (Figure 9) [46].

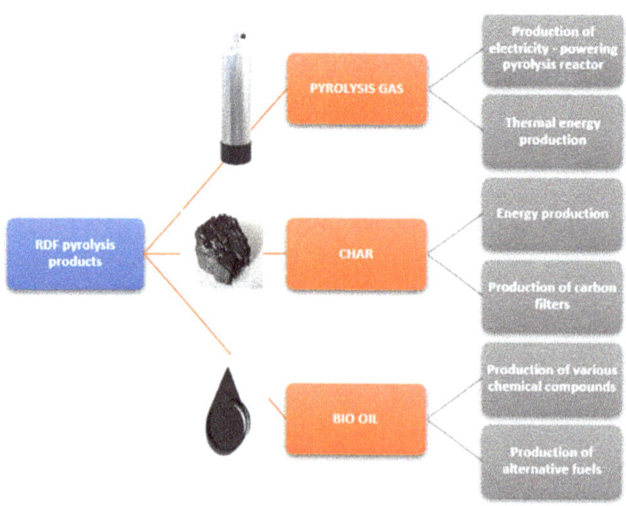

Figure 9. Uses of RDF pyrolysis products [24,24,62,68–70].

The gas fraction produced at higher temperatures, without a catalyst, is characterized by a high calorific value of about 28 MJ/m^3. So high calorific value makes this gas an attractive fuel which can be used not only in industrial boilers, but also in gas engines and turbines, enabling the production of electricity and heat [71]. At the same time, gas obtained after the catalytic process can be used to produce methanol, hydrocarbons, ammonia or liquid fuels by Fischer–Tropsch synthesis. Pyrolysis gas may also be a potential source of hydrogen due to its significant content [72]. This aspect can have a significant impact on the environment, as about 95% of H_2 comes from fossil fuels, and the share of this product in the energy market is increasing due to the growing demand for zero fuel emissions. Therefore, hydrogen obtained in this way can play an important role in the petrochemical, electronic, metallurgical and transportation industries. In addition, mention should be made of the potential use of pyrolysis gas as a fuel in integrated gasification combined cycle (IGCC) boilers, where its environmentally friendly nature could contribute to reducing the cost of gas purging equipment [73,74].

Biochar and pyrolysis oil can be used as energy raw materials or as base substances for the production of various chemicals [75–77]. Oil produced by RDF pyrolysis can be a good base for the production of substances such as styrene, benzene, toluene, xylenes (BTX) and naphthalene derivatives [24,60,69–71]. Biochar, which is a solid carbonized product of thermochemical conversion of municipal waste, is currently a popular alternative to expensive activated carbon due to its properties, especially in adsorption applications [42]. At the same time, this solid product from the pyrolysis of municipal waste can be a valuable fuel due to its relatively high calorific value (18–30 MJ/kg) (Figure 10) [42,78].

Taking into account the calorific values (Figure 10) featured by the individual RDF pyrolysis products, it can be seen that this factor determines their use in the industry [10,74,75]. The energy crisis caused by the armed conflict between Russia and Ukraine, the increase in fuel prices and the diversification of fuel suppliers are very important factors that affect the management of pyrolysis gas, biochar and bio-oil [76].

The products of pyrolysis vary in calorific value and composition due to different parameters [79], including feedstock type, reactor system, gas dwell time, heating rate, exposure time, temperature, pressure ranges, catalyst activity (if any), and the presence of hydrogen gas or hydrogen compounds [80].

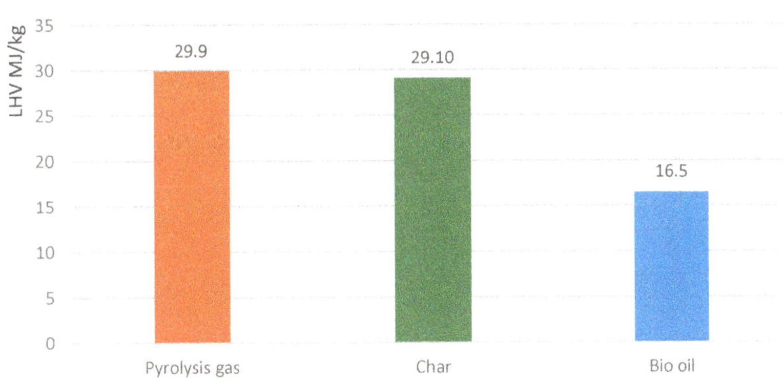

Figure 10. Comparison of calorific values of the individual RDF pyrolysis products [10,81,82].

The metallurgical industry is one of the key industries in developed countries [83]. Iron and steel production plays a key role in the global market economy. The world leader in iron and steel production is China, which supplies more than 50% of all raw materials in the metallurgical industry to the world market [84]. Metal products are the leading industrial material in almost every industry. Steel is of great importance for industry, urbanization and transport [34]. At the same time, the metallurgical industry is one of the most energy-intensive and polluting industries in the world [85]. These problems are met by the idea of a circular economy and the EU's "Green Steel for Europe" program [86], the aim of which is to decarbonize the steel industry [87], improve energy efficiency [88], achieve carbon neutrality and implement sustainability [89,90]. The transformation of the iron and steel industries is a long and complex process (Figure 11) [91]. Each mill has its own specificity of production, which involves individual adaptation of solutions to unique, already operating, process lines [92].

The metallurgical industry is currently intensifying its efforts to reduce CO_2 emissions [93]. At the same time, large-scale research is being carried out to reduce the use of fossil fuels or eliminate them altogether [94].

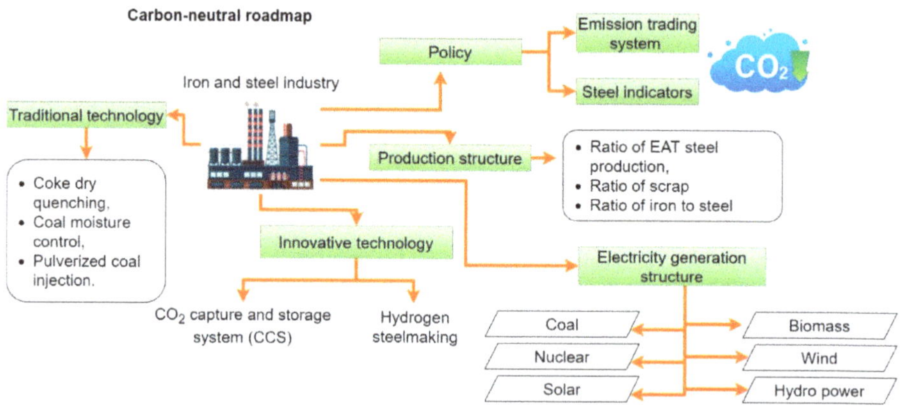

Figure 11. The path to decarbonization of the metallurgical industry [95].

The products of RDF pyrolysis show great potential for substituting conventional fuels. Countries such as Brazil show that it is possible to use biochar in the metallurgical industry. As much as 75% of total production of biochar in these countries goes to the iron and steel industries. In order for biochar to be used in metallurgy, it must meet strictly defined requirements, similar to those that characterize conventional fuels [96].

Biochar can replace conventional fuels in several stages of iron and steel production. Figure 12 shows the potential pathways for biochar to be introduced into the iron and steel production process [43]. Biochar can replace bituminous coal and coke in, among others:

- coke ovens;
- sintering processes;
- blast furnace processes;
- carbonization of liquid steel.

The metallurgical industry, as one of the most energy-intensive industries, consumes huge amounts of gaseous fuels. Just one of the mills operating in Poland consumes 2,500,000 Nm^3 of natural gas per month. The monthly costs incurred by the company are EUR 5,575,000. In 2022, during the deepest energy crisis in recent years, the price of natural gas reached a staggering price of EUR 2.63 (the mean EUR: PLN exchange rate in 2022 was 1:4.68). Steelworks are trying to cut production costs by switching to coke oven gas supplied from coking plants. Unfortunately, the price of this gas has also increased almost 3 times (0.21 EUR). Taking into account the calorific value of gas from the pyrolysis of RDF, which varies from 20 to even 30 MJ/m^3 [10,26,44,97], one can consider the use of pyrolysis gas for pre-heating furnaces before the plastic processing of metals [20]. Figure 13 compares the calorific values of conventional gaseous fuels with those of pyrolysis gas obtained from the pyrolysis of RDF.

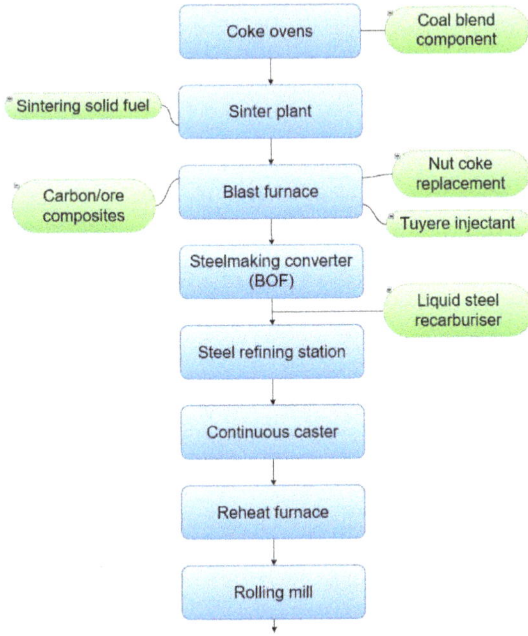

Figure 12. Biochar applications in metallurgy [95].

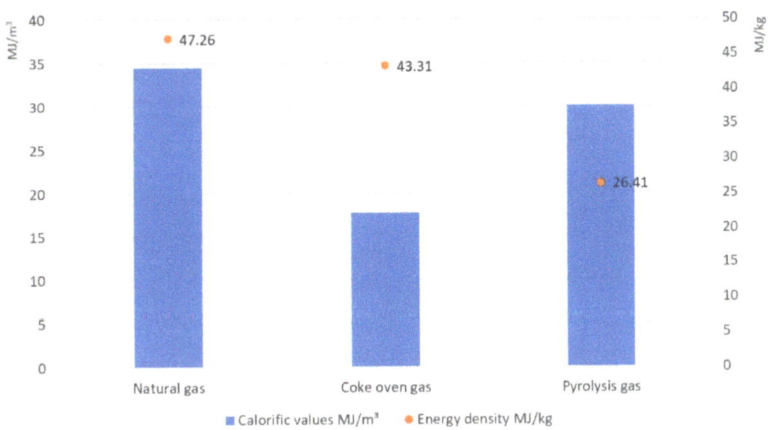

Figure 13. Comparison of the calorific values of conventional gaseous fuels with those of pyrolysis gas [9,10].

According to Szwaja et al. [67], it is possible to use gas obtained from thermal conversion (in this case from the torrefaction of biomass) for energy purposes in the metallurgical industry. The authors [20] proposed a concept of using torgas for firing pusher furnaces in the process of co-combustion with natural and coke-oven gases. The paper has shown that, despite low calorific value of torgas (6 MJ/Nm^3), its energy use offers economic benefits and, additionally, biochar is produced during the process, which can be used at other stages of iron and steel production. The use of gases from the thermal conversion of waste, for example pyrolysis, where the calorific value is much higher than in the case of torgas, may contribute to reducing the consumption of natural gas, which will ultimately allow the diversification of fuel and energy sources, while bringing measurable economic and environmental benefits [20,67].

4. Analysis of Strengths and Weaknesses of Generating Energy from RDF Pyrolysis Products

As suggested by numerous literature sources, the pyrolysis of RDF seems to be a promising solution [9,58]. The management of surplus RDF production has economic, environmental and social benefits. The use of RDF pyrolysis products for energy purposes, like any process, shows not only advantages but also disadvantages [47,98,99]. An analysis of these strengths and weaknesses is presented in Table 7.

Table 7. Strengths and weaknesses of using RDF pyrolysis products as energy sources [47,98,99].

Strengths	Weaknesses
(1) The generation of energy from the pyrolysis products offers environmental benefits: • More waste managed instead of landfilled; • Lower emissions of pollutants compared to conventional combustion; • Reduced production and use of conventional fossil fuels. (2) Pyrolysis is a very flexible process. By controlling the parameters of the process, you can get larger quantities of a more desirable product. (3) Pyrolysis is also suitable for the disposal of various types of waste, which is very beneficial considering diverse morphological compositions of RDF. (4) The pyrolysis plant does not require a large area, which enables integrating it into the existing infrastructure of the steelworks or other plant. (5) The conversion of RDF is in line with the ideas of circular economy. In addition, the products of RDF pyrolysis can be further used for generation of energy, which brings the metallurgical industry closer to carbon neutrality. (6) Continuous and intensive research on the pyrolysis process and the possibilities of managing its products enrich the world of science and bring the economy and industry closer to the introduction of pyrolysis for use on an industrial scale. (7) Finding new applications for pyrolysis products. Proposal to use pyrolysis gas and biochar for energy purposes, reducing the consumption of conventional fuels. (8) Possibility of co-financing investments from domestic sources (the National Fund for Environmental Protection and Water Management may co-finance up to 50% of eligible investment costs—up to PLN 50 million of non-returnable aid). Significant reduction of investment costs, which has a positive effect on the willingness to manage post-process products of the RDF pyrolysis. (9) The calorific values of pyrolysis gas and biochar are sufficient to treat them as attractive alternative fuels that can reduce process costs and bring the steel industry closer to carbon neutrality.	(1) RDF consists of a processed over-sieve fraction from the municipal waste stream. This involves additional processes (grinding, drying, pelletization) that the waste must undergo before it is delivered to the pyrolysis reactor. (2) The chlorine and sulfur contents in RDF, under certain process conditions, may corrode the plant, which may shorten the lifetime of the reactor, which will result in greater expenditures on maintenance. (3) Unknown investment costs: due to the lack of a pyrolysis plant operating on an industrial scale, investment and maintenance, and also profits, are just estimates at this stage, which discourages local governments and investors from building new pyrolysis systems. (4) Insufficient processing capacities of thermal waste disposal plans and lack of investment financing from the EU. (5) Lack of knowledge of the technology by the public, which may result in social reluctance at the stage of consultations or plans for the construction of infrastructure. Public opposition may effectively block or delay the finalization of the investment. (6) Low environmental awareness of the society, which may result in a lack of understanding of the need to obtain new sources of alternative fuels (a perception of no need to build new infrastructure for thermal conversion of waste). (7) Frequent incomprehensible changes in the law, too short time to implement changes, very high environmental standards to meet in a short time—all this gives rise to uncertainty that discourages companies from investing in new, prospective technologies.

5. Summary and Conclusions

The energy transformation taking place in Europe, additionally stimulated by the energy crisis resulting from the geopolitical situation and Russia's military operation against Ukraine, is a platform for the implementation of new solutions and technologies. One such solution is to replace fossil fuels with alternative ones. The use of RDF is in this case an example of a solution that allows for efficient use of industrial and municipal waste, thus reducing the amount of waste deposited in landfills and at the same time reducing the demand for fossil fuels. This approach meets global trends in this area and is in line with the policy of sustainable development. RDF produced from the over-sieve fraction of waste can be used in many industries. The metallurgical industry looks particularly attractive in this case, as it seems to be the most absorbent industry for alternative fuels in the coming years. This is due to both technological conditions, production costs incurred by metallurgical companies, as well as the volume of production and the volume of demand for fuels. The use of biochar or, for example, gases from the thermal conversion of waste makes economic sense in this case [20]. Despite the investment and adaptation costs that the company must incur in the initial phase of technological transformation, in the long run it may bring measurable benefits to the steel industry, which will significantly compensate for the costs incurred. Already a 20% share of energy produced from biochar would give measurable effects in the form of a significant reduction in the relationship between the amount of carbon dioxide from non-renewable fossil fuels and the amount of carbon dioxide from alternative sources. These are arguments for the implementation of new technologies and the use of alternative fuels in the metallurgical industry. Reducing CO_2 emissions will have a positive impact on climate change by limiting the greenhouse effect, which will bring measurable benefits to the entire society [100].

Author Contributions: Conceptualization, M.S. (Magdalena Skrzyniarz), M.S. (Marcin Sajdak), M.Z., A.B.-P., J.I. and A.S.; writing—original draft preparation, M.S. (Magdalena Skrzyniarz), M.S. (Marcin Sajdak) and M.Z.; writing—review and editing, A.B.-P., J.I. and A.S.; supervision, M.S. (Marcin Sajdak) and M.Z.; project administration, M.S. (Marcin Sajdak) and A.B.-P. All authors have read and agreed to the published version of the manuscript.

Funding: This research received no external funding.

Data Availability Statement: Publicly available datasets were analyzed in this study.

Conflicts of Interest: The authors declare no conflict of interest.

References

1. Quek, A.; Balasubramanian, R. Life cycle assessment of energy and energy carriers from waste matter—A review. *J. Clean. Prod.* **2014**, *79*, 18–31. [CrossRef]
2. Singh, A. Managing the uncertainty problems of municipal solid waste disposal. *J. Environ. Manag.* **2019**, *240*, 259–265. [CrossRef] [PubMed]
3. Di Giacomo, G.; Romano, P. Evolution and Prospects in Managing Sewage Sludge Resulting from Municipal Wastewater Purification. *Energies* **2022**, *15*, 5633. [CrossRef]
4. Kundariya, N.; Mohanty, S.S.; Varjani, S.; Hao Ngo, H.; Wong, J.W.C.; Taherzadeh, M.J.; Chang, J.S.; Yong Ng, H.; Kim, S.H.; Bui, X.T. A review on integrated approaches for municipal solid waste for environmental and economical relevance: Monitoring tools, technologies, and strategic innovations. *Bioresour. Technol.* **2021**, *342*, 125982. [CrossRef]
5. Das, A.K.; Islam, M.N.; Billah, M.M.; Sarker, A. COVID-19 and municipal solid waste (MSW) management: A review. *Environ. Sci. Pollut. Res.* **2021**, *28*, 28993–29008. [CrossRef]
6. Ng, K.S.; Phan, A.N. Evaluating the Techno-economic Potential of an Integrated Material Recovery and Waste-to-Hydrogen System. *Resour. Conserv. Recycl.* **2021**, *167*, 105392. [CrossRef]
7. Maj, I.; Matus, K. Aluminosilicate Clay Minerals: Kaolin, Bentonite, and Halloysite as Fuel Additives for Thermal Conversion of Biomass and Waste. *Energies* **2023**, *16*, 4359. [CrossRef]
8. Chand Malav, L.; Yadav, K.K.; Gupta, N.; Kumar, S.; Sharma, G.K.; Krishnan, S.; Rezania, S.; Kamyab, H.; Pham, Q.B.; Yadav, S.; et al. A review on municipal solid waste as a renewable source for waste-to-energy project in India: Current practices, challenges, and future opportunities. *J. Clean. Prod.* **2020**, *277*, 123227. [CrossRef]
9. Sieradzka, M.; Rajca, P.; Zajemska, M.; Mlonka-Mędrala, A.; Magdziarz, A. Prediction of gaseous products from refuse derived fuel pyrolysis using chemical modelling software—Ansys Chemkin-Pro. *J. Clean. Prod.* **2020**, *248*, 119277. [CrossRef]
10. Rajca, P.; Poskart, A.; Chrubasik, M.; Sajdak, M.; Zajemska, M.; Skibiński, A.; Korombel, A. Technological and economic aspect of Refuse Derived Fuel pyrolysis. *Renew. Energy* **2020**, *161*, 482–494. [CrossRef]
11. Wielgosiński, G. Review of waste thermal treatment technology. *New Energy* **2011**, *1*, 55–67.
12. Xie, Y.; Wang, J.; Cai, X.; Huang, Z. Self-acceleration of cellular flames and laminar flame speed of syngas/air mixtures at elevated pressures. *Int. J. Hydrogen Energy* **2016**, *41*, 18250–18258. [CrossRef]
13. Javed, N.; Muhammad, S.; Iram, S.; Ramay, M.W.; Jaffri, S.B.; Damak, M.; Fekete, G.; Varga, Z.; Székács, A.; Aleksza, L. Analysis of Fuel Alternative Products Obtained by the Pyrolysis of Diverse Types of Plastic Materials Isolated from a Dumpsite Origin in Pakistan. *Polymers* **2023**, *15*, 24. [CrossRef] [PubMed]

14. Sanchis-Sebastiá, M.; Ruuth, E.; Stigsson, L.; Galbe, M.; Wallberg, O. Novel sustainable alternatives for the fashion industry: A method of chemically recycling waste textiles via acid hydrolysis. *Waste Manag.* **2021**, *121*, 248–254. [CrossRef] [PubMed]
15. Rollinson, A.N.; Oladejo, J.M. 'Patented blunderings' efficiency awareness, and self-sustainability claims in the pyrolysis energy from waste sector. *Resour. Conserv. Recycl.* **2019**, *141*, 233–242. [CrossRef]
16. Peng, N.; Li, Y.; Liu, Z.; Liu, T.; Gai, C. Emission, distribution and toxicity of polycyclic aromatic hydrocarbons (PAHs) during municipal solid waste (MSW) and coal co-combustion. *Sci. Total Environ.* **2016**, *565*, 1201–1207. [CrossRef]
17. Jouhara, H.; Olabi, A.G. Editorial: Industrial waste heat recovery. *Energy* **2018**, *160*, 1–2. [CrossRef]
18. Sun, Y.; Seetharaman, S.; Liu, Q.; Zhang, Z.; Liu, L.; Wang, X. Integrated biomass gasification using the waste heat from hot slags: Control of syngas and polluting gas releases. *Energy* **2016**, *114*, 165–176. [CrossRef]
19. Perondi, D.; Poletto, P.; Restelatto, D.; Manera, C.; Silva, J.P.; Junges, J.; Collazzo, G.C.; Dettmer, A.; Godinho, M.; Vilela, A.C.F. Steam gasification of poultry litter biochar for bio-syngas production. *Process Saf. Environ. Prot.* **2017**, *109*, 478–488. [CrossRef]
20. Zajemska, M.; Sajdak, M.; Iwaszko, J.; Skrzyniarz, M.; Biniek-Poskart, A.; Skibi, A.; Maroszek. A. The role of calorific waste in transformation of iron and steel industry towards sustainable production. *Resour. Conserv. Recycl.* **2023**, *191*, 2022–2024. [CrossRef]
21. Násner, A.M.L.; Lora, E.E.S.; Palacio, J.C.E.; Rocha, M.H.; Restrepo, J.C.; Venturini, O.J.; Ratner, A. Refuse Derived Fuel (RDF) production and gasification in a pilot plant integrated with an Otto cycle ICE through Aspen plus™ modelling: Thermodynamic and economic viability. *Waste Manag.* **2017**, *69*, 187–201. [CrossRef] [PubMed]
22. Bensidhom, G.; Ben Hassen-Trabelsi, A.; Alper, K.; Sghairoun, M.; Zaafouri, K.; Trabelsi, I. Pyrolysis of Date palm waste in a fixed-bed reactor: Characterization of pyrolytic products. *Bioresour. Technol.* **2018**, *247*, 363–369. [CrossRef] [PubMed]
23. Wu, K.T.; Lee, H.T.; Juch, C.I.; Wan, H.P.; Shim, H.S.; Adams, B.R.; Chen, S.L. Study of syngas co-firing and reburning in a coal fired boiler. *Fuel* **2004**, *83*, 1991–2000. [CrossRef]
24. Malinowski, A.; Chwiałkowski, W. Energy recycling of RDF fraction of municipal solid waste by continuous pyrolysis. *Chem. Environ. Biotechnol.* **2017**, *20*, 27–33. [CrossRef]
25. Gerassimidou, S.; Velis, C.A.; Williams, P.T.; Komilis, D. Characterisation and composition identification of waste-derived fuels obtained from municipal solid waste using thermogravimetry: A review. *Waste Manag. Res.* **2020**, *38*, 942–965. [CrossRef]
26. Rajca, P.; Skibiński, A.; Biniek-Poskart, A.; Zajemska, M. Review of Selected Determinants Affecting Use of Municipal Waste for Energy Purposes. *Energies* **2022**, *15*, 57. [CrossRef]
27. Variny, M.; Varga, A.; Rimár, M.; Janošovský, J.; Kizek, J.; Lukáč, L.; Jablonský, G.; Mierka, O. Advances in biomass co-combustion with fossil fuels in the European context: A review. *Processes* **2021**, *9*, 100. [CrossRef]
28. Rozporządzenie Ministra Gospodarki z dnia 16 lipca 2015 r. w Sprawie Dopuszczania Odpadów do Składowania na Składowiskach, Dz.U. 2015 poz. 1277. Available online: https://isap.sejm.gov.pl/isap.nsf/download.xsp/WDU20150001277/O/D20151277.pdf (accessed on 15 July 2023).
29. Berardi, P.; Almeida, M.F.; de Lopes, M.; Maia Dias, J. Analysis of Portugal's refuse derived fuel strategy, with particular focus on the northern region. *J. Clean. Prod.* **2020**, *277*, 123262. [CrossRef]
30. Hussain, S.A.; Razi, F.; Hewage, K.; Sadiq, R. The perspective of energy poverty and 1st energy crisis of green transition. *Energy* **2023**, *275*, 127487. [CrossRef]
31. Ochoa-Barragán, R.; Munguía-López, A.d.C.; Ponce-Ortega, J.M. A hybrid machine learning-mathematical programming optimization approach for municipal solid waste management during the pandemic. *Environ. Dev. Sustain.* **2023**. [CrossRef]
32. Durak, H. Comprehensive Assessment of Thermochemical Processes for Sustainable Waste Management and Resource Recovery. *Processes* **2023**, *11*, 2092. [CrossRef]
33. Gabbar, H.A.; Aboughaly, M.; Ayoub, N. Comparative study of MSW heat treatment processes and electricity generation. *J. Energy Inst.* **2018**, *91*, 481–488. [CrossRef]
34. Zhang, S.; Yi, B.W.; Worrell, E.; Wagner, F.; Crijns-Graus, W.; Purohit, P.; Wada, Y.; Varis, O. Integrated assessment of resource-energy-environment nexus in China's iron and steel industry. *J. Clean. Prod.* **2019**, *232*, 235–249. [CrossRef]
35. Malinauskaite, J.; Jouhara, H.; Czajczyńska, D.; Stanchev, P.; Katsou, E.; Rostkowski, P.; Thorne, R.J.; Colón, J.; Ponsá, S.; Al-Mansour, F.; et al. Municipal solid waste management and waste-to-energy in the context of a circular economy and energy recycling in Europe. *Energy* **2017**, *141*, 2013–2044. [CrossRef]
36. Wasielewski, R.; Bałazińska, M. Energy recovery from waste in the aspect of qualificationsof electricity and heat as coming from renewable energysources and to participate in the emissions trading system. *Energy Policy J.* **2018**, *21*, 129–142.
37. Kumar, A.; Agrawal, A. Recent trends in solid waste management status, challenges, and potential for the future Indian cities—A review. *Curr. Res. Environ. Sustain.* **2020**, *2*, 100011. [CrossRef]
38. Dong, J.; Tang, Y.; Nzihou, A.; Chi, Y.; Weiss-Hortala, E.; Ni, M. Life cycle assessment of pyrolysis, gasification and incineration waste-to-energy technologies: Theoretical analysis and case study of commercial plants. *Sci. Total Environ.* **2018**, *626*, 744–753. [CrossRef]
39. Chen, D.; Yin, L.; Wang, H.; He, P. Pyrolysis technologies for municipal solid waste: A review. *Waste Manag.* **2014**, *34*, 2466–2486. [CrossRef]
40. Qureshi, M.S.; Oasmaa, A.; Pihkola, H.; Deviatkin, I.; Tenhunen, A.; Mannila, J.; Minkkinen, H.; Pohjakallio, M.; Laine-Ylijoki, J. Pyrolysis of plastic waste: Opportunities and challenges. *J. Anal. Appl. Pyrolysis* **2020**, *152*, 104804. [CrossRef]

41. Primus, A.; Rosik-Dulewska, C. Energy production in low-power cogeneration systems usingthe gasification technology of post-municipal waste.The legal and economic conditions. *Energy Policy J.* **2017**, *20*, 79–92.
42. Sajdak, M.; Muzyka, R.; Gałko, G.; Ksepko, E.; Zajemska, M.; Sobek, S.; Tercki, D. Actual Trends in the Usability of Biochar as a High-Value Product of Biomass Obtained through Pyrolysis. *Energies* **2023**, *16*, 355. [CrossRef]
43. Mousa, E.; Wang, C.; Riesbeck, J.; Larsson, M. Biomass applications in iron and steel industry: An overview of challenges and opportunities. *Renew. Sustain. Energy Rev.* **2016**, *65*, 1247–1266. [CrossRef]
44. Edo, M.; Skoglund, N.; Gao, Q.; Persson, P.E.; Jansson, S. Fate of metals and emissions of organic pollutants from torrefaction of waste wood, MSW, and RDF. *Waste Manag.* **2017**, *68*, 646–652. [CrossRef] [PubMed]
45. Malindzakova, M.; Strąka, M.; Rosova, A.; Kanuchova, M.; Trebuna, P. Modeling the process for incineration of municipal waste. *Chem. Ind.* **2015**, *94*, 1260–1264. [CrossRef]
46. Dong, J.; Tang, Y.; Nzihou, A.; Chi, Y. Key factors influencing the environmental performance of pyrolysis, gasification and incineration Waste-to-Energy technologies. *Energy Convers. Manag.* **2019**, *196*, 497–512. [CrossRef]
47. Korombel, A.; Ławińska, O.; Zajemska, M. Pyrolysis-Based Municipal Solid Waste Management in Poland—SWOT Analysis. *Energies* **2022**, *15*, 510.
48. Zajemska, M.; Rajca, P.; Szwaja, S.; Morel, S. The chemical mechanism of the HCl formation in the pyrolysis process of selected wastes. *Przemysł Chem.* **2019**, *98*, 907–910.
49. Czajczyńska, D.; Anguilano, L.; Ghazal, H.; Krzyżyńska, R.; Reynolds, A.J.; Spencer, N.; Jouhara, H. Potential of pyrolysis processes in the waste management sector. *Therm. Sci. Eng. Prog.* **2017**, *3*, 171–197. [CrossRef]
50. Skrzyniarz, M. Directions of plastic waste management in the era of COVID-19. *Mater. Econ. Logist. J.* **2021**, *08*, 25–31. [CrossRef]
51. Piechocki, J.; Sołowiej, P.; Neugebauer, M. Gasification of waste miomass from agricultural production. *Agric. Eng.* **2010**, *5*, 219–224.
52. Białowiec, A.; Pulka, J.; Stępień, P.; Manczarski, P.; Gołaszewski, J. The RDF/SRF torrefaction: An effect of temperature on characterization of the product—Carbonized Refuse Derived Fuel. *Waste Manag.* **2017**, *70*, 91–100. [CrossRef] [PubMed]
53. Nobre, C.; Vilarinho, C.; Alves, O.; Mendes, B.; Gonçalves, M. Upgrading of refuse derived fuel through torrefaction and carbonization: Evaluation of RDF char fuel properties. *Energy* **2019**, *181*, 66–76. [CrossRef]
54. Nobre, C.; Alves, O.; Longo, A.; Vilarinho, C.; Gonçalves, M. Torrefaction and carbonization of refuse derived fuel: Char characterization and evaluation of gaseous and liquid emissions. *Bioresour. Technol.* **2019**, *285*, 121325. [CrossRef] [PubMed]
55. Chen, Z.; Liao, Y.; Chen, Y.; Ma, X. Insight into the gas pollutants emission of rural solid waste during the gasification-combustion process: Influencing factors and mechanisms. *Fuel* **2024**, *355*, 129510. [CrossRef]
56. Mentes, D.; Nagy, G.; Szabó, T.J.; Hornyák-Mester, E.; Fiser, B.; Viskolcz, B.; Póliska, C. Combustion behaviour of plastic waste—A case study of PP, HDPE, PET, and mixed PES-EL. *J. Clean. Prod.* **2023**, *402*, 136850. [CrossRef]
57. Chen, P.; Wang, Z.; Zhang, Y.; Guo, T.; Li, Y.; Hopke, P.K.; Li, X. Volatility distribution of primary organic aerosol emissions from household crop waste combustion in China. *Environ. Pollut.* **2023**, *323*, 121353. [CrossRef]
58. Jagodzińska, K.; Zaini, I.N.; Svanberg, R.; Yang, W.; Jönsson, P.G. Pyrolysis of excavated waste from landfill mining: Characterisation of the process products. *J. Clean. Prod.* **2021**, *279*, 123541. [CrossRef]
59. Kijo-Kleczkowska, A.; Gnatowski, A. Recycling of PlasticWaste, with Particular Emphasis on Thermal Methods—Review. *Energies* **2022**, *15*, 2114. [CrossRef]
60. Gałko, G.; Mazur, I.; Rejdak, M.; Jagustyn, B.; Hrabak, J.; Ouadi, M.; Jahangiri, H.; Sajdak, M. Evaluation of alternative refuse-derived fuel use as a valuable resource in various valorised applications. *Energy* **2023**, *263*, 125920. [CrossRef]
61. Simshauser, P. The 2022 energy crisis. Fuel poverty and the impact of policy interventions in Australia's National Electricity Market. *Energy Econ.* **2023**, *121*, 106660. [CrossRef]
62. Sikarwar, V.S.; Masláni, A.; Hlína, M.; Fathi, J.; Mates, T.; Pohorelý, M.; Meers, E.; Syc, M.; Jeremiás, M. Thermal plasma assisted pyrolysis and gasification of RDF by utilizing sequestered CO_2 as gasifying agent. *J. CO2 Util.* **2022**, *66*, 102275. [CrossRef]
63. Ooi, T.C.; Thompson, D.; Anderson, D.R.; Fisher, R.; Fray, T.; Zandi, M. The effect of charcoal combustion on iron-ore sintering performance and emission of persistent organic pollutants. *Combust. Flame* **2011**, *158*, 979–987. [CrossRef]
64. Zajemska, M.; Magdziarz, A.; Iwaszko, J.; Skrzyniarz, M.; Poskart, A. Numerical and experimental analysis of pyrolysis process of RDF containing a high percentage of plastic waste. *Fuel* **2022**, *320*, 123981. [CrossRef]
65. Smoliński, A.; Wojtacha-Rychter, K.; Król, M.; Magdziarczyk, M.; Polański, J.; Howaniec, N. Co-gasification of refuse-derived fuels and bituminous coal with oxygen/steam blend to hydrogen rich gas. *Energy* **2022**, *254*, 124210. [CrossRef]
66. Gałko, G.; Rejdak, M.; Tercki, D.; Bogacka, M.; Sajdak, M. Evaluation of the applicability of polymeric materials to BTEX and fine product transformation by catalytic and non-catalytic pyrolysis as a part of the closed loop material economy. *J. Anal. Appl. Pyrolysis* **2021**, *154*, 105017. [CrossRef]
67. Szwaja, S.; Zajemska, M.; Szwaja, M.; Maroszek, A. Integration of waste biomass thermal processing technology with a metallurgical furnace to improve its efficiency and economic benefit. *Clean Technol. Environ. Policy* **2023**, *25*, 577–587. [CrossRef]
68. Manyà, J.J.; García-Ceballos, F.; Azuara, M.; Latorre, N.; Royo, C. Pyrolysis and char reactivity of a poor-quality refuse-derived fuel (RDF) from municipal solid waste. *Fuel Process. Technol.* **2015**, *140*, 276–281. [CrossRef]
69. Chavando, J.A.M.; de Matos, E.C.J.; Silva, V.B.; Tarelho, L.A.C.; Cardoso, J.S. Pyrolysis characteristics of RDF and HPDE blends with biomass. *Int. J. Hydrogen Energy* **2022**, *47*, 19901–19915. [CrossRef]

70. Zaini, I.N.; García López, C.; Pretz, T.; Yang, W.; Jönsson, P.G. Characterization of pyrolysis products of high-ash excavated-waste and its char gasification reactivity and kinetics under a steam atmosphere. *Waste Manag.* **2019**, *97*, 149–163. [CrossRef]
71. Czajczyńska, D.; Krzyżyńska, R.; Jouhara, H.; Spencer, N. Use of pyrolytic gas from waste tire as a fuel: A review. *Energy* **2017**, *134*, 1121–1131. [CrossRef]
72. Abbas-Abadi, M.S.; Haghighi, M.N.; Yeganeh, H.; McDonald, A.G. Evaluation of pyrolysis process parameters on polypropylene degradation products. *J. Anal. Appl. Pyrolysis* **2014**, *109*, 272–277. [CrossRef]
73. Veses, A.; Sanahuja-Parejo, O.; Callén, M.S.; Murillo, R.; García, T. A combined two-stage process of pyrolysis and catalytic cracking of municipal solid waste for the production of syngas and solid refuse-derived fuels. *Waste Manag.* **2020**, *101*, 171–179. [CrossRef] [PubMed]
74. Lu, J.S.; Chang, Y.; Poon, C.S.; Lee, D.J. Slow pyrolysis of municipal solid waste (MSW): A review. *Bioresour. Technol.* **2020**, *312*, 123615. [CrossRef]
75. Pressley, P.N.; Aziz, T.N.; Decarolis, J.F.; Barlaz, M.A.; He, F.; Li, F.; Damgaard, A. Municipal solid waste conversion to transportation fuels: A life-cycle estimation of global warming potential and energy consumption. *J. Clean. Prod.* **2014**, *70*, 145–153. [CrossRef]
76. Czajczyńska, D.; Nannou, T.; Anguilano, L.; Krzyzyńska, R.; Ghazal, H.; Spencer, N.; Jouhara, H. Potentials of pyrolysis processes in the waste management sector. *Energy Procedia* **2017**, *123*, 387–394. [CrossRef]
77. Ouadi, M.; Jaeger, N.; Greenhalf, C.; Santos, J.; Conti, R.; Hornung, A. Thermo-Catalytic Reforming of municipal solid waste. *Waste Manag.* **2017**, *68*, 198–206. [CrossRef]
78. Zhang, H.L. *Pyrolysis Properties of Municipal Organic Wastes and Study on the Improvement Effects of Biochar on Soil*; Chongqing University: Chongqing, China, 2012.
79. Fombu, A.H.; Ochonogor, A.E. Design and Construction of a Semi-batch Pyrolysis Reactor for the Production of Biofuel. *IOP Conf. Ser. Earth Environ. Sci.* **2021**, *730*, 012041. [CrossRef]
80. Ślefarski, R.; Jójka, J.; Czyżewski, P.; Gołębiewski, M.; Jankowski, R.; Markowski, J.; Magdziarz, A. Experimental and Numerical-Driven Prediction of Automotive Shredder Residue Pyrolysis Pathways toward Gaseous Products. *Energies* **2021**, *14*, 1779. [CrossRef]
81. Xue, Y.; Zhou, S.; Brown, R.C.; Kelkar, A.; Bai, X. Fast pyrolysis of biomass and waste plastic in a fluidized bed reactor. *Fuel* **2015**, *156*, 40–46. [CrossRef]
82. Eke, J.; Onwudili, J.A.; Bridgwater, A.V. Influence of Moisture Contents on the Fast Pyrolysis of Trommel Fines in a Bubbling Fluidized Bed Reactor. *Waste Biomass Valorization* **2019**, *11*, 3711–3722. [CrossRef]
83. Fan, Z.; Friedmann, S.J. Low-carbon production of iron and steel: Technology options, economic assessment, and policy. *Joule* **2021**, *5*, 829–862. [CrossRef]
84. Li, Z.; Hanaoka, T. Plant-level mitigation strategies could enable carbon neutrality by 2060 and reduce non-CO_2 emissions in China's iron and steel sector. *One Earth* **2022**, *5*, 932–943. [CrossRef]
85. Khanna, R.; Li, K.; Wang, Z.; Sun, M.; Zhang, J.; Mukherjee, P.S. *Biochars in Iron and Steel Industries*; Elsevier: Amsterdam, The Netherlands, 2019; Volume 2017, pp. 429–446.
86. Griffin, P.W.; Hammond, G.P. The prospects for 'green steel' making in a net-zero economy: A UK perspective. *Glob. Transit.* **2021**, *3*, 72–86. [CrossRef]
87. Somers, J.; Moya, J. Decarbonisation of industrial heat: The iron and steel sector. *Ee-IpOrg* **2020**, *4*. Available online: https://setis.ec.europa.eu/decarbonisation-industrial-heat-iron-and-steel-sector_en (accessed on 15 July 2023).
88. Zhang, S.; Worrell, E.; Crijns-Graus, W.; Wagner, F.; Cofala, J. Co-benefits of energy efficiency improvement and air pollution abatement in the Chinese iron and steel industry. *Energy* **2014**, *78*, 333–345. [CrossRef]
89. Bailera, M.; Lisbona, P.; Peña, B.; Romeo, L.M. A review on CO_2 mitigation in the Iron and Steel industry through Power to X processes. *J. CO2 Util.* **2021**, *46*, 101456. [CrossRef]
90. Shukla, I. Potential of renewable agricultural wastes in the smart and sustainable steelmaking process. *J. Clean. Prod.* **2022**, *370*, 133422. [CrossRef]
91. Mallett, A.; Pal, P. Green transformation in the iron and steel industry in India: Rethinking patterns of innovation. *Energy Strategy Rev.* **2022**, *44*, 100968. [CrossRef]
92. Ren, L.; Zhou, S.; Peng, T.; Ou, X. A review of CO_2 emissions reduction technologies and low-carbon development in the iron and steel industry focusing on China. *Renew. Sustain. Energy Rev.* **2021**, *143*, 110846. [CrossRef]
93. Lukáč, L.; Kizek, J.; Jablonský, G.; Karakash, Y. Defining the Mathematical Dependencies of NOx and CO Emission Generation after Biomass Combustion in Low-Power Boiler. *Civ. Environ. Eng. Rep.* **2019**, *29*, 153–163. [CrossRef]
94. Wen, Z.; Wang, Y.; Li, H.; Tao, Y.; De Clercq, D. Quantitative analysis of the precise energy conservation and emission reduction path in China's iron and steel industry. *J. Environ. Manag.* **2019**, *246*, 717–729. [CrossRef] [PubMed]
95. Shen, J.; Zhang, Q.; Xu, L.; Tian, S.; Wang, P. Future CO_2 emission trends and radical decarbonization path of iron and steel industry in China. *J. Clean. Prod.* **2021**, *326*, 129354. [CrossRef]
96. Rousset, P.; Figueiredo, C.; De Souza, M.; Quirino, W. Pressure effect on the quality of eucalyptus wood charcoal for the steel industry: A statistical analysis approach. *Fuel Process. Technol.* **2011**, *92*, 1890–1897. [CrossRef]

97. Chan, W.P.; Veksha, A.; Lei, J.; Oh, W.-D.; Dou, X.; Giannis, A.; Lisak, G.; Lim, T.T. A novel real-time monitoring and control system for waste-to-energy gasification process employing differential temperature profiling of a downdraft gasifier. *J. Environ. Manag.* **2019**, *234*, 65–74. [CrossRef]
98. Li, F.; He, X.; Shoemaker, C.A.; Wang, C.H. Experimental and numerical study of biomass catalytic pyrolysis using Ni_2P-loaded zeolite: Product distribution, characterization and overall benefit. *Energy Convers. Manag.* **2020**, *208*, 112581. [CrossRef]
99. Liu, Y.; Xue, L.; Ma, J.; Peng, C.; Bai, F.; Li, Y.; Zhao, J. Three-dimensional numerical simulation, energy efficiency and economic benefit estimation of oil shale in situ pyrolysis process. *Geoenergy Sci. Eng.* **2023**, *227*, 211804. [CrossRef]
100. Škrbić, S.; Ašonja, A.; Prodanović, R.; Ristić, V.; Stevanović, G.; Vulić, M.; Janković, Z.; Radosavac, A.; Igić, S. Analysis of plant-production-obtained biomass in function of sustainable energy. *Sustainability* **2020**, *12*, 5486. [CrossRef]

energies

MDPI

Brief Report

Hydrothermal Carbonisation as Treatment for Effective Moisture Removal from Digestate—Mechanical Dewatering, Flashing-Off, and Condensates' Processing

Halina Pawlak-Kruczek [1,*], Agnieszka Urbanowska [2,*], Lukasz Niedzwiecki [1,3], Michał Czerep [1], Marcin Baranowski [1], Christian Aragon-Briceño [4,5], Małgorzata Kabsch-Korbutowicz [2], Amit Arora [6], Przemysław Seruga [7,8], Mateusz Wnukowski [1], Jakub Mularski [1], Eddy Bramer [4], Gerrit Brem [4] and Artur Pożarlik [4]

1 Department of Energy Conversion Engineering, Faculty of Mechanical and Power Engineering, Wrocław University of Science and Technology, Wyb. Wyspiańskiego 27, 50-370 Wrocław, Poland; lukasz.niedzwiecki@pwr.edu.pl (L.N.); michal.czerep@pwr.edu.pl (M.C.); marcin.baranowski@pwr.edu.pl (M.B.); mateusz.wnukowski@pwr.edu.pl (M.W.); jakub.mularski@pwr.edu.pl (J.M.)
2 Department of Environment Protection Engineering, Faculty of Environmental Engineering, Wrocław University of Science and Technology, Wyb. Wyspiańskiego 27, 50-370 Wrocław, Poland; malgorzata.kabsch-korbutowicz@pwr.edu.pl
3 Energy Research Centre, Centre for Energy and Environmental Technologies, VŠB—Technical University of Ostrava, 17. Listopadu 2172/15, 708 00 Ostrava, Czech Republic
4 Department of Thermal and Fluid Engineering, University of Twente, Drienerlolaan 5, 7522 NB Enschede, The Netherlands; caragon@fcirce.es (C.A.-B.); e.a.bramer@utwente.nl (E.B.); g.brem@utwente.nl (G.B.); a.k.pozarlik@utwente.nl (A.P.)
5 Fundacion Circe, Parque Empresarial Dinamiza, Avda. Ranillas 3D, 1ª Planta, 50018 Zaragoza, Spain
6 Department of Chemical Engineering, Shaheed Bhagat Singh State University, Ferozepur 152004, India; amitarora@sbsstc.ac.in or aroraamitlse@yahoo.com
7 Zaklad Gospodarowania Odpadami Gac Sp. z o.o, Gac 90, 55-200 Olawa, Poland; przemyslaw.seruga@zgo.org.pl
8 Department of Bioprocess Engineering, Wroclaw University of Economics, Komandorska 118/120, 53-345 Wroclaw, Poland
* Correspondence: halina.pawlak@pwr.edu.pl (H.P.-K.); agnieszka.urbanowska@pwr.edu.pl (A.U.)

Citation: Pawlak-Kruczek, H.; Urbanowska, A.; Niedzwiecki, L.; Czerep, M.; Baranowski, M.; Aragon-Briceño, C.; Kabsch-Korbutowicz, M.; Arora, A.; Seruga, P.; Wnukowski, M.; et al. Hydrothermal Carbonisation as Treatment for Effective Moisture Removal from Digestate—Mechanical Dewatering, Flashing-Off, and Condensates' Processing. *Energies* **2023**, *16*, 5102. https://doi.org/10.3390/en16135102

Academic Editor: Alok Kumar Patel

Received: 2 May 2023
Revised: 5 June 2023
Accepted: 28 June 2023
Published: 1 July 2023

Abstract: One of the processes that can serve to valorise low-quality biomass and organic waste is hydrothermal carbonization (HTC). It is a thermochemical process that transpires in the presence of water and uses heat to convert wet feedstocks into hydrochar (the solid product of hydrothermal carbonization). In the present experimental study, an improvement consisting of an increased hydrophobic character of HTC-treated biomass is demonstrated through the presentation of enhanced mechanical dewatering at different pressures due to HTC valorisation. As part of this work's scope, flashing-off of low-quality steam is additionally explored, allowing for the recovery of the physical enthalpy of hot hydrochar slurry. The flashing-off vapours, apart from steam, contain condensable hydrocarbons. Accordingly, a membrane system that purifies such effluent and the subsequent recovery of chemical energy from the retentate are taken into account. Moreover, the biomethane potential is calculated for the condensates, presenting the possibility for the chemical energy recovery of the condensates.

Keywords: digestate; hydrothermal carbonisation; nanofiltration; mechanical dewatering

1. Introduction

Hydrothermal carbonisation (HTC) is promising in terms of the valorisation of various types of wet biomass [1–3]. HTC is a thermal valorisation process that transpires in subcritical water at elevated temperatures (typically 170 to 260 °C) [4–6] with a residence time ranging between 30 min to 2 h [7–11] and at an autogenic pressure, which is higher than the saturation pressure of water. During the HTC process, complex reaction pathways are carried out, with different reactions proceeding in parallel, including hydrolysis, decarboxylation, and dehydration as well as aldol condensation and polymerisation [12,13]. Dehydration is important from the perspective of the global processing efficiency of installations processing wet biomass, as it decreases the number of hydroxyl groups (OH) in the process [14]. From a practical perspective, a loss of hydroxyl groups makes hydrochars more hydrophobic both in terms of decreased equilibrium moisture content [15] and with respect to facilitating mechanical dewatering [14]. Some of the advantages offered by the process include the improved grindability of the hydrochars [16,17], an enhancement in pelletising [16], an increased heating value of the valorised biomass [18,19], and improved sorption capacity for some compounds [20], among others. A positive influence across the whole value chain of wet biomass has been shown using LCA [21–24]. The HTC process produces liquid by-products, which contain some chemical energy that could be used for the production of biogas [25–27].

Ahmed et al. [26] evaluated the influence of HTC process conditions on Capillary Suction Time (CST) and dewaterability with a centrifuge for sewage sludge samples obtained from a wastewater treatment plant in Trento, Italy. HTC at 190 °C with a residence time of 30 min resulted in the CST decreasing from 2.78 s/g/L for raw sewage sludge to 2.67 s/g/L for the HTC-treated material [26]. A further increase in the HTC residence time resulted in a significant decrease in CST, i.e., HTC residence times of one, two, and three hours resulted in CST values as low as 0.38 s/g/L, 0.37 s/g/L, and 0.27 s/g/L, respectively [26]. This is in apparent contradiction to the results reported by Wang et al. [28], who observed an increase in CST for hydrothermally treated samples of sewage sludge from a wastewater treatment plant in Hefei, China, treated at temperatures ranging between 50 °C and 170 °C with a residence time of 30 min. Wang et al. [28] attributed this to the disintegration of flocs caused by the thermal treatment. It seems plausible to suspect that the relatively low temperature and residence time of the hydrothermal treatment applied by Wang et al. [28] was enough to disintegrate the flocs but not enough to decrease the content of oxygenated functional groups (i.e., OH groups) forming on the surface of the hydrochars, which has a maximum at a certain hydrothermal treatment temperature and then decreases with a further temperature increase, as reported by Jain et al. [29]. Gao et al. [30] reported the benefits of the in situ mechanical dewatering of sewage sludge in an HTC reactor, resulting in a reduction of 27.7–59.6% of the moisture content, depending on the HTC severity. Wang et al. [31] also reported a decrease in the moisture content of sewage sludge using in situ dewatering in an HTC reactor and demonstrated that the dewatering performance is significantly better in hot conditions in comparison to dewatering performed after cooling the products to ambient temperature for HTC performed at 180 °C with residence times ranging between 10 and 90 min. Aragon-Briceño et al. [32] demonstrated the positive influence of HTC treatment on mechanical dewatering for the digestate from the anaerobic digestion of the wet fraction of municipal solid waste treated at temperatures ranging from 180 °C to 230 °C with residence times of 30–120 min. Wilk et al. [33] treated effluents after the HTC of sewage sludge with vacuum distillation and demonstrated that the chemical oxygen demand (COD) of the filtrate decreased 32-fold during the process. Czerwińska, Śliz, and Wilk [34] reported that distillation at atmospheric pressure caused a 95% decrease in COD and total organic carbon (TOC). Czerwińska et al. [35] performed a double nanofiltration of effluents after HTC, achieving an 84.5% decrease in COD.

Lisseth et al. [36] proposed recovering heat by using flash vapours as the heat source in an installation with HTC. Such a generation of vapours would also entail the removal of moisture from the hydrochars. This waste stream, i.e., the condensate obtained in the distillation process, can be both a source of water and valuable substances. However, this requires its proper processing. To recover water and high-value substances, membrane pressure-driven processes can be used, among which nanofiltration seems to be the most useful method.

The aims of this work are as follows:

- The optimisation of the dewatering pressure of municipal solid waste digestate after HTC;
- The determination of a possible means of purifying the condensate with nanofiltration membranes after the condensing of vapours flashed for heat recovery after HTC;
- The evaluation of the biomethane potential of flashed vapours.

2. Materials and Methods

Samples of the digestate, produced using the organic fraction of municipal solid waste, were taken from a biogas plant located at the premises of ZGO Gać near Oława in Lower Silesia, Poland. The diagram of the experimental setup for hydrothermal carbonisation (Figure 1) given below shows the autoclave rig. The autoclave was filled with 3.0 litres of wet digestate, which had solid content of 30%. Moisture content was determined with Radwag MA.X2.A (Radom, Poland)at 105 °C.

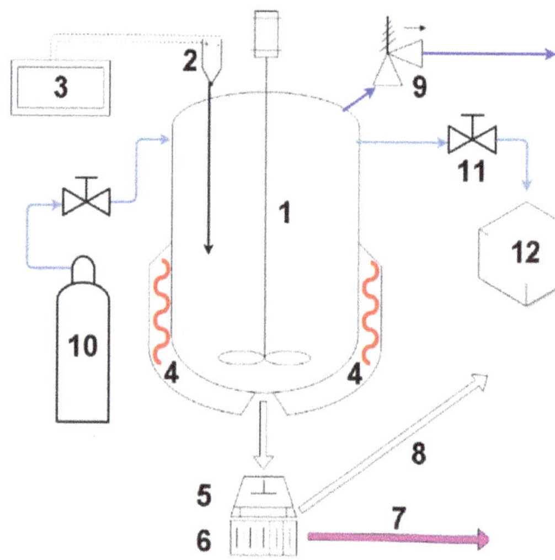

Figure 1. Diagram of the HTC rig (1—vessel; 2—thermocouple; 3—PLC; 4—heaters; 5—cotton filter; 6—filter's base; 7—effluent; 8—separated hydrochar; 9—safety valve; 10—purge gas; 11—flashing-off valve; 12—laboratory cooler).

HTC temperature of 200 °C was chosen based on the range between 200 °C and 260 °C [37,38] specified in the literature and by taking into account the design preference in industrial-scale HTC installations for a lower range of pressure, which allows for comparably lower thickness of a reactor's walls. After heating the biomass with a heating rate of 1.57 °C/min, the biomass was kept in the reactor for 120 min. Subsequently, the heating was halted, and the rig was allowed to cool overnight.

Flashing-off was performed, after the separation of wet hydrochars from liquid effluent, at a temperature of 110 °C. When this temperature had been achieved in the autoclave, the flashing-off valve was slightly opened, and vapours were released into an Allihn-type glass cooler with 1 m of effective length and cooled with tap water. The opening of the valve was kept at a level allowing for a sufficient pressure drop in order to prevent the glass cooler from shattering due to vapour pressure. Condensate was collected for 2 h into 5 containers, with each container being changed after approx. 25 min. Characterisation of each condensate was carried out according to the Standard Methods for the Examination of Water and Wastewater, 23rd Edition. Chemical Oxygen Demand (COD), Biological Oxygen Demand (BOD_5), Dissolved Organic Carbon (DOC), pH, conductivity, dry mass, alkaline metal content (Na^+, Mg^{2+}, Ca^{2+}, and K^+), phosphorus content (total P and $PO4^{3-}$), nitrogen content (total nitrogen, NO_3^-, and NH_4^+), sulphate content (SO_3^{2-}), and halogen content

(Cl^-, Br^-, and F^-) were determined. The characteristics of the test solutions (K0–K4) are presented in Table 1.

Table 1. Properties of the test solutions of condensate.

Parameter	Test Solution				
	K0	K1	K2	K3	K4
pH	8.12	10.46	10.88	9.98	9.61
conductivity, mS/cm	32.8	8.23	5.74	1.841	1.042
dry mass, mg/dm^3	27,050	790	850	680	570
COD, mg O_2/dm^3	30,940	2670	2130	1410	840
BOD_5, mg O_2/dm^3	6200	484	448	464	460
DOC, mg C/dm^3	7970	1250	940	660	370
N-NH_4^+, mg/dm^3	3026	2364	2396	327	218
N-NO_3^-, mg/dm^3	0	0	0	17.6	18
N, mg/dm^3	3280	2880	2640	389	261
Na^+, mg/dm^3	2300	1720	1290	630	460
K^+, mg/dm^3	2150	1580	1030	550	390
Mg^{2+}, mg/dm^3	285	41	11.8	10.3	10.9
Ca^{2+}, mg/dm^3	440	95.2	22.1	20.8	21.3
F^-, mg/dm^3	82.5	8.14	2.83	11	11.3
Cl^-, mg/dm^3	5965	35	50.2	14.6	14.1
Br^-, mg/dm^3	14.6	1.14	1.14	0.25	1.24
P-PO_4^{3-}, mg/dm^3	30.3	5.52	2.54	2.87	3
SO_4^{2-}, mg/dm^3	1240	83.3	45	35.8	35.4
P, mg/dm^3	31.8	6.2	5.3	5.6	4.9

Treatment of individual condensates was carried out using NPO10P and NPO30P flat nanofiltration membranes from Mann + Hummel Water & Fluid Solutions (Ludwigsburg, Germany). Their characteristics are shown in Table 2.

Table 2. NF membranes used in the experiments [39].

Membrane Type	Membrane Material	Na_2SO_4 Retention	MWCO, kDa	Contact Angle	Max Temp., °C	pH Range	Effective Filtration Surface, cm^2
NP010P NP030P	polyethersulfone	25–40% 80–95%	1040–1400 520–700	62.5°	95	0–14	45.3

The membranes were conditioned before the actual membrane filtration process by filtering the redistilled water through the membranes successively under different transmembrane pressures from 0.1 to 0.4 MPa until constant water flux values were obtained.

After each experiment, the membranes were cleaned (chemically regenerated) with 0.1 mol/dm^3 of NaOH solution (Avantor Performance Materials Poland S.A., Gliwice, Poland) and rinsed with redistilled water until the initial values of permeate flux were obtained.

The nanofiltration process was carried out on a test stand equipped with an Amicon 8400 cell produced by Millipore (Figure 2). This cell allows for a dead-end filtration process and is designed to work with flat sheet membranes. The volume of the Amicon 8400 cell is 400 cm^3, and the membrane diameter is 76 mm. The cell was placed on an ARE magnetic stirrer produced by OMC Envag (Warsaw, Poland) so that the contaminant concentration was uniform throughout the solution volume. The transmembrane pressure used in this study was 0.3 MPa.

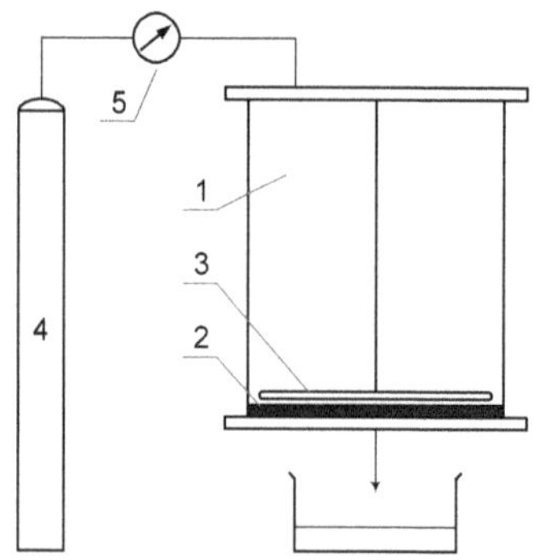

Figure 2. Amicon 8400 dead-end membrane system (1—filtration cell, 2—membrane, 3—stirrer, 4—pressurized nitrogen cylinder, and 5—pressure valve).

The separation efficiency was evaluated by determining the concentration of impurities in the treated solution and in the purified solution and by determining the value of the reduction factor (removal/retention), R, using the following formula:

$$R = \left(1 - c_p/c_0\right) \cdot 100,\ \% \tag{1}$$

where:

c_p—concentration of impurities in the treated solution, g/m^3;

c_0—initial concentration of impurities in the condensate, g/m^3.

Equation (2) presents the formula used for the calculation of the theoretical biomethane potential (BMP) in the condensates.

$$BMP_{Th} = 0.39 \cdot COD \cdot \frac{DOC}{COD} \cdot 0.9,\ {dm^3_{CH4}}\Big/{dm^3_{liquid}} \tag{2}$$

To assess the theoretical BMP of the condensates, the stochiometric Formula (3) of methane oxidation was used. This formula allows for the calculation of the potential amount of methane produced based on the COD balance of a sample [40].

The COD conversion to methane at 35 °C is 0.39 dm^3 of CH_4 per gram of COD. Furthermore, a second correction factor was applied to render the prediction more accurate. The first correction factor was the DOC-to-COD ratio, which was used since the COD refers to all the organics and inorganics that can be oxidized, while the DOC only refers to the organic carbon compounds that can potentially be converted into methane. The second correction factor was the biodegradability of HTC process water, which was 90%. This factor was based on a comparison of the real BMP reported in previous studies versus the theoretical BMP value from the normal conversion of the stochiometric formula [25,41].

$$CH_4 + 2O_2 \rightarrow CO_2 + 2H_2O \tag{3}$$

3. Results

The obtained results (Table 3) show that HTC resulted in a reduction in the moisture content of the digestate, as moisture content values of the dewatered hydrochars were much lower in comparison to the moisture content of the raw digestate (75.7%$_{w.b.}$) and the digestate after dewatering at 3 MPa (65.3%). Furthermore, the increase in dewatering pressure allowed for a further reduction in the moisture content, which allowed achieving 37.1% moisture content after dewatering at a pressure of 10 MPa.

Table 3. Moisture content of hydrochars after dewatering at different pressures (w.b.—wet basis).

Sample	$p_{dewatering}$, MPa	Moisture Content, $\%_{w.b.}$
Raw digestate	-	75.7
Dewatered digestate	3	65.3
Dewatered hydrochars	3	45.4
	5	42.9
	7	41.4
	10	37.1

The results of the tests for determining the suitability of nanofiltration membranes for the purification of condensates obtained in the distillation process of the liquid fraction of municipal post-HTC digestate are presented in Figure 3. The effect of the membrane cut-off (MWCO) value on the separation efficiency of the organic compounds from the analysed solutions was analysed. Upon comparing the results obtained, it can be seen that both of the tested nanofiltration membranes can be applied in the treatment of condensates, although a deterioration in the quality of the permeate was observed as the cut-off value increased. In the nanofiltration process, the separation effect of organic macromolecules is determined by the sieve mechanism as well as the solution–diffusion mechanism and electrostatic interactions between the membrane and the solution components. Purification with the NPO10P membrane allowed for the acquisition of the retention coefficients of biological oxygen demand (BOD_5), COD, and DOC at levels of up to 62, 47, and 50% and, for the NPO30P membrane, 73, 61, and 82%, respectively. The better separation properties of the NPO30P membrane may be due to its dense structure. According to Kovacs et al. [42], the cut-off value of the NPO10P membrane is higher and is in the range of 1010–1400 Da (with a pore diameter of 0.80–1.29 nm), while that for the NPO30P membrane is 500–700 Da (with a pore diameter of 0.57–0.93 nm).

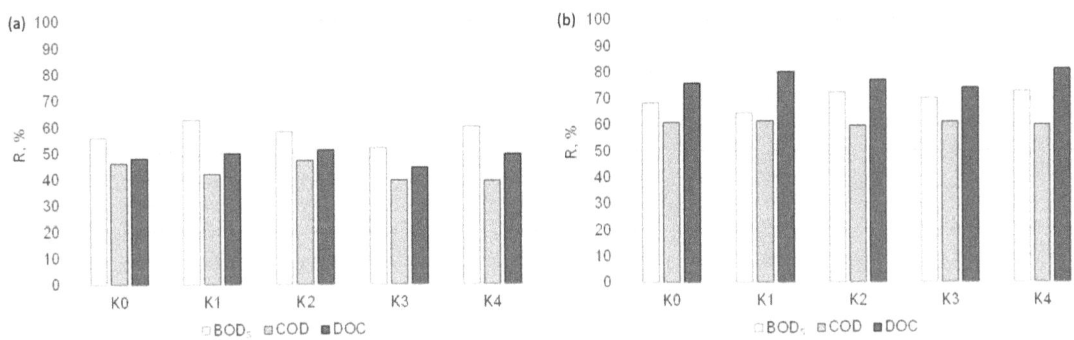

Figure 3. Effect of nanofiltration membrane type, i.e., (**a**) NPO10P and (**b**) NPO30P, on the efficiency of the separation of organic pollutants from condensates obtained from the liquid fraction of municipal post-HTC digestate (Δp = 0.3 MPa).

Upon analysing the results obtained for the individual condensates, it was found that regardless of the type of nanofiltration membrane, the retention coefficients of BOD_5, COD, and DOC remained at a relatively constant level, which may mean that the duration of the distillation process had no significant effect on the efficiency of organic compound removal from the analysed liquids.

In this study, the potential for methane production from the retentate through cascade membrane filtration was considered for a more circular approach. This approach has been adopted in some studies that have used the retentates from the membrane filtration of wastewater for anaerobic digestion to produce biogas. Luo et al. [43] suggested this approach in their study, where they applied an ultrafiltration/nanofiltration system in a dairy wastewater model. They did not conduct experimental work but highlighted this approach nonetheless, as the retentate from membrane filtration system contained a high concentration of organics. Chen et al. [44] integrated isoelectric precipitation, nanofiltration, and anaerobic digestion into a cascade, using a model dairy wastewater. The aim of the study was to increase the production of biogas (focused on hydrogen) through concentrating the short chain organics in the retentate and, at the same time, reduce the fouling of the membrane. The result was an increase in hydrogen in the biogas from 32.4% to 58.8%. Campell et al. [45] used HTC-processed water from spent coffee grounds and applied nanofiltration and reverse osmosis to treat the process water and reduce its COD. Both retentates were mixed and subjected to anaerobic digestion, obtaining yields of 0.21 dm^3/g_{COD}. However, one of the main points to consider is that the retentates can contain high concentrations of metals and nutrients (phosphorus and nitrogen) that can inhibit the anaerobic digestion process [46].

Table 4 shows the biomethane potential of the five different retentates. The first sample of condensate (K0) presented the highest and most significant BMP value compared with those of the subsequent samples, which gradually decreased. However, an alternative approach should be considered since nanofiltration has shown significant COD reduction potential. Therefore, it might be beneficial to perform the anaerobic digestion of the retentates remaining after membrane purification.

Table 4. Biomethane potential (BMP) for subsequently taken condensate samples.

Parameter	Test Solution				
	K_0	K_1	K_2	K_3	K_4
BMP, dm^3_{CH4}/dm^3_{liquid}	2.80	0.44	0.33	0.23	0.13

4. Conclusions

This research shows that it is beneficial to increase the dewatering pressure in the dewatering of hydrochars after hydrothermal carbonisation. However, technical difficulties related to increased problems with the strength of the materials used in the construction of dewatering presses should also be considered when selecting dewatering parameters. Flashing-off a share of the liquid by-products seems to be a feasible way to recover the sensible heat of hydrochars after the HTC process. In this study, the condensates had some biomethane potential, which could be utilized to further improve the energy balance of such installations. The nanofiltration membranes, which were made of polyethersulfone, demonstrated good COD removal rates, proving their suitability for the purification of condensates originating from flashed vapours. Overall, the existing possibilities for the purification of condensates make flashing-off a viable dewatering option. We recommend that further research is conducted on the combination of novel in situ dewatering methods presented in the literature with the use of flashing as a heat recovery option in HTC installations.

Author Contributions: Conceptualisation: H.P.-K., A.U. and L.N.; methodology: A.U., C.A.-B. and L.N.; validation: M.K.-K., A.A. and P.S.; formal analysis: A.U., C.A.-B., L.N. and M.W.; investigation: A.U., M.C., M.B. and M.W.; resources: H.P.-K., M.K.-K., E.B. and G.B.; data curation: P.S., M.W., J.M. and A.A.; writing—original draft preparation: A.U., L.N. and C.A.-B.; writing—review and editing: L.N., H.P.-K., M.K.-K., E.B., G.B. and A.P.; visualisation: J.M. and P.S; supervision: H.P.-K., M.K.-K., E.B., G.B. and A.P.; project administration: H.P.-K., A.P. and P.S.; funding acquisition: H.P.-K., A.P., E.B., G.B., P.S. and A.U. All authors have read and agreed to the published version of the manuscript.

Funding: The authors would like to thank the European Commission, the National Centre for Research and Development (Poland), Nederlandse Organisatie Voor Wetenschappelijk Onderzoek (Netherlands), and the Swedish Research Council Formas for providing funding through the framework of the collaborative international consortium (RECOWATDIG) financed under the 2018 Joint call of the WaterWorks 2017 ERA-NET Cofund. This ERA-NET is an integral part of the activities developed by the Water JPI. National Centre for Research and Development agreement number WATERWORKS2017/I/RECOWATDIG/01/2019.

Data Availability Statement: Data are available on request.

Conflicts of Interest: The authors declare no conflict of interest. The funders had no role in the design of the study; in the collection, analyses, or interpretation of data; in the writing of the manuscript; or in the decision to publish the results.

References

1. Aragón-Briceño, C.I.; Ross, A.B.; Camargo-Valero, M.A. Mass and energy integration study of hydrothermal carbonization with anaerobic digestion of sewage sludge. *Renew. Energy* **2021**, *167*, 473–483. [CrossRef]
2. Shi, W.; Fenton, O.; Ashekuzzaman, S.M.; Daly, K.; Leahy, J.J.; Khalaf, N.; Hu, Y.; Chojnacka, K.; Numviyimana, C.; Healy, M.G. An examination of maximum legal application rates of dairy processing and associated STRUBIAS fertilising products in agriculture. *J. Environ. Manag.* **2022**, *301*, 113880. [CrossRef]
3. Numviyimana, C.; Warchoł, J.; Khalaf, N.; Leahy, J.J.; Chojnacka, K. Phosphorus recovery as struvite from hydrothermal carbonization liquor of chemically produced dairy sludge by extraction and precipitation. *J. Environ. Chem. Eng.* **2022**, *10*, 106947. [CrossRef]
4. Śliz, M.; Wilk, M. A comprehensive investigation of hydrothermal carbonization: Energy potential of hydrochar derived from Virginia mallow. *Renew. Energy* **2020**, *156*, 942–950. [CrossRef]
5. Wilk, M.; Magdziarz, A.; Kalemba-Rec, I.; Szymańska-Chargot, M. Upgrading of green waste into carbon-rich solid biofuel by hydrothermal carbonization: The effect of process parameters on hydrochar derived from acacia. *Energy* **2020**, *202*, 117717. [CrossRef]
6. Magdziarz, A.; Wilk, M.; Wądrzyk, M. Pyrolysis of hydrochar derived from biomass—Experimental investigation. *Fuel* **2020**, *267*, 117246. [CrossRef]
7. Kumar, N.; Weldon, R.; Lynam, J.G. Hydrothermal carbonization of coffee silverskins. *Biocatal. Agric. Biotechnol.* **2021**, *36*, 102145. [CrossRef]
8. Sobek, S.; Tran, Q.-K.; Junga, R.; Werle, S. Hydrothermal carbonization of the waste straw: A study of the biomass transient heating behavior and solid products combustion kinetics. *Fuel* **2021**, *314*, 122725. [CrossRef]
9. Lühmann, T.; Wirth, B. Sewage Sludge Valorization via Hydrothermal Carbonization: Optimizing Dewaterability and Phosphorus Release. *Energies* **2020**, *13*, 4417. [CrossRef]
10. Román, S.; Libra, J.; Berge, N.; Sabio, E.; Ro, K.; Li, L.; Ledesma, B.; Álvarez, A.; Bae, S. Hydrothermal carbonization: Modeling, final properties design and applications: A review. *Energies* **2018**, *11*, 216. [CrossRef]
11. Smith, A.M.; Ekpo, U.; Ross, A.B. The influence of pH on the combustion properties of bio-coal following hydrothermal treatment of swine manure. *Energies* **2020**, *13*, 331. [CrossRef]
12. Djandja, O.S.; Yin, L.-X.; Wang, Z.-C.; Duan, P.-G. From wastewater treatment to resources recovery through hydrothermal treatments of municipal sewage sludge: A critical review. *Process Saf. Environ. Prot.* **2021**, *151*, 101–127. [CrossRef]
13. Picone, A.; Volpe, M.; Messineo, A. Process Water Recirculation during Hydrothermal Carbonization of Waste Biomass: Current Knowledge and Challenges. *Energies* **2021**, *14*, 2962. [CrossRef]
14. Funke, A.; Ziegler, F. Hydrothermal carbonisation of biomass: A summary and discussion of chemical mechanisms for process engineering. *Biofuels Bioprod. Biorefining* **2010**, *4*, 160–177. [CrossRef]
15. Acharjee, T.C.; Coronella, C.J.; Vasquez, V.R. Effect of thermal pretreatment on equilibrium moisture content of lignocellulosic biomass. *Bioresour. Technol.* **2011**, *102*, 4849–4854. [CrossRef] [PubMed]
16. Sharma, H.B.; Panigrahi, S.; Dubey, B.K. Hydrothermal carbonization of yard waste for solid bio-fuel production: Study on combustion kinetic, energy properties, grindability and flowability of hydrochar. *Waste Manag.* **2019**, *91*, 108–119. [CrossRef]
17. Kambo, H.S.; Dutta, A. Comparative evaluation of torrefaction and hydrothermal carbonization of lignocellulosic biomass for the production of solid biofuel. *Energy Convers. Manag.* **2015**, *105*, 746–755. [CrossRef]
18. Czerwińska, K.; Śliz, M.; Wilk, M. Hydrothermal carbonization process: Fundamentals, main parameter characteristics and possible applications including an effective method of SARS-CoV-2 mitigation in sewage sludge. A review. *Renew. Sustain. Energy Rev.* **2022**, *154*, 111873. [CrossRef]
19. Mlonka-Mędrala, A.; Sieradzka, M.; Magdziarz, A. Thermal upgrading of hydrochar from anaerobic digestion of municipal solid waste organic fraction. *Fuel* **2022**, *324*, 124435. [CrossRef]
20. Mokrzycki, J.; Lorenc-Grabowska, E.; Kordek-Khalil, K.; Rutkowski, P. Hydrothermal and pyrolytic biochars from waste milk thistle (Silybum marianum) extrudates as precursors for production of effective isoproturon adsorbents. *J. Water Process Eng.* **2020**, *37*, 101459. [CrossRef]
21. Mayer, F.; Bhandari, R.; Gäth, S.A. Life cycle assessment on the treatment of organic waste streams by anaerobic digestion, hydrothermal carbonization and incineration. *Waste Manag.* **2021**, *130*, 93–106. [CrossRef]
22. Stobernack, N.; Mayer, F.; Malek, C.; Bhandari, R. Evaluation of the energetic and environmental potential of the hydrothermal carbonization of biowaste: Modeling of the entire process chain. *Bioresour. Technol.* **2020**, *318*, 124038. [CrossRef]

23. Medina-Martos, E.; Istrate, I.R.; Villamil, J.A.; Gálvez-Martos, J.L.; Dufour, J.; Mohedano, Á.F. Techno-economic and life cycle assessment of an integrated hydrothermal carbonization system for sewage sludge. *J. Clean. Prod.* **2020**, 277. [CrossRef]
24. Mendecka, B.; Lombardi, L.; Micali, F.; de Risi, A. Energy Recovery from Olive Pomace by Hydrothermal Carbonization on Hypothetical Industrial Scale: A LCA Perspective. *Waste Biomass Valorization* **2020**, *11*, 5503–5519. [CrossRef]
25. Aragón-Briceño, C.; Ross, A.B.B.; Camargo-Valero, M.A.A. Evaluation and comparison of product yields and bio-methane potential in sewage digestate following hydrothermal treatment. *Appl. Energy* **2017**, *208*, 1357–1369. [CrossRef]
26. Ahmed, M.; Andreottola, G.; Elagroudy, S.; Negm, M.S.; Fiori, L. Coupling hydrothermal carbonization and anaerobic digestion for sewage digestate management: Influence of hydrothermal treatment time on dewaterability and bio-methane production. *J. Environ. Manag.* **2021**, *281*, 111910. [CrossRef] [PubMed]
27. Cao, Z.; Hülsemann, B.; Wüst, D.; Oechsner, H.; Lautenbach, A.; Kruse, A. Effect of residence time during hydrothermal carbonization of biogas digestate on the combustion characteristics of hydrochar and the biogas production of process water. *Bioresour. Technol.* **2021**, *333*, 125110. [CrossRef] [PubMed]
28. Wang, L.-F.; Qian, C.; Jiang, J.-K.; Ye, X.-D.; Yu, H.-Q. Response of extracellular polymeric substances to thermal treatment in sludge dewatering process. *Environ. Pollut.* **2017**, *231*, 1388–1392. [CrossRef]
29. Jain, A.; Balasubramanian, R.; Srinivasan, M.P. Hydrothermal conversion of biomass waste to activated carbon with high porosity: A review. *Chem. Eng. J.* **2016**, *283*, 789–805. [CrossRef]
30. Gao, N.; Li, Z.; Quan, C.; Miskolczi, N.; Egedy, A. A new method combining hydrothermal carbonization and mechanical compression in-situ for sewage sludge dewatering: Bench-scale verification. *J. Anal. Appl. Pyrolysis* **2019**, *139*, 187–195. [CrossRef]
31. Wang, L.; Zhang, L.; Li, A. Hydrothermal treatment coupled with mechanical expression at increased temperature for excess sludge dewatering: Influence of operating conditions and the process energetics. *Water Res.* **2014**, *65*, 85–97. [CrossRef] [PubMed]
32. Aragon-Briceño, C.; Pożarlik, A.; Bramer, E.; Brem, G.; Wang, S.; Wen, Y.; Yang, W.; Pawlak-Kruczek, H.; Niedźwiecki, Ł.; Urbanowska, A.; et al. Integration of hydrothermal carbonization treatment for water and energy recovery from organic fraction of municipal solid waste digestate. *Renew. Energy* **2022**, *184*, 577–591. [CrossRef]
33. Wilk, M.; Czerwińska, K.; Śliz, M.; Imbierowicz, M. Hydrothermal carbonization of sewage sludge: Hydrochar properties and processing water treatment by distillation and wet oxidation. *Energy Rep.* **2023**, *9*, 39–58. [CrossRef]
34. Czerwińska, K.; Śliz, M.; Wilk, M. Thermal Disposal of Post-processing Water Derived from the Hydrothermal Carbonization Process of Sewage Sludge. *Waste Biomass Valorization* **2023**, 1–10. [CrossRef]
35. Czerwińska, K.; Marszałek, A.; Kudlek, E.; Śliz, M.; Dudziak, M.; Wilk, M. The treatment of post-processing liquid from the hydrothermal carbonization of sewage sludge. *Sci. Total Environ.* **2023**, *885*, 163858. [CrossRef]
36. Lisseth, C.; Martinez, M.; Sermyagina, E.; Saari, J.; Silva, M.; Jesus, D.; Cardoso, M.; de Almeida, G.M.; Vakkilainen, E. Biomass and Bioenergy Hydrothermal carbonization of lignocellulosic agro-forest based biomass residues. *Biomass Bioenergy* **2021**, *147*, 106004. [CrossRef]
37. Wnukowski, M.; Owczarek, P.; Niedźwiecki, Ł. Wet Torrefaction of Miscanthus—Characterization of Hydrochars in View of Handling, Storage and Combustion Properties. *J. Ecol. Eng.* **2015**, *16*, 161–167. [CrossRef]
38. Yan, W.; Hastings, J.T.; Acharjee, T.C.; Coronella, C.J.; Vásquez, V.R. Mass and energy balances of wet torrefaction of lignocellulosic biomass. *Energy Fuels* **2010**, *24*, 4738–4742. [CrossRef]
39. MICRODYN-NADIR. Flat Sheet Membrane Data Sheets—MICRODYN-NADIR n.d. Available online: https://www.microdyn-nadir.com/flat-sheet-membrane-data-sheets/ (accessed on 15 February 2021).
40. Tauber, J.; Parravicini, V.; Svardal, K.; Krampe, J. Quantifying methane emissions from anaerobic digesters. *Water Sci. Technol.* **2019**, *80*, 1654–1661. [CrossRef]
41. Aragón-Briceño, C.I.; Grasham, O.; Ross, A.B.; Dupont, V.; Camargo-Valero, M.A. Hydrothermal carbonization of sewage digestate at wastewater treatment works: Influence of solid loading on characteristics of hydrochar, process water and plant energetics. *Renew. Energy* **2020**, *157*, 959–973. [CrossRef]
42. Kovacs, Z.; Samhaber, W. Characterization of nanofiltration membranes with uncharged solutes. *Membrantechnika* **2008**, *12*, 22–36.
43. Luo, J.; Ding, L.; Qi, B.; Jaffrin, M.Y.; Wan, Y. A two-stage ultrafiltration and nanofiltration process for recycling dairy wastewater. *Bioresour. Technol.* **2011**, *102*, 7437–7442. [CrossRef] [PubMed]
44. Chen, Z.; Luo, J.; Chen, X.; Hang, X.; Shen, F.; Wan, Y. Fully recycling dairy wastewater by an integrated isoelectric precipitation–nanofiltration–anaerobic fermentation process. *Chem. Eng. J.* **2016**, *283*, 476–485. [CrossRef]
45. Campbell, B.S.; Thorpe, R.B.; Peus, D.; Lee, J. Anaerobic digestion of untreated and treated process water from the hydrothermal carbonisation of spent coffee grounds. *Chemosphere* **2022**, *293*, 133529. [CrossRef] [PubMed]
46. Zacharof, M.-P.; Mandale, S.J.; Oatley-Radcliffe, D.; Lovitt, R.W. Nutrient recovery and fractionation of anaerobic digester effluents employing pilot scale membrane technology. *J. Water Process Eng.* **2019**, *31*, 100846. [CrossRef]

MDPI AG
Grosspeteranlage 5
4052 Basel
Switzerland
Tel.: +41 61 683 77 34

Energies Editorial Office
E-mail: energies@mdpi.com
www.mdpi.com/journal/energies

www.ingramcontent.com/pod-product-compliance
Lightning Source LLC
Chambersburg PA
CBHW041949260926
51850CB00019B/2092